AF531366

LIPID
[illegible]OTECHNOLOGY

By

Dr. P.R. Yadav
Lecturer
Department of Zoology
D.A.V. College
Muzaffarnagar (U.P.)

&

Dr. Rajiv Tyagi
Department of Zoology
M.M. College
Modi Nagar (U.P.)

DISCOVERY PUBLISHING HOUSE
NEW DELHI-110002

First Published-2006

ISBN 81-8356-106-3

Published by

DISCOVERY PUBLISHING HOUSE
4831/24, Ansari Road, Prahlad Street,
Darya Ganj, New Delhi-110002 (India)
Phone: 23279245 • Fax: 91-11-23253475
E-mail:dphtemp@indiatimes.com

Printed at
Arora Offset Press
Laxmi Nagar, Delhi–92

Contents

Chapter 1

Introduction

Lipids occur throughout the living world in micro-organisms, higher plants and animals. They occur in all cell types and contribute to cellular structure, provide stored fuel and participate in many biological processes ranging from transcription of the genetic code to regulation of vital metabolic pathways and physiological responses. In this book, they will be described mainly in terms of their functions, although from time to time it will be convenient, even necessary, to deal with lipid classes based on their chemical structures and properties.

DEFINITIONS

Lipids are defined on the basis of their solubility properties, not primarily their chemical structure.

The word 'lipid' (in older literature spelled also as lipide or lipoid) is used by chemists to denote a chemically heterogeneous group of substances having in common the property of insolubility in water, but solubility in non-aqueous solvents such as chloroform, hydrocarbons or alcohols. It is necessary to use this definition based on physical properties since there may be little or no chemical relationship between the numerous compounds now classified as lipids, many of which are described in this book. It is not always possible to discern a clear distinction between the terms fat and lipid. The term fat is more familiar to the layman and brings to mind substances that are clearly fatty in nature, greasy in texture cmd immiscible with water. Familiar examples are butter and the fatty parts of meats. Fats are thought of as solid in texture as distinct from oils which are liquid at ambient temperatures. Chemically, however, there is little distinction between a fat and an oil, since the substances that the layman thinks of as edible fats and oils are composed predominantly of esters of glycerol with fatty acids. These are called triacylglycerols and are chemically quite distinct from the oils used in the petroleum industry, which are generally hydrocarbons. The term 'lipid' to the chemist embraces the wider range of fatty substances that are described in this book.

STRUCTURAL CHEMISTRY AND NOMENCLATURE

Naming systems are complex and have to be learned. The naming of lipids often poses problems. When the subject was in its infancy, research workers gave names to substances that they had newly discovered. Often, these substances would turn out to be impure mixtures and as the chemical structures of individual lipids became established, rather more systematic naming systems came into being. Later, these were further formalized under naming conventions laid down by the International Union of Pure and Applied Chemistry (IUPAC) and the International Union of Biochemistry (IUB). Thus, triacylglycerol is now preferred to triglyceride, but the latter is still frequently used especially by nutritionists and clinicians and you will need to learn both. Likewise, outdated names for phospholipids: 'lecithin' (for phosphatidylcholine) and 'cephalin' (for an ill-defined mixture of phosphatidylethanolamine and phosphatidylserine) will be avoided in this book but you should be aware of their existence. Their main structural features are their chain lengths, the presence of unsaturation (double bonds) and of substituent groups. In regard to chain length, it is cumbersome to have to say every time: 'a chain length of ten carbon atoms' and we shall, therefore, refer to a 'IOC fatty acid'.

If we wish to refer to a specific carbon atom in a chain, we shall write, for example: 'the substituent at C10'. The numbering of fatty acid carbon atoms is done from the carboxyl end of the chain with the carboxyl carbon as C1. An old system of identifying carbon atoms was to give them Greek letters. Thus, C2 was the C1-carbon, C3 the B carbon and so on, ending with the Co-carbon as the last in the chain, furthest from the carboxyl carbon. Remnants of this system still survive and can sometimes be discerned in this book. However, we will routinely use the numbering, rather than the Greek lettering system. Hence, we will use 3 hydroxybutyrate, not B-hydroxybutyrate etc.

An important aspect of unsaturated fatty acids is the opportunity for isomerism, which may be either positional or geometric. Positional isomers occur when double bonds are located at different positions in the carbon chain. Thus, for example, a 16C monounsaturated fatty acid may have positional isomeric forms with double bonds at C_7 or C_9, sometimes written $\Delta 7$ or $\Delta 9$. (The position of unsaturation is numbered with reference to the first of the pair of carbon atoms between which the double bond occurs.) Two positional isomers of an 18C diunsaturated acid are illustrated in Fig. 1.1(c) and (d). Geometric isomerism refers to the possibility that the configuration at the double bond can be *cis* or *trans*. (Although the convention Z/E is now preferred by chemists instead of *cis/ trans,* we shall use the more traditional and more common *cis/ trans* nomenclature throughout this book.) In the *cis* form, the two hydrogen substituents are on the same side of the molecule, while in the *trans* form they are on opposite sides.[Fig. 1.1(a) and (b)]

Another important feature of biological molecules is their stereochemistry. In lipids based on glycerol, for example, there is an inherent asymmetry at the central carbon atom of glycerol. Thus, chemical synthesis of phosphoglycerides yields an equal mixture of two stereoisomeric forms, whereas almost all naturally occurring phosphoglycerides have a single stereochemical configuration, much in the same way as most natural amino

H H
C=C
$CH_3(CH_2)_yCH_2$ $CH_2(CH_2)_xCOOH$

(a) *cis* (Z)

H $CH_2(CH_2)_xCOOH$
C=C
$CH_3(CH_2)_yCH_2$ H

(b) *trans* (E)

$CH_3(CH_2)_4\overset{12}{C}H{=}\overset{}{C}H{\cdot}CH_2{\cdot}CH{=}\overset{9}{C}H(CH_2)_7COOH$

(c) cis, cis –9, 12–octadecadienoic acid

$CH_3(CH_2)_7CH{=}\overset{9}{C}H{\cdot}CH_2{\cdot}CH{=}\overset{6}{C}H(CH_2)_4COOH$

(d) cis, cis –6, 9–octadecadienoic acid

Fig. 1.1 Isomerism in unsaturated fatty acids.

acids are of the L (or S) series. In the past, naturally occurring compounds were designated L-α- and represented by a Fisher projection (see Fig. 1.2). The glycerol derivative was put into the same category as that glyceraldehyde into which it would be transformed by oxidation, without any alteration or removal of substituents. Phosphatidylcholine was therefore named: L-α-phospha tidylcholine.The IUPAC-IUB convention has now abolished the DL (or even the more recent *RS)* terminology and has provided rules for the unambiguous numbering of the glycerol carbon atoms. Under this system, phosphatidylcholine becomes, 1,2-diacyl-*sn*-glycero-3 phosphorylcholine or more shortly, 3-*sn*-phosphatidylcholine. The letters *sn* denote the *stereochemical numbering* and indicate that this system is being used. The stereochemical numbering system is too cumbersome to use routinely in a book of this type and, therefore, we shall normally use the terms 'phosphatidylcholine' etc., but introduce the more precise name when necessary.

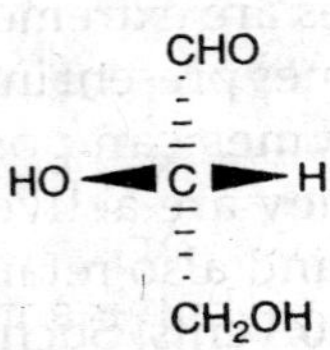

L–Glyceraldehyde

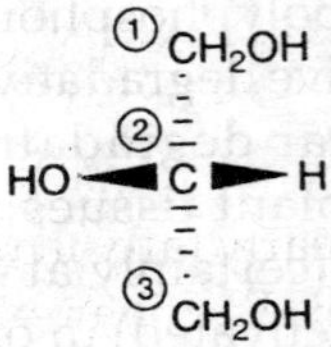

Glycerol

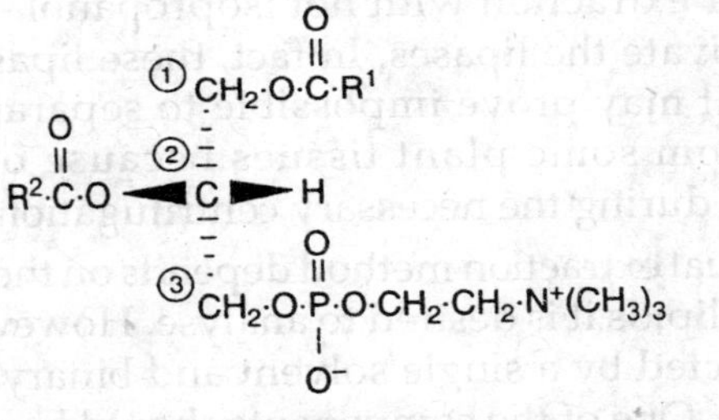

1, 2–Diacyl–*sn*–glycero–3–phosphorylcholine

Fig. 1.2 The stereochemical numbering of lipids derived from glycerol.

Another field in which nomenclature has grown up haphazardly is that of the enzymes of lipid metabolism. This has now been formalized to some extent under the Enzyme Commission (EC) nomenclature. The system is incomplete and not all lipid enzymes have EC names and numbers. Moreover, the system is very cumbersome for routine use and we have decided not to use it here.

EXTRACTION OF LIPIDS FROM NATURAL SAMPLES

Since lipids are characterized generally by their ability to dissolve in water-immiscible organic solvents, advantage is taken of this property at many stages of analysis.

Extraction of lipids from natural samples, like that of many biological molecules, is best accomplished as soon as possible to minimize the degradative changes which would otherwise take place. When storage has to take place it is best done at as Iowa temperature as possible (say -20°C or less) and under an inert nitrogen atmosphere to minimize oxidation of groups such as double bonds. If great care is not taken during the extraction process, many lipids will be partly or completely lost. For example, the polyphosphoinositides are extremely labile, due to the very active degradative enzymes present in many animal tissues. Similar degradative enzymes can pose particular problems in plant tissues since they are active at very low temperatures (certainly at –20°C) and also retain activity (or may even be activated) in organic solvents! Such enzymes are obviously best inactivated as quickly as possible by, for example, brief exposure of the tissue to steam or boiling water or by a prior extraction with hot isopropanol - all measures which inactivate the lipases. In fact, these lipases may be so active that it may prove impossible to separate membrane fractions from some plant tissues because of continuous degradation during the necessary centrifugation steps!

The actual extraction method depends on the type of tissue and also the lipids it is desired to analyse. However, few lipids can be extracted by a single solvent and binary mixtures are usually used. One of the components should have some water solubility and hydrogen bonding ability because lipid-protein complexes, such as those encountered in membranes, have to be split.

Common methods are often based on the original procedure of Bligh and Dyer. The latter used a mixture of chloroform and methanol in a ratio with tissue water (1:2:0.4) to form a one-phase system. Homogenization of tissues in this mixture efficiently extracts most lipids. More chloroform and methanol are then added to give two phases, the upper (aqueous) one containing non-lipid impurities; the lower (chloroform) phase can then be removed, washed with fresh upper phase and finally evaporated to dryness. Residual water can be removed with anhydrous sodium sulphate or by filtration through Sephadex columns. In some cases it may be desirable to use salt or dilute acid solutions in the upper phase to prevent losses of polar lipids. Finally, after removing solvent by vacuum evaporation, the crude lipid residue should be protected from oxidation by inert nitrogen gas. In fact, before storage for any length of time, lipids are best re-dissolved in a small amount of solvent containing an antioxidant such as 3, 5-di-*t*-butyl-4 hydroxytoluene (BHT) before storage at – 20°C or less in the dark under nitrogen.

LIKELY COMPONENTS OF THE CRUDE LIPID EXRACT

Since the initial extraction has been based on solubility properties, the crude lipid extract will contain any molecule which also dissolves preferentially in the organic solvents used. Thus, significant quantities of non-lipids, e.g. hydrophobic proteins, may be present at this stage. The mixture of lipids (Table 1.1) will depend on the nature of the tissue extracted.

GENERAL FEATURES OF LIPIDS IMPORTANT FOR THEIR ANALYSIS

Characterization of a lipid requires separation from other components and detection based on chemical and physical features.

Most of the lipid classes cited as major components in different tissues (Table 1.1) contain esterified fatty acids. They are thus termed acyl lipids. This is important during analysis for several reasons:

(1) The nature of the fatty acids determines much of the physical and biological properties of the lipid and, therefore, it is of importance to analyse their properties.
(2) The type of fatty acid influences the stability of the sample.
(3) Because mixtures of fatty acids are found in any given lipid class, the latter contains a number of molecular species. Thus, for example, when one refers to the triacylglycerol content of a certain tissue one should say triacylglycerols to show that a mixture of molecular species is present. Sphingolipids and sterols are present in leaves bur are minor components. Sterols in animals are predominantly cholesterol.

Total fatty acid analysis is usually conducted by forming volatile derivatives (such as methyl esters) for gas chromatography. Special techniques have to be used occasionally to avoid destroying unusual functional groups such as cyclopropene rings. Moreover, the danger of auto-oxidation means that analysis of samples containing polyunsaturated fatty acids has to be especially careful. For complete identification of individual acyl groups, degradation or derivatization usually has to be employed—for example, when double bond positions have to be assigned.

Table 1.1 Major components of typical lipid extracts from different tissues

Erythrocytes	Liver	Leaves	Cyanobacteria	Gram negative bacteria
Phosphoglycerides	Phosphoglycerides	Phosphoglycerides	Phosphoglycerides	Phosphoglycerides
Sphingolipids	Sphingolipids	-	-	-
-	-	Glycosy-glycerides	Glycosyl-glycerides	-
Sterols	Sterols	-	-	-
-	Triacylglycerols	-	-	-
		Others[a,b]	Others[a]	Others[c]

[a] Pigments.
[b] Waxes, cutin.
[c] Lipopolysaccharide.

In order to determine the positional distribution of acyl groups on, say, the glycerol backbone, enzymic cleavage is usually utilized. For example, phospholipases are available which have a specificity for either the *sn*-1 or the *sn*-2 position. Use of such enzymes will release fatty acids from one position and these can be separated from the partly deacylated product. Analysis of fatty acids and deacylated lipid will then reveal which fatty acids are present at each glycerol carbon.

Molecular species of lipids can be separated on the basis of size and/or unsaturation. In the past it has often been necessary to derivatize or remove the polar part of the lipid making analyses time-consuming. For example, because the charge on the head-group of phosphatidylcholine is large in relation to differences in acyl unsaturation, it was usually necessary to degrade such phosphoglycerides to diacylgly-cerols before analysis by chromatography. Modem methods of *High Performance Liquid Chromatography* (HPLC) have, however, rendered such methods unnecessary provided that adequate methods of detection are available.

Single lipid classes can be separated from each other by methods which make use of differences in their size and charge. They can often be provisionally identified by co-chromatography with authentic standards in various systems. Important constituent groups will be revealed by spectroscopic techniques or with specific colour reagents. However, unambiguous identification may require that the various products of hydrolysis be isolated, characterized and quantified. When enzymes are used to cause hydrolysis, their action (or otherwise) may also provide information about the stereochemistry of particular linkages.

CHROMATOGRAPHIC TECHNIQUES FOR SEPARATING LIPIDS

The principles of chromatography are based on distribution between two phases, one moving, the other stationary.

The two phases can be arranged in a variety of ways

A chromatogram (so-named by its Polish inventor, Tswett, because he used the technique to separate plant pigments) consists of two immiscible phases. One phase is kept stationary by either being held on an inert microporous support or being itself a microporous or particulate adsorbent solid; the other phase is percolated continuously through the stationary phase. Various phase pairs are possible although liquid-solid and gas-liquid are the most common.

If we take any single substance and mix it with any of these phase pairs, it will distribute itself between the two phases, the ratio of the concentrations in the two phases (at equilibrium) being known as the partition coefficient.

The partition coefficient is a physical constant dependent on the nature and magnitude of solute-solvent interactions in the two phases. Let us consider two substances A and B and imagine that, at equilibrium, substance A distributes itself between the two phases so that 90% is in the stationary phase and 10% in the moving phase. Substance B, however, distributes so that 10% is in the stationary phase and 90% in the moving phase. Then a mixture of A and B dissolved in a small volume of moving phase and applied to the chromatogram will begin to separate when the moving phase is added and washes them through the system. They will distribute themselves independently of one another: B will move as a zone at 9/10ths of the velocity of the moving phase and A at 1/10th of the velocity. Clearly the two substances will rapidly move apart. After separation, the substances can either be visualized directly on the system by colour sprays or be eluted as pure components.

There are basically two types of chromatogram geometry:

(1) The *column* consists of a metal, glass or even plastic tube with a ratio of length to diameter of at least 10:1 and is packed with either an absorbent solid (silica gel, alumina etc.) or an inert solid, such as Kieselguhr, of large surface area, that can hold, by surface tension, a liquid as one member of the phase pair. Gas chromatograms can afford a much greater ratio of length to diameter because of their inherently lower resistance to flow than can liquid columns.

(2) The *plate* or *strip* consists of the stationary phase support arranged as a flat surface. Mixtures can be spotted and dried on the surface and when the bottom of the plate is immersed in the moving phase in a closed vessel, capillarity ensures that the liquid will move through the porous material (paper, or porous solid held to a glass or metal surface). The fact that both phases exist as relatively thin films means that the solutes have only short distances to move as they pass from phase to phase. Very refined separations can thus be obtained rapidly. The thin layer chromatogram is particularly useful for lipid separations.

In general, compounds are not eluted from flat plate chromatograms during development (i.e. movement of the liquid phase). Instead, the development is stopped just before the solvent front has reached the end of the strip which is dried to remove solvent and the position of the zones revealed by spraying. The sprays can be of a destructive type such as dilute sulphuric acid followed by heating (this produces black spots by carbonization where there is an organic material) or a non-destructive type such as dichloro- or dibromo-fluoresceins that show a changed fluorescence where there is a zone. In the latter case, the lipids can be recovered by scraping the adsorbent from the plate (wherever the spray has indicated that a compound is present) and extracting with a suitable solvent. Under standard conditions it will be found that a given substance will move relative to a standard substance to a constant ratio or relative. This relative Rf is a useful confirmation of structure of an unknown substance, but it is unwise to use it as an absolute indicator.

Only infrequently can every component of a complex mixture be resolved by one solvent system. However, by using two-dimensional development refined separations can be achieved. These involve carrying out a separation with one solvent, drying the plate, turning it at right angles and using a second solvent system for development.

Gas-Liquid Chromatography

With this method the moving phase is a permanent gas and columns contain either an inert support (e.g. Celite) on the surface of which is the stationary liquid phase or else are themselves narrow tubes on the wall of which is a thin layer of the stationary phase. The latter are known as wall-coated-open tubular (WCOT) or capillary columns. Because no inert support is used in these columns, the flow of gas is relatively unimpeded and very long thin columns (say 100 m long and only 0.25 mm internal diameter) capable of quite remarkable separations are possible (see Fig. 1.3). Columns containing an inert support may be either packed or support coated-open-tubular (SCOT). Packed columns are the 'work horses' of gas chromatography and glass or stainless steel columns are usual. Glass columns are better because they are almost completely inert and any breaks or deterioration in the packing material can be seen easily. Of course, they are fragile – as many students know to their cost!

In a typical gas chromatograph the column is held in an oven either at a fixed temperature (isothermal operation) or at a temperature which increases during the separation (temperature programming). The latter method is particularly useful when the mixture to be analysed contains components with a wide range of molecular weights (Fig. 1.3). Located at the end of the column is the detector which quantifies the eluted components. For lipids the flame ionization detector is usually used. Eluates are burned in a flame of hydrogen and air to produce ions which are detected and measured by an electrical device capable of detecting about 10^{-12} gml^{-1}. Flame ionization detectors respond to almost all organic compounds, have a good signal to-noise ratio and are robust. If required, a T-piece can be used just before the detector so that fractions can be collected or separate radioactivity measurements (in, for example, a gas-flow proportional counter) and mass spectrometric analysis undertaken.

Individual components separated at a fixed temperature (isothermal use) and with constant gas flow can be provisionally identified by reference to standard mixtures run under the same conditions. However, because of the difficulties of keeping conditions exactly identical and also because of slow changes in the column's characteristics, it is better to use *relative retention time* to provisionally identify particular components. The comparison is usually made with a known component in the mixture or with an internal standard (which can also be used for quantitation and which, itself, is a compound not present in the original mixture). By the use of different column packings, confirmation of these identities can be made although, for complete analysis, some degradative methods must be used also. A rather convenient technique uses a combination of gas chromatography and mass spectrometry, where the eluate is split so that part goes to the detector and the remainder enters a mass spectrometer for (fragmentation) analysis.

Gas-liquid chromatography (GLC) can, of course, only be used for compounds which can be volatilized without decomposition. In practice this normally limits its application to molecules with molecular mass of less than 800 Da. Even so many lipids, such as phosphoglycerides, are not amenable to analysis as intact molecules.

Perhaps the most important use for the lipid biochemist is in the analysis of the fatty acid composition of lipids or tissues. In addition, gas chromatography may be used for steroid, glycerol, sphingosine, inositol and carbohydrate analysis. Molecular species of partial glycerides or, on highly thermostable silicone liquid phases, even triacylglycerols may be analysed also.

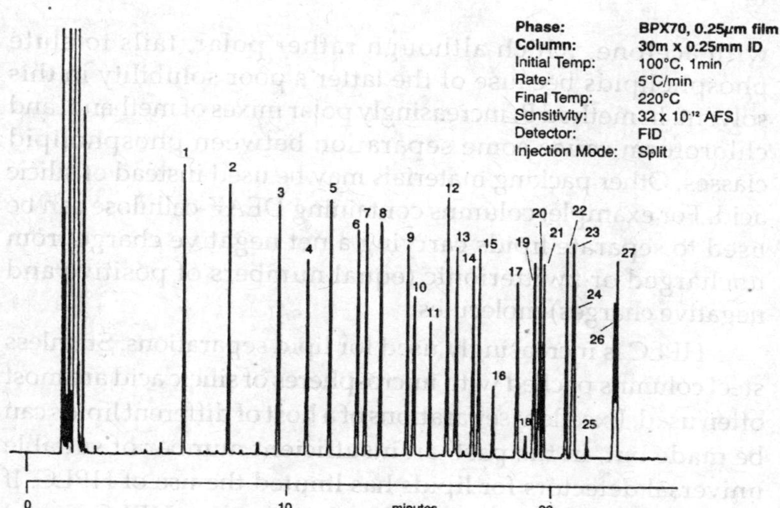

1. C11:0; 2. C12:0; 3. C13:0; 4. C10:0 2-OH; 5. C14;6. C15;0-*i*; 7. C15:0-a; 8. C15:0; 9. C12:0 2-OH; 10. C16:0-*i*; 11. C16:0; 12. C15:1n7*cis*; 13. C12:0 3-OH; 14. C17:0-a; 15. C17:0; 16. C9. 10-methylene 16:0; 17. C14:0 2-OH; 18. C18:0; 19. C18:1n9-*trans*; 20. C18:1n9-*cis*: 21. C18:1n7-*cis*; 22. C18:2n6-*cis*; 23. C19:0; 24. C16:0 2-OH; 25. C9. 10-methylene 18:0; 26. C20:0; 27 C18:0 2-OH. a = anteiso-branched; i = iso-branched.

Fig. 1.3. Separation of fatty acid methyl esters by gas-liquid chromatography on a capillary column. The column was 30 m long and only 0.25 mm internal diameter. Initial temperature was 100°C and this was then increased at 5°C min^{-1} to 220°C. Note that some very small differences in compounds can be detected. For example, peaks 19-21 are all of octadecenoic acids that only differ in the position or configuration of their single double bond.

Adsorption Column Chromatography

In this technique a moving liquid phase elutes lipids selectively from a solid support. Traditionally the most useful columns for lipid separations have been those containing silicic acid. Polar lipids tend to be tightly adsorbed onto these and can be separated easily from less polar materials such as sterol (-esters) or triacylglycerols. By an appropriate choice of elution solvents, good separations may also be obtained within a single lipid class – say phospholipids. Two simple methods are summarized in Table 1.2. In method A, glycolipids are eluted with acetone, which although rather polar, fails to elute phospholipids because of the latter's poor solubility in this solvent. In method B, increasingly polar mixes of methanol and chloroform cause some separation between phospholipid classes. Other packing materials may be used instead of silicic acid. For example, columns containing DEAE-cellulose can be used to separate lipids carrying a net negative charge from uncharged or zwitterionic (equal numbers of positive and negative charges) molecules.

HPLC is increasingly used for lipid separations. Stainless steel columns packed with microspheres of silicic acid are most often used. Excellent separations of a host of different lipids can be made but, in the past, an insufficient number of suitable universal detectors for lipids has limited the use of HPLC. If only lipids had convenient absorbances then HPLC would become the separation method of choice for practically every occasion!

There are two solutions to the detection problem—either to use a detector which does not detect compounds by their absorbances or to derivatize the lipids for analysis so that they absorb in the UV or visible region and can, therefore, be readily measured. An example of derivatization is to convert fatty acids to their phenacyl derivatives. Fluorescent derivatives (e.g. anthrylmethyl esters) give a further increase in the sensitivity of detection.

Two general detectors are useful for lipids. The first is a differential refractometer, which senses minute differences in the refractive index of an eluant brought about by material eluting from the column. However, it can only be used easily with a solvent of constant composition. A second method is to use an evaporative light-scattering detector. In this technique the eluate is evaporated in a jet of air to produce a 'mist' of (lipid) particles which scatter light. The method is sensitive and can be used with almost any lipid. Unlike the differential refractometer, it is not affected by solvent gradients although it does need solvents that are volatile.

Table 1.2 Separation of complex lipids on silicic acid columns

Fraction	Compound eluted	Solvent	Column volume
Method A			
1	Simple lipids	Chloroform	10
2	Glycolipids (and traces of acidic phospholipids)	Acetone	40
3	Phospholipids	Methanol	10
Method B (for fraction 3 separated above)			
1[a]	DPG + PA	Chloroform/ methanol (95:5)	10
2[b]	PE + PS	Chloroform/ methanol (80:20)	20
3[c]	PI+PC	Chloroform/ methanol (50:50)	20
4	Sphingomyelin + lyso PC	Methanol	20

[a] DPG = diphosphatidylglycerol; PA = phosphatidic acid.
[b] PE = phosphatidylethanolamine; PS = phosphatidylserine.
[c] PI = phosphatidylinositol; PC = phosphatidylcholine.

Thin layer adsorption chromatography

In this technique a thin (usually 0.25 mm) layer is spread on a glass plate and separations are achieved by allowing appropriate solvents to rise up the plate by capillary action. Silica gel is the most usual adsorbent and samples, which are applied close to the edge of the plate, are fractionated due to their different adsorptions to the support phase and/or their solubility in the solvent. Very good separations can be achieved although the thickness of the layer means that only small amounts of lipids are usually analysed. When the plates have been run, they are dried and can be sprayed with various reagents to reveal individual lipid classes. If the spray reagent is non-destructive (e.g. 2',7'-dichlorofluorescein, which causes all lipids to fluoresce) then the lipids can be recovered by scraping the appropriate area off the plate and eluting with a suitable solvent.

Complex lipid mixtures cannot always be separated by one-dimensional chromatography. In that case, either a preliminary fractionation on a column (to give say, neutral, glycolipid and phospholipid fractions) or two-dimensional thin layer chromatography (TLC) can be used.

TLC offers a number of advantages over column chromatography. It is more rapid and sensitive, gives better resolution and is usually much quicker. Moreover, the apparatus required is minimal and, especially if plates are made in the laboratory, the technique is inexpensive. By incorporating various chemicals in the thin layer, special lipid separations can be made. For example, silver nitrate allows fatty acids (or more complex lipids) to be separated on the basis of their unsaturation. Silicone oil-silica gel TLC works on the basis of reverse-phase separation and can be used to fractionate fatty acid mixtures based on their hydrophobicity, with shorter chain or unsaturated components migrating faster. Boric acid impregnation allows separation of *threo-* or *erythro-*iso mers of vicinal diols or fractionation of molecular species of ceramides.

A number of useful chemical methods is available for detecting or quantifying lipid classes. In some cases, the specificity of the colours developed can be used to help with identifications. Also when a non-destructive fluorescent reagent is used, lipids can be quantified on the plates by a scanning fluorimeter—though, because different lipids have different fluorescence values, calibration curves must be made.

Specific chemical methods can be used to reveal phospholipids, glycolipids, sterols or their esters as well as compounds with quaternary nitrogens or vicinal diols. Particularly useful for many membrane extracts is the reaction of ammonium molybdate with inorganic phosphate released from phospholipids. The phosphomolybdic acid so produced is then reduced to give an intense blue colour. The method can be adapted to a spray reagent or, more often, used to detect as little as 1 μg of phosphorus in scraped samples.

OTHER USEFUL METHODS

A detailed discussion of the wide variety of methods that are available to the lipid biochemist for identification and analysis is beyond the scope of this book. However, a few of the more widespread techniques are listed in Table 1.3.

SUMMARY

In contrast to carbohydrates, proteins and nucleic acids, lipids are defined on the basis of their physical properties (insolubility in water) rather than on the basis of consistent chemical features. For this reason, the student will need to learn and remember a wide range of different chemical types and their rather complex nomenclature.

Lipids can usually be extracted easily from tissues by making use of their hydrophobic characteristics. However, such extractions yield a complex mixture of different lipid classes which have to be purified further for quantitative analysis. Moreover, the crude lipid extract may be contaminated by other hydrophobic molecules, e.g. by intrinsic membrane proteins.

Table 1.3 Some other methods for lipid analysis

Method	Use
IR spectoscopy	Identification of organic bases in phospholipids, or *trans*-double bonds.
NMR spectoscopy	Widely used for lipid structure determination particularly identification and location of double bonds in fatty acids, functional groups (e.g. hydroxyl) on fatty adds and preliminary identification of glycerides, glycolipids and phospholipids. ^{1}H spectra give better quantitative data but ^{13}C spectra provide more information on functional groups. ^{31}p spectra are used often to assess solidl liquid phase ratios of lipids.
UV spectoscopy	Analysis of conjugated double bond systems, especially those formed by oxidation reactions.
Supercritical fluid chromatography	Supercritical fluid chromatography uses a compressed gas (often C_2) above its critical temperature and pressure to elute compounds from a column. Can be used at lower temperatures than GLC and more detectors are available than for HPLC.
Mass spectrometry	Often linked to a gas chromatogram or HPLC. Has a wide variety of uses from the estimation

	of molecular weights (e.g. use of thermospray mass spectrometry on-line with an HPLC to quantify molecular species) to the location of double bonds in aliphatic chains or identification of functional groups. Different sample inlet systems (direct insertion, fast atom bombardment) and ion sources (electron or chemical ionization) are used depending on the lability of the sample and whether structural elucidation or detection and quantitation are the main aims.
Stable isotopes	This technique makes use of stable isotopes of H, C and 0 and mass spectrometry. It can be utilized for quantification of substances in mixtures (by isotope dilution), in metabolic experiments, to detect food adulteration and to determine sources of lipids in geological sediments or ecological samples.
Enzyme degradation	Used to show nature of specific bonds and substituents in lipids. For example, the positional distribution of acyl groups can be determined or the presence of individual sugars or bases revealed. The stereo-specificity of most enzymes also allows the configuration of the target linkage to be demonstrated.

Of the various types of separation, thin layer and column chromatography are most useful for intact lipids. High performance liquid chromatography (HPLC) is becoming increasingly used. With the development of the evaporative light-scattering detector (sometimes called the 'universal' detector) its utility has been considerably enhanced and HPLC has the twin advantages that it is usually used at room temperature and that lipid oxidation is much reduced compared to thin-layer methods.

A powerful tool for quantitation of the majority of lipids is gas-liquid chromatography (GLC). The method is very sensitive and, if adapted with capillary columns, can provide information with regard to such subtle features as the position or configuration of substituents along acyl chains. By using light-scattering detectors with HPLC, this method can also be made quantitative. By coupling GLC or HPLC to a radioactivity detector, then the techniques are also very useful for metabolic measurements.

Although research laboratories use generally sophisticated analytical methods such as GLC to analyse and quantify lipid samples, chemical derivatizations are often used in hospitals. For these methods, the lipid samples are derivatized to yield a product which can be measured simply and accurately - usually by colour. Thus, total triacylglycerol, cholesterol or phospholipid-phosphorus can be quantified conveniently without bothering with the extra information of molecular species etc. which might be determined by more thorough analyses.

Chapter 2

Dietary Lipids

The main emphasis of this chapter is on lipids in the human diet, their metabolism in the healthy body and their relationship to disease processes. Lipid metabolism in other animals will be touched on, however, for two main reasons. First, a great deal of the food we eat is supplied by farm animals and the types of lipids we consume from this source are determined by the animals' own metabolism, their diets and by farming practices. Second, rigorously controlled studies of human metabolism are difficult and expensive to achieve and scientists are, therefore, frequently obliged to make inferences about the metabolic effects of lipids in man from research with experimental animals, generally laboratory rodents.

LIPIDS IN FOOD

Whereas food lipids originate from biological matter, they may be modified by processing before they are eaten.

The fats in foods are derived from the structural and storage fats of animals and plants

With few exceptions, all food originates from living matter; therefore the lipids in the diet of humans are the structural and storage lipids of animals and plants. Animal storage fats (mainly triacylglycerols with some dissolved cholesterol and fat-soluble vitamins) are present in meat that retains parts of the adipose tissue. Lard, for example, is a cooking fat derived from the adipose tissue of the pig. Milk fat occurs in full fat milk, cheese, cream, full fat yoghurt, butter and low fat dairy spreads. Plant storage fats (mainly triacylglycerols with some fat soluble vitamins) are present in their original form in nuts (e.g. peanuts), cereal grains and some fruits such as the avocado. They are also familiar in cooking oils, salad oils and mayonnaise. Both animal and plant storage fats are used in the manufacture of spreading fat. Eggs are rich sources of fat. The lipid is present as lipoproteins in which triacylglycerols, cholesterol and phospholipids are associated together with proteins. The lipoproteins are either in a high-density form ('lipovitellin', 78% protein, 12% phospholipid, 9% triacylglycerol) or a low-density form (18% protein, 22% phospholipid, 58% triacylglycerol). They are somewhat analogous to the plasma lipoproteins.

The predominance of fat in foods like cooking oils, margarines, butter and meat fat is obvious: we call these the visible fats. Where fat is incorporated into the structure of the food, either naturally (e.g. in peanuts or cheese), or when it is added in the cooking or manufacturing process (e.g. in cakes, pastries, meats), it is less obvious to the consumer and we call these hidden fats.

Because most food derives from biological tissues and because all tissues contain biological membranes, all foods contain, to a greater or lesser extent, structural fats. These are predominantly phospholipids and glycolipids with, in animal tissues, cholesterol and in plant tissues, the plant sterols. Thus, the lean part of meat contains the structural lipids, phospholipids and cholesterol of the muscle membranes and minor quantities of glycolipids. Consumption of tissues like brain would introduce larger quantities of animal sphingolipids into the diet. Dairy products contribute small amounts of structural lipids because of the presence of the milk fat globule membrane. The structural lipids of plants that are important in the diet are those in green leafy vegetables, which are predominantly galactolipids and the membrane lipids of cereal grains, vegetables and fruits.

The fatty acid composition of dietary lipids depends on the relative contributions of animal and plant structural or storage lipids

The most significant aspect of lipids in the human diet is their content of different types of fatty acids: saturated, monounsaturated and polyunsaturated. It is important to mention at the outset that all natural fats contain complex mixtures of all three types of fatty acids. It is incorrect, therefore, to describe a fat as saturated or polyunsaturated: only the constituent fatty acids can be so described. Yet fats in which the saturated acids form the largest single fraction are frequently categorized as saturated fats to contrast them with fats in which polyunsaturated fatty acids (PUFA) predominate. As an example of where this can be misleading to a layman, pig fat (lard) is frequently categorized as a saturated fat, yet over half (about 55%) of the fatty acids are unsaturated. Similarly, among vegetable oils, palm oil is frequently categorized as 'saturated' even though 50% of its fatty acids are unsaturated.

In general, the animal storage fats are characterized by a predominance of saturated and monounsaturated fatty acids. Storage fats from ruminant animals tend to have a greater proportion of saturated fatty acids because of the extensive hydrogenation of the animal's dietary fats in the rumen. Nevertheless, ruminant fats contain a substantial proportion of monounsaturated fatty acids because of the presence in mammary gland and in adipose tissue of desaturases. Milk fat is unusual among animal fats in containing significant amounts of saturated fatty acids with chain lengths of 12 carbon atoms or less. The proportions of 4:0, 6:0, 8:0, 10:0 and 12:0 are highly dependent on species (Table). These acids are synthesized in the mammary gland itself. The fatty acid composition of simple-stomached animals is more dependent on their diet since they do not extensively hydrogenate the dietary unsaturated fatty acids, but can incorporate them directly into the adipose tissue. Thus pig adipose tissue can be significantly enriched in linoleic acid, simply by feeding the animals rations supplemented with soybean oil. Even ruminant fats can be enriched with polyunsaturated fatty acids, but in order to do so the dietary PUFA have to be protected from hydrogenation in the rumen. This can be achieved by treating the oilseeds used to supplement the feed with formaldehyde, which cross links the protein and renders the seed oil unavail able to the rumen micro-organisms. When the food reaches the acid environment of the true stomach, the cross-links are digested and the PUFA are available for absorption just as in simple-stom ached animals.

Such feeding methods are theoretically useful for modifying the fatty acid composition of food products, if higher contents of polyunsaturated fatty acids are required. Polyunsaturated ruminant products are not generally available, however: they are expensive and subject to oxidative deterioration, which may result in poor taste and appearance. The wide range of fatty acid compositions that can occur in animal storage fats and the changes in PUFA content that can be achieved by dietary manipulation. Examples of animal storage fats naturally rich in PUFA of the n-3 family are those of many species of fish and marine animals.

Seed oils contain a wide variety of fatty acids, the composition of which is characteristic of the family to which the plant belongs. Generally one fatty acid predominates. It may either be one of the normal fatty acids, palmitic, oleic, or linoleic, as exemplified by palm oil, olive oil and sunflower seed oil or it may be an unusual acid, for example erucic acid in older varieties of rapeseed oil. Coconut oil and palm kernel oil are unusual among seed oils in having a preponderance of medium chain length saturated fatty acids. It is therefore an unjustified generalization to characterize all vegetable oils as unsaturated.

The fatty acid composition of the structural lipids of the animal tissues used as food is remarkably uniform irrespective of the animal's diet or whether it is simple-stomached or a ruminant (Table 2.1), although there are clearly some species differences, and diet can influence membrane composition in subtle ways. An important PUFA in animal structural lipids is arachidonic acid, a member of the n-6 family, and meat provides most of our dietary supply of this fatty acid. The flesh of fish, however, contains relatively less arachidonic acid and more of the longer chain PUFA of the *n*-3 family. Plant structural fats are similarly dominated by PUFA and supply mainly (α-linolenic acid, a member of the *n*-3 family.

Table 2.1 The fatty acid composition of some structural lipids important in foods

	Fatty acid (g per 100 g total fatty acids)*								
	16:0	16:1	18:0	18:1	18:2	18:3	20:4	LCPUFA[a]	total
Foodstuff									
Beef: muscle	16	2	11	20	26	1	13		89
Lamb: muscle	22	2	13	30	18	4	7		96
Lamb: brain	22	1	18	28	1		4	14	88
Chicken: muscle	23	6	12	33	18	1	6		99
Chicken: liver	25	3	17	26	15	1	6		99
Chicken: egg yolk	29	4	9	43	11				96
Pork: muscle	19	2	12	19	26		8		86
Green leaves	13	3		7	16	56			95

[a] Longer chain polyunsaturates, mainly 20:5*n*-3 and 22:6*n*-3.

* The figures are rounded to the nearest whole number. Therefore the content of an individual fatty acid of less than 0.5% is not recorded. The figures across a row rarely sum to 100% because each sample contains a large number of components each contributing less than 1%. Thus the samples for ruminants contain a wide variety of odd and branched-chain fatty acids.

Industrial processing influences

The fatty acid composition of animal food lipids can be influenced by animal feeding practices as discussed. Likewise, the fatty acid composition of plant fats can be modified by breeding. The best example of this is the change that was brought about by breeding the erucic acid out of rapeseed in exchange for its metabolic precursor, oleic acid. By far the greatest influence on food fatty acid composition, however, is industrial processing.

Catalytic hydrogenation

The objective of industrial hydrogenation is to reduce the degree of unsaturation in fatty acids. There are two main reasons for this. The first is to improve oxidative stability. The more highly unsaturated a fat, the greater is its susceptibility to oxidation, resulting in poor flavour and colour and even generating toxic compounds.

The second is to improve the physical properties of the fat. Highly unsaturated oils have low melting points and as such are unsuitable for many food uses where a solid texture is desired. Industrial hydrogenation (hardening) raises the melting point, thereby giving better textural properties to many food fats. By careful choice of catalyst and temperature, the oil can be selectively hydrogenated so as to achieve the desired characteristics. Indeed, the process is seldom taken to completion, since fully saturated fats, especially those that would be derived from the very long chain fatty acids of fish oils, would have melting points that were too high.

Table 2.2 Changes in the fatty acid composition of a vegetable oil during hydrogenation (g per 100 g total fatty acids)

Fatty acid		Soybean oil	Hydrogenated soybean oil
16:0		11	11
18:0		4	7
18:1	*cis*	22	33
	trans	0	12
18:2	*cis*-9, *cis*-12	54	22
	cis-9, *trans*-12 / *trans*-9, *cis*-12	0	8
	cis-9, *trans*-12 / *trans*-8, *cis*-12	0	4
18:3	all isomers	8	2

Hydrogenation is carried out in an enclosed tank in the presence of 0.05-0.20% of a finely powdered catalyst (usually nickel) at temperatures up to 180°C after which all traces of the catalyst are removed by filtration. Chemically, there are three main results of hydrogenation:

(1) the total number of double bonds in the fatty acid molecule is reduced;

(2) some of the double bonds that are present in the original oil in the *cis* geometric configuration are isomerized to the *trans* form; and

(3) the double bonds may be shifted along the hydrocarbon chain from their original positions in the natural fat to yield a range of positional isomers (Table 2.1).

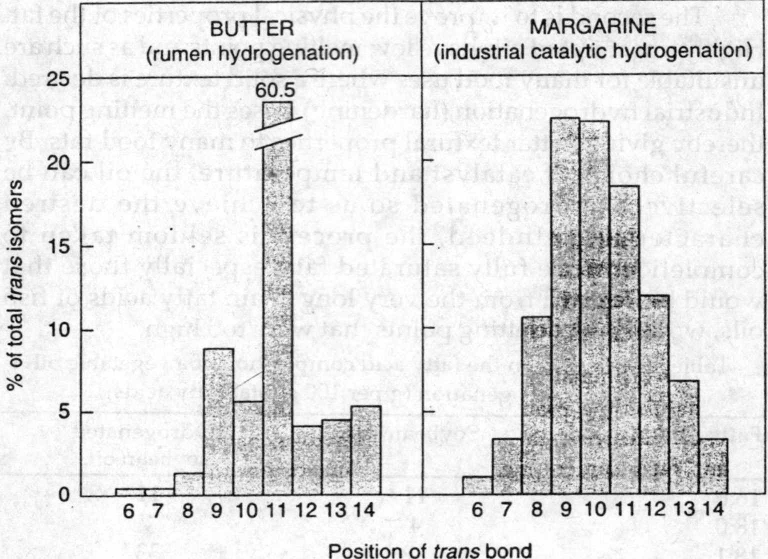

Fig. 2.1 The positional distribution of *trans* double bonds in butter and margarine. Note the sharp peak at position 11 for butter, resulting from biohydrogeneration by anaerobic rumen bacteria, compared with the broad distribution in the industrially hydrogenated product.

It is important to note that *trans* double bonds do occur in natural fats, as well as in industrially processed fats, but generally much less abundantly than *cis* bonds. Thus some seed oils have a significant content of fatty acids with *trans* unsaturation although these are not important sources of fat in the human diet. All green plants contain small quantities of *trans*-3-hexadecenoic acid. The most important naturally occurring isomeric fatty acids in human foods are those in ruminant fats. In principle, the chemical outcome of rumen biohydrogenation is similar to that of industrial hydrogenation. However, whereas in ruminant fats there is a preponderance of the monoene with the *trans* bond at position 11 (*trans*-vaccenic acid), in industrially hydrogenated fats, monoenes with double bonds at positions 8, 9, 10, 11 and 12 make similar contributions (Fig. 2.1). Whereas *trans*-monoenes usually predominate, PUFA with *trans* unsaturation also occur in both natural and processed fats and these can have *cis* and *trans* bonds within the same molecule. Since the early 1990s, interest has been growing in a group of such isomers given the general description 'conjugated linoleic acid' (CLA). It was discovered that some components of the mixture, notably the *cis*-9, *trans*-11 isomer, could inhibit the growth of cancer cells in vivo or in vitro.

The fatty acid structural changes that occur as a result of industrial hydrogenation may have nutritional significance.

Heating

Either industrially or in the home, heating may be used to make food palatable and to make it safe by killing micro-organisms. Risk of reduction of heatlabile and oxygen-sensitive nutrients requires a balance to be struck between microbiological safety and nutritive value. Heating fat that has little contact with air (e.g. in a deep fat fryer) gives rise to gradual accumulation of polymeric products. The first substances to be formed are cyclic monomers of triacylglycerols followed by polymeric products. As long as oils are not reused excessively, such polymeric material does not significantly reduce the functional properties of the oil, nor is it digested and absorbed.

More important changes in fats are brought about by heating in the presence of oxygen and trace metal catalysts (e.g. iron, copper). The major pathway is attack by reactive oxygen radicals at the double bonds of unsaturated fatty acids with the initial formation of a lipid peroxide. The more highly unsaturated the fatty acid, the greater the susceptibility to oxidation. If allowed to go too far, lipid peroxidation can significantly reduce concentrations of essential fatty acids, vitamin A and vitamin E in foods and may also result in damage to proteins (including enzymes) and DNA. Some products may be carcinogenic. Lipid peroxidation may be held in check by the presence of lipid-soluble antioxidants, either natural (e.g. vitamin E, carotenoids) or synthetic (butylated hydroxytoluene, BHT, or butylated hydroxyanisole, BHA). Cholesterol in foods can also auto-oxidize, a process that is accelerated in heating or drying. Thus, 25-dihydroxycholesterol, 7 -ketocholesterol, 7-hydroxycholesterol, 5- and 7-hydroperoxycholesterol and cholesterol-3,5,6-triol have been detected in products such as dried egg powder. These components are known to initiate arterial disease in rabbits when they are given in low quantities in the diet, where cholesterol itself has no effect.

Irradiation

Irradiation, used to kill pathogens in some types of foods, may generate lipid radicals and thus cause peroxidative damage. Vitamins E and K are particularly susceptible to irradiation damage but not, surprisingly, carotenes. In practice, however, very little direct nutritional damage occurs as a result of food irradiation because of its limited applications.

Interesterification

The process of interesterification involves treatment of a fat or fat blend with a catalyst (e.g. sodium methoxide) at an elevated temperature so as to randomize the fatty acids among the triacylglycerol molecules, as distinct from their natural stereo-specific distribution. Interesterificationis used to change physical properties of lipids (crystal structure, melting point) resulting in a changed food texture. The process is now being used more extensively as an alternative to catalytic hydrogenation because of concerns about potential adverse metabolic effects of overconsuming *trans*-unsaturated fatty acids. Some evidence has been presented that interesterified fats are metabolized differently from their natural counterparts. For example, work in the 1960s showed that butter that had been subjected to interesterification to randomize its fatty acids had little influence on the concentration of cholesterol in plasma when given in the diet, in contrast to the generally observed cholesterol-raising effect of natural butter. However, these and other similar observations have not been confirmed by more recent research.

Fractionation

All natural fats and oils are mixtures of triacylglycerols

with different fatty acid distributions and it is possible to effect partial or almost complete separations (fractionation) with various physical methods. Thus fractions enriched in more highly saturated or unsaturated molecular species can be obtained, yielding products with desirable physical, textural and metabolic properties. For example, fractions of triacylglycerols enriched in medium-chain fatty acids (8:0, 10:0) can be obtained from coconut oil. These find clinical use for patients with fat malabsorption. A recent development has been the production of spreadable butters by selective removal of higher melting triacylglycerols. Reduced cholesterol foods can now be produced by differential extraction with, for example, carbon dioxide. (Alternatively, the same result has also been achieved by treatment with cholesterol degrading enzymes.)

Structured fats

'Structured fats' is the name now generally given to semi-synthetic mixed-acid triacylglycerolscontaining both very long and very short chain fatty acids. It is proposed that they may find applications in reduced energy foods. There are two main principles here. First, the short-chain acids have an intrinsically lower energy value than the longer chain fatty acids and are, in any case, metabolized in the liver in such a way that they are not deposited in adipose tissue. Second, the very long chain acids are less efficiently absorbed. The combination of reduced gross energy and poorer absorption reduces the overall metabolizable energy of the fat. This is a relatively new concept and there is as yet no convincing evidence that such products contribute effectively to weight-reducing diets in the long term. Another type of structured triacylglycerol, which contains palmitic acid at position *sn*-2 to mimic the structure of a major molecular species in human milk triacylglycerols, has been introduced commercially for use in infant feeds. Molecules with this structure are digested and absorbed more efficiently by the human infant.

Other products that are aimed at reducing dietary energy, including sucrose polyesters, are also now approved in the USA for limited food use. These are not triacylglycerols,but have fat-like texture. They are not attacked by pancreatic lipase and so are not absorbed and contribute no energy value to the diet. Much discussion has taken place about potentially harmful effects of ingesting such compounds. These may include significant reductions in the absorption of fat-soluble vitainins (which can be overcome by fortification) or gastrointestinal discomfort and fatty stools.

Cyclopropenes

Of the fatty acids, those containing a cyclopropene ring have been considered toxic as a result of their ability to inhibit the Δ9-desaturase. One result of this is to alter membrane permeability as seen in 'pink-white disease'. If cyclopropene fatty acids are present in the diet of laying hens, the permeability of the membrane surrounding the yolk is increased, allowing release of pigments into the yolk. Rats given diets containing 5% of dietary energy as sterculic acid died within a few weeks and at the 2% level, the reproductive performance of females was completely inhibited. Cottonseed oil is the only important oil in the human diet that contains cyclopropene fatty acids. However, their concentration in the natural oil is low (0.6-1.2%) and is reduced still further to harmless levels (0.1-0.5%) by processing. There has been no evidence that consumption of cottonseed oil in manufactured products has had any adverse nutritional effects.

Long-chain monoenes

The long-chain monoenoic acid, erucic acid is present in high concentration (up to 45% of total fatty acids) in the seed oil of older varieties of rape *(Brassica napus)*. When young rats were given diets containing more than 5% of dietary energy as high erucic rapeseed oil, their heart muscle became infiltrated with triacylglycerols. After about a week on the diet, the hearts contained three to four times as much lipid as normal hearts and although, with continued feeding, the size of the fat deposits (referred to as a lipidosis) decreased, other pathological changes were noticeable. These included the formation of fibrous tissue in the heart muscle. The biochemistry of the heart muscle was also affected. Mitochondrial oxidation of substrates, such as glutamate, was reduced and the rate of ATP biosynthesis was impaired. The degradation of triacylglycerols that contain erucic acid is somewhat slower than with fatty acids of normal chain length and this may have contributed to the accumulation of lipid deposits.

Despite lack of evidence for harmful effects in man, an extensive breeding programme was undertaken to replace older varieties of rape with new 'zero-erucic' varieties. The use of low erucic varieties in manufactured products is now mandatory in most industrial countries, but high erucic varieties continue to be used in countries such as China.

High concentrations of long-chain monoenoic fatty acids also occur in some fish oils and therefore contribute to the diets of people consuming these fish, as well as to diets containing certain fat spreads that incorporate hardened fish oils. The nutritional and toxicological consequences of long-term consumption of marine long-chain monoenes have been less extensively studied than have those of high erucic acid-containing rapeseed oil.

Trans-*unsaturated fatty acids*

Concerns about possible toxic effects of fatty acids with *trans*-unsaturation began with the publication of results of experiments with pigs given diets containing hydrogenated vegetable fat for 8 months. They had more extensive arterial disease than those given otherwise equivalent diets devoid of *trans*-unsaturated fatty acids. Subsequently, numerous animal feeding trials, epidemiological studies of human populations and controlled dietary experiments with human subjects have been reported. Some, but not all, epidemiological studies have shown an apparent statistically significant correlation between intkes of certain types of *trans* unsaturated fatty acids and increased risk of coronary heart disease. Critical evaluation of all such studies leads to the conclusion that an association between *trans*-unsaturated fatty acid consumption and coronary heart disease has not been demonstrated; nor is there any reliable evidence linking *trans*-unsaturated fatty acid intake with cancer or other chronic diseases.

A theoretical basis for some of the claimed effects of *trans*-unsaturated fatty acids on heart disease has been discussed in terms of their influence on plasma lipoproteins and on components of the blood-clotting system. While there is no conclusive evidence for the latter, there is an effect of *trans*unsaturated fatty acids in raising LDL-cholesterol and decreasing HDL-cholesterol concentrations in the blood.

High concentrations of *trans*-unsaturated fatty acids may adversely affect the metabolism of the parent essential fatty acids, linoleic and (α-linolenic acids to their long-chain metabolites by inhibiting desaturase activity or competing with normal substrates for desaturases. In this way, they may raise the dietary requirement for essential fatty acids as explained later. This kind of 'nutritional imbalance' must be distinguished, however, from direct and specific toxic effects. It is worth noting that the hydrogenated fats used in many experiments designed to investigate potential adverse effects of dietary *trans*-unsaturated fatty acids, contain many unusual *cis* as well as *trans* isomers and their potential toxicity has been less extensively researched.

Lipid peroxides

Lipid peroxides may be formed from poly unsaturated fatty acids when they are in contact with oxygen. The reaction is accelerated in the presence of catalysts such as transition metal ions or haem compounds and by heating. When lipid hydroperoxides are ingested, they are rapidly metabolized in the mucosal cells of the small intestine to various oxyacids that are then rapidly oxidized to carbon dioxide. High concentrations of lipid peroxides in the gut may damage the mucosa and potentiate the growth of tumours. However, there is little evidence for the absorption of unchanged hydroperoxides nor for their incorporation into tissue lipids. Hydroperoxyalkenals, lower molecular weight breakdown products of lipid hydroperoxides, are absorbed and may be toxic. Rats given diets enriched in these compounds manifested increased liver weight, increased concentrations of malondialdehyde, peroxides and other carbonyl compounds and decreased concentrations of (α-tocopherol and linoleic acid in tissues.

ROLES OF DIETARY LIPIDS

Lipids contribute to intakes of energy, structural components of cells, essential fatty acids and fat-soluble vitamins as well as enhancing the palatability of foods.

Triacylglycerols provide a major source of metabolic energy especially in affluent countries

More than 90% of dietary lipids are triacylglycerols, which originate from the adipose tissue or milk of animals or from plant seed oils, mainly in the form of manufactured products. Some 35-45% of dietary energy in industrialized countries comes from triacylglycerols. World-wide, however, the contribution that fat makes to the diet can vary enormously. For example, the Ho tribe in India makes little or no use of fat and their fat intake has been estimated to be no more than 2-4 g day or 2% of dietary energy. Certain Inuit communities, by contrast, may consume as much as 80% of their energy as fat. Whereas the presence of fat can contribute substantially to the palatability of the diet, the main nutritional contribution of triacylglycerols is to supply metabolic energy. The useful energy available to the body (metabolizable energy, ME) depends mainly on the digestibility of the fat and the chain length of the constituent fatty acids. Most common food fats are efficiently digested by most people with insignificant differences in their digestible energies. As discussed some practical use has been made of differences in chain length as reflected in their metabolism. Useful metabolic energy is derived from triacylglycerols mainly through β-oxidation of the constituent fatty acids. At each step of the cycle, acetyl-CoA is generated and fed into the triocarboxylic acid cycle, where the end-products of oxidation are carbon dioxide and water. The reduced pyridine nucleotides generated in the triocarboxylic acid cycle enter the mitochondrial electron transport chain to fuel the generation of ATP as a source of metabolic energy. One mole of palmitic acid (16:0) yields 129 moles of ATP (Table 2.3).

Table 2.3 The A TP yield of palmitate oxidation

Reaction	ATP yield
A. Beta-oxidation	
$CH_3(CH_2)_{14}COSCoA + 7FAD + 7I\text{-}H_2O + 7NAD^+ + 7CoA$	
$\longrightarrow 7FADH_2 + 7NADH + 7H^+ + 8CH_3COSCoA$	
(B) (C) (D)	
B. Oxidation of $FADH_2$ in the mitochondrial electron transport chain	
$14H^+ + 7FADH_2 + 14Pi + 14ADP + 3.5O_2$	
$\longrightarrow 7FAD + 21H_2O + 14ATP$	14
C. Oxidation of NADH in the mitochondrial electron transport chain	
$7NADH + 28H^+ + 21Pi + 21ADP + 3.5O_2$	
$\longrightarrow 7NAD^+ + 28H_2O + 21ATP$	21
D. Oxidation of acetyl-CoA in the tricarboxylic acid cycle	
$96H^+ + 8CH_3COSCoA + 16O_2 + 96Pi + 96ADP$	
$\longrightarrow 8CoA + 104H_2O + 16CO_2 + 96ATP$	96
Overall equation	
$131H^+ + CH_3(CH_2)_{14}COSCoA + 131Pi + 131ADP + 23O_2$	
$\longrightarrow 131ATP + 16CO_2 + 146H_2O + CoA$	131
Subtract 2ATP required for initial activation of palmitate	-2
(formation of CoA thiolester)	129

Components Supplied by Lipids

Some lipids that are needed for tissue growth and development can be synthesized by the cells themselves. However, the fatty acids linoleic (18:2 *n*-6) and α-linolenic (18:3*n*-3) cannot be made by animal cells and these so-called 'essential fatty acids' must be supplied by the diet. Once incorporated into cells they may be further elongated and desaturated to longer chain polyunsaturated products. Will describe the nature and function of essential fatty acids and their dietary requirements, but first it will be useful to discuss more general needs for lipids in growth.

Foetal growth

From conception, the cells of the growing foetus need to incorporate lipids into their rapidly proliferating membranes.

The foetus is totally dependent on the placental transfer of substrates from the mother's circulation, which are then elaborated into lipids. Glucose is a major substrate for the foetus, which possesses the enzymic activities for its conversion into (non-essential) fatty acids and for the production of the glycerol moiety of glycerides via glycerol phosphate. A crucial question in developmental biology is how essential fatty acids are acquired and metabolized at this early stage.

Placental transfer of fatty acids

Figure 2.2 illustrates several pathways by which the foetus could obtain the polyunsaturated fatty acids that it requires for growth. The placenta of most mammals, including human beings, is permeable to non-esterified fatty acids although there are large differences between animal species in rates of transfer. There are also large species differences in the lipid biosynthetic capacity of the placenta and of foetal tissues, so that generalizations cannot be made from research on any one species. It is well established that the amounts of 18:2*n*-6 and 18:3*n*-3 and of their elongation/desaturation products in the mother's diet influences the fatty acid composition of the developing foetus. Desaturation activity is present in developing foetal brain and liver (Fig. 2.2 reaction 7), so that the foetus may not be totally dependent on a supply of long-chain PUFA (LCPUFA) from the mother (Fig. 2.2 reaction 6). There is, however, little evidence for placental desaturase activity in simple-stomached animals, so that reaction 5 in Fig. 2.2 is unlikely to provide significant amounts of LCPUFA.

Concentrations of all lipoprotein classes increase in the maternal circulation during pregnancy, a process that is mediated by the sex hormones. A lipoprotein lipase has been discovered in human placenta. This is consistent with the hypothesis that during hyperlipidaemia of pregnancy, release of fatty acids by placental lipoprotein lipase could generate substrates for synthesis of lipids by the foetus. The transfer of intact lipoproteins by a receptor-mediated pathway has not been demonstrated but cannot be ruled out. Placental lipoprotein lipase is present on the maternal, but not on the foetal,

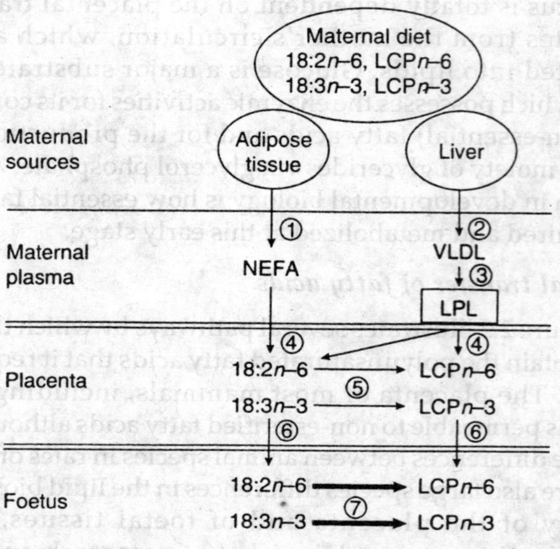

Fig. 2.2 Origins and Metabolism of polyunsaturated fatty acids in the foetus. The maternal diet contains a mixture of n–6 and n–3 parent and long-chain polyunsaturated fatty acids (LCP), which will vary widely among species and individuals. After digestion and absorption of the dietary lipids, the re-esterified fatty acids are transported inthe maternal plasma as lipoproteins and incorporated into maternal tissues, e.g. adipose tissue and liver. Non-esterified fatty acids (NEFA) are released from adipose tissue stores (stage 1 in the figure). The liver exports fatty acids as triacylglycerols in very low density lipoproteins (VLDL; stage 2 in figure) into the maternal plasma. The maternal face of the placenta contains a lipoprotein lipase (LPL) (stage 3), which releases NEFA from VLDL. The placenta is permeable to NEFA. The human placenta contains minimal desaturase activity and stage 5 is probably not a significant source of LCP. Parent polyunsaturates and LCP are transported to the foetal circulation (stage 6) and the foetus can convert a proportion of the parent polyunsaturates into LCP (stage 7).

side of the placenta and hydrolyses very low density lipoproteins, but apparently not chylomicrons. Perhaps this constitutes a mechanism for ensuring a flow of fatty acids from mother to foetus, rather than in the reverse direction. It may also allow the mother to control the fatty acids available to the placenta since VLDL are supplied by synthesis in the mother's liver, whereas chylomicrons derive from dietary fat.

In simple-stomached animals, arachidonic and some other LCPUFA are present in higher concentration in foetal than in maternal plasma and in foetal tissues than in foetal plasma. Arachidonic acid seems to be selectively incorporated and trapped into placental phospholipids for export to the foetal circulation. The term 'biomagnification' has been coined for a process in which the proportion of long-chain polyunsaturates increases in phospholipids progressing from maternal blood, to cord blood, foetal liver and foetal brain. Such a biomagnification process probably incorporates the combined effects of placental fatty acid uptake, selective protection of specific fatty acids against β–oxidation and selective direction of fatty acids into membrane or storage lipid biosynthesis.

The study of foetal conservation of essential fatty acids in ruminants is extremely rewarding since the availability of essential fatty acids to the mother is severely limited by rumen hydrogenation. The ratio of 20:3*n*-9/20:4*n*-6 in foetal lamb tissues is 1.6. In simple-stomached animals, a value of this ratio above 0.4 is a biochemical marker of essential fatty acid deficiency. By 10 days after birth, the ratio falls to 0.4 and by 30 days to 0.1. These values are well within the normal range despite the extremely low concentration of linoleic acid in ewe's milk (0.5% of energy). Ruminants are, therefore, able to conserve linoleic acid with supreme efficiency. Sheep placenta transfers linoleic acid at a relatively slow rate, but has a very high Δ6-desaturase activity by comparison with non-ruminants, providing the major source of arachidonic acid for the foetus. This metabolite is concentrated into glycerophospholipids whereas the linoleic acid precursor is in higher concentration in the triacylglycerols. This molecular compartmentation has the effect of conserving arachidonic acid and directing it into membranes.

Storage in adipose tissue

In some species, reserves of fatty acids are already being built up in the foetus. For example, the development of human fat cells begins in the last third of gestation and at birth a baby weighing 3.5 kg has, on average, 560 g of adipose tissue. Guinea pigs are also born with a large amount of adipose tissue whereas, in contrast, pigs, cats and rats are born with little or none. There is a remarkable parallelism between the placental permeability to non-esterified fatty acids and the tendency to accumulate adipose tissue by the foetus, suggesting that an important source of foetal fat reserves is derived from circulating maternal lipids. The adipose tissue composition of the foetal and new-born fat will, therefore, reflect the fatty acid composition of the maternal diet as illustrated in Table 2.4.

Brain development

A large proportion of the long-chain polyunsaturated fatty acids synthesized or accumulated during the perinatal period is destined for the growth of the brain, of which 50% of the acyl groups may consist of 20:4*n*-6, 22:4*n*-6, 22:5*n*-6 and 22:6*n*-3. As with adipose tissue, there are large species differences in the time at which birth occurs in relation to the extent of brain development. The peak rate of brain development occurs in guinea pigs in foetal life; in the rat, post-natally; while in man and pig it reaches a peak in late gestation and continues after birth. It has been suggested that transfer from the placenta is the major source of LCPUFA for the human foetal brain, but the experimental difficulties of demonstrating this in man are enormous. Because of their remarkable similarity to man in the timing of brain development and its lipid composition, pigs have been used as models. Long-chain derivatives of linoleic acid increase in brain from mid-gestation to term. Little linoleic acid, however, accumulates until birth, when its concentration

Table 2.4 Influence of the fatty acid composition of the mother's dietary fat on the fatty acid composition of adipose tissue in new-born guinea-pigs (g per 100 g total fatty acids)

	Maize oil		Beef tallow	
Fatty acid	Maternal diet	Adipose tissue	Maternal diet	Adipose tissue
16:0	13	23	25	26
18:0	3	6	18	9
18:1	26	28	36	42
18:2	54	33	7	16
18:3	2	5	2	6
Others	2	5	12	1

increases threefold while the concentrations of its elongation/ when its concentration increases threefold while the desaturation products remain constant. By labelling experiments with [1-^{14}C]linoleic acid in vivo, it has been shown that linoleic acid is metabolized to LCPUFA by piglet brain and liver throughout the perinatal period. The contribution of the liver was many-fold greater than the brain at all stages. Whether foetal tissues can supply all the needs of the nervous system for LCPUFA without the need for the maternal transfer of intact LCPUFA is still to be resolved. In regard to maternal transfer, it has been proposed that an intracellular a-fetoprotein, derived from maternal plasma a-fetoprotein may play a role in delivering LCPUFA to foetal brain. It binds fatty acids avidly, especially the LCPUFA such as 20:4*n*-6 and 22:6*n*-3. Indeed plasma non-esterified 22:6*n*-3 seems to be almost exclusively transported on a-fetoprotein despite the quantitative predominance of albumin.

An outstanding feature of the composition of brain phospholipid is its remarkable consistency, irrespective of species or diet. The concentrations of the parent essential fatty acids are extremely low (18:2*n*-6, 0.1-1.5% and 18:3*n*-3, 0.1-1.0%) while arachidonic (20:4*n*-6) and docosahexaenoic (22:6*n*3) acids predominate at 8-170% and 13-29%, respectively in all species (Table 2.5).

Liver lipids

In contrast to the conservative lipid composition of brain, the liver lipids exhibit much greater variation. The parent essential fatty acids are present in much greater concentrations than they are in brain and there are major differences in the elongation/ desaturation products. For example, 22:5 is the major *n*-3 fatty acid in the liver lipids of ruminants and other herbivores while 22:6*n*-3 predominates in the carnivores and omnivores. Fatty acids of the *n*-6 family usually predominate in liver phosphoglycerides, even when the dietary intake is in favour of the *n*-3 fatty acids. Thus, zebra and dolphin, both species that have an overwhelming excess of *n*-3 fatty acids in the diet, attain a preponderance of *n*-6 acids in the liver phosphoglycerides (Table 2.5).

Table 2.5 Principal polyunsaturated fatty acids of the liver and brain phospholipids of zebra and dolphin

	Zebra	Dolphin
Food fatty acids		
n-6/*n*-3	1:3	1:20
Liver fatty acids (g per 100 g total fatty acids)		
18:2*n*-6	47	2
20:4*n*-6	4	13
18:3*n*-3	2	
20:5*n*-3	1	8
22:6*n*-3		11
Brain fatty acids (g per 100 g total fatty acids)		
18:2*n*-6	1	
20:4*n*-6	10	7
18:3*n*-3		
20:5*n*-3		1
22:6*n*-3	18	27

The zebra's food polyunsaturates, both *n*-6 and *n*-3 are mainly 18C and the dolphin's 20C and 22C. Of the liver and brain fatty acids, only the principal polyunsaturates are shown for clarity. Figures are rounded to the nearest whole number so that components contributing less than 0.5% of total are not recorded. Note the preponderance of *n*-6 polyunsaturates in the liver lipids of both species despite the dietary excess of *n*-3 polyunsaturates and a preponderance of long-chain *n*-3 polyunsaturates in the brain, with virtually none of the parent acids (18:2*n*-6, 18:3*n*-3).

Post-natal growth

At birth, quite large changes in lipid metabolism occur. Whereas the foetus had relied extensively on glucose from which to synthesize (non-essential) fatty acids and triacylglycerols, the sole source of nutrition for the new-born is milk from which about 50% of energy comes from fat. The enzymes of fatty acid synthesis are suppressed and the baby's metabolism becomes geared to using fat directly from the diet.

Milk composition

The lipid composition of human milk is therefore the main factor that determines the availability of lipids for the breast-fed baby's development, especially in respect of the essential fatty acids. Triacylglycerols comprise about 98% of the total lipids in milk and provide most of the required linoleic and α-linolenic acids. Although the glycerophospholipids represent only about 1% of total milk lipids, they provide about 50% of the long-chain *n*-3 and *n*-6 PUFA in milk. The fatty acid composition of human milk (and indeed that of any simple stomached animal) is highly variable since it is strongly influenced by the fatty acid composition of the mother's diet. However, the milk and dietary fatty acid compositions are not related in a linear manner because the maternal adipose tissue stores also make a contribution to milk fatty acid composition. Table 2.6 illustrates this variability and also compares the composition of human milk with that of cow's milk, which provides the base for most commercial infant formulas. Women eating a vegetarian diet tend to have higher concentrations of 18:2*n*-6 and 18:3*n*-3 in their milk. Concentrations of 20:5*n*-3 and

Table 2.6 The fatty acid composition of human and cow's milk fat (g per 100 g total fatty acids)

	Cow's milk	Human milk	
		Vegans	Omnivores
Fatty acid			
4:0	3		
6:0	2		
8:0	1		
10:0	3		
12:0	4	4	3
14:0	12	7	8
16:0	26	16	28
16:1	3	1	4
18:0	11	5	11
18:1	28	31	35
18:2	2	32	7
18:3	1	1	1
20:4		1	1
Total	96	98	98

Figures are rounded to the nearest whole number. Note the high proportion of saturated fatty acids in cow's milk fat and especially those with short- and medium-chain lengths. The components not listed are mainly odd-chain and branched-chain fatty acids, individually in very small amounts. Note the high proportion of linoleic acid in the milk of human vegans, which mainly replaces palmitic acid. The components not listed are mainly longer chain polyunsaturated fatty acids of the *n*-3 and *n*-6 families each in very small amounts.

22:6*n*-3 are increased many-fold when the diet is supplemented with fish oils. The increase in *n*-3 LCPUFA is not, however, accompanied by a reduction in 20:4*n*-6.

The fatty acid composition of the milk feed given to newborn infants is quickly reflected in the composition of the adipose tissue, as illustrated in Table 2.7.

Brain development

A dietary supply of 18:2*n*-6, 18:3*n*-3 and their long chain metabolites is important because, as discussed earlier, human brain development continues after birth. Data on the sequential development of human brain and nervous system are difficult to collect, but such data as exist, together with information from pigs, suggest a preferential accumulation of *n*-6 fatty acids in early foetal development, followed by an accumulation of *n*-3 acids later in gestation. The deposition of myelin occurs late in the development of the central nervous system, predominantly after birth in human babies. The fatty acid composition of immature myelin reflects the composition of the plasma membrane of the oligodendrocyte from which it is made. The high lipid content of the myelin sheath and the rapid rate of myelin synthesis in early post-natal life put great demands on the supply of milk LCPUFA. Later, there is a marked decrease in the LCPUFA content of myelin and a concomitant increase in monoenoic and saturated fatty acids, thus reducing the demands on the milk PUFA supply.

Infant formulas traditionally contained cow's milk fat having a low concentration of essential fatty acids. A major development was replacement of the cow's milk fat by vegetable oils providing high intakes of 18:2*n*-6. More recent developments have been to supply a higher content of *n*-3 PUFA, but the optimal ratio of 18:3*n*-3 to its longer chain metabolites is still debated. Limited data suggest that a dietary supply of preformed long-chain *n*-3 PUFA results in significantly higher deposition of these acids in developing organs than when dietary 18:3*n*-3 alone is supplied and the practice of adding fish oils to infant formulas is becoming more widespread. Reasons for the superiority of the LCPUFA may be the significantly higher β-oxidation of 18:3*n*-3 compared with

LCPUFA and limited activity in the elongation/ desaturation pathway.

Table 2.7 Influence of the type of fat in infant formulas on adipose tissue composition of babies (g per 100 g total fatty acids)

	British[a]			Dutch[b]		
Fatty acid	Fat in formula	Adipose tissue		Fat in formula	Adipose tissue	
		Birth	6-12 months		Birth	6-12 months
14:0	11.5	3.8	8.5	0	3.3	4.8
16:0	30.0	48.9	32.0	10.7	48.5	29.7
18:0	14.3	4.1	5.6	2.0	3.8	2.9
16:1	2.0	12.6	7.0	0	15.2	13.8
18:1	31.1	29.6	43.4	27.2	29.0	40.8
18:2	1.8	1.0	3.5	58.2	2.9	8.0

[a] British babies were given a formula in which the fat component came from cow's miik.

[b] Dutch babies were given a formula in which the fat component was provided by a vegetable oil.

An adequate supply of LCPUFA in milk is even more important in infants born prematurely, since these acids would normally have been supplied to the foetus from the mother's blood or metabolized from parent acids in foetal tissues. There is limited evidence that giving formula enriched with fish oil improves the 22:6*n*-3 status and visual function of premature babies, but their 20:4*n*-6 status deteriorates. Thus, the optimum amounts of parental essential fatty acids and their elongation/ desaturation products, and the optimal ratios between *n*-3 and *n*-6 families remain crucial topics in paediatric nutrition research.Dietary lipids supply essential fatty acids that are essential to life but cannot be made in the animal body.

HISTORICAL BACKGROUND

In 1929 the Americans Burr and Burr described how acute deficiency states could be produced in rats by feeding fat-free diets and that these deficiencies could be eliminated by adding only certain specific fatty acids to the diet. Until that time it had been thought that fat was important only in so far as its energy contribution was concerned. It was shown that linoleic and later arachidonic acid were responsible for this effect and the term vitamin F was coined for them. They are now known universally as 'essential fatty acids' (EFA).

EFA deficiency can be produced in a variety of animals, including man, but data for the rat are the best documented (Table 2.8). The disease is characterized by skin symptoms, such as dermatosis and the skin becomes more 'leaky' to water. Growth is retarded, reproduction is impaired and there is degeneration or impairment of function in many organs of the body. Biochemically, EFA deficiency is characterized by changes in the fatty acid composition of many tissues, especially their biological membranes, whose function is impaired and, in the mitochondria, the efficiency of oxidative phosphorylation is much reduced. Well-documented EFA deficiency in man is rare, but was first seen in children fed virtually fat-free diets. Four hundred infants were fed milk formulas containing different amounts of linoleic acid. When the formulas contained less than 0.1% of the dietary energy as linoleic acid, clinical and chemical signs of EFA deficiency ensued. The skin abnormalities were very similar to those described in rats and these and other signs of EFA deficiency disappeared when linoleic acid was added to the diet.

In 1971, the first unequivocal case of EFA deficiency in an adult was reported. The patient, a man of 44, had all but 60 cm of his bowel surgically removed. He was then given intravenous feeding only, with preparations containing no fat, and after 100 days he developed scaly dermatitis. In a London hospital in 1974, three patients with chronic disease of the small bowel, who had been treated with low fat diets, but not given intravenous feeding, were suspected of having EFA deficiency. They responded successfully to the application to the skin of lipids containing a high proportion of linoleic acid, demonstrating that EFA need not necessarily be absorbed through the conventional route to be effective.

Biochemical basis for EFA deficiency

Once EFA deficiency had been demonstrated, the hunt was on to discover which fatty acids had EFA activity. The effects of different acids were first compared by measurement of growth

Table 2.8 Major effects of *n*-6 essential fatty acid deficiency in rats

Skin	Dermatosis, water permeability increased Sebum secretion decreased Epithelial hyperplasia
Body weight	Decreased
Circulation	Heart enlargement Capillary resistance decreased
Kidney	Enlargement; intertubular haemorrhage
Lung	Cholesterol accumulation
Endocrine glands	Adrenals: weight decreased in females, increased in males Thyroid: weight increased
Reproduction	Females: irregular oestrus; impaired lactation, reproduction Males: degeneration of the seminiferous tubules
Metabolism	Changes in tissue fatty acid composition (see Table 2.9) Reduced cholesterol concentration in plasma Increased cholesterol concentration in liver, adrenals and skin Mitochondrial swelling and uncoupled oxidative phosphorylation Increased triacylglycerol output by liver

response. Later, the team at the Unilever Laboratories in The Netherlands used a test based on disturbances in water metabolism, since an important result of EFA deficiency is an altered permeability of biological membranes. In this way they were able to rank fatty acids according to their 'EFA activity' and to correlate such activity with molecular structure. Linoleic (all-*cis*-9,12-18:2), γ-linolenic (all-*cis*-6,9,12-18:3) and arachidonic (all-CIS-5,8,11,14-20:4) acids (all in the *n*-6 family) had a similar order of activity whereas that of α-linolenic acid (all-*cis*-9,12,15-18:3, *n*-3) was much lower. Work with a wide range of synthetic unsaturated fatty acids demonstrated that virtually all acids with significant activity were polyunsaturated with a *cis, cis* methylene-interrupted double bond system.

An important observation that gave further clues about which fatty acids were 'essential' was that in the tissues of animals that developed EFA deficiency, there was an accumulation of a PUFA that was either present in extremely small concentration or undetectable. This fatty acid is all-*cis*-5,8,11-eicosatrienoic acid (20:3*n*-9). Among the first biochemists to appreciate its significance was the American, James Mead, and it is now frequently referred to as the 'Mead acid'. Since that time, the ratio of the concentration of the Mead acid to arachidonic acid in tissues or plasma phospholipids (the 'triene/tetraene ratio') has been used as a biochemical index of essential fatty acid deficiency. In health, the ratio is about 0.1 or less, rising to 1.0 in severe EFA deficiency. A ratio of 0.4 has been used conventionally as indicating EFA deficiency, although more recently some authors have argued that it might be prudent to revise this figure downwards to nearer 0.2. The relationship between the proportion of linoleic acid in the diet and the magnitude of the triene/tetraene ratio in plasma phospholipids is similar in all species and all tissues examined (Fig. 2.3). It is a sensitive indicator since it increases significantly before overt EFA deficiency signs are apparent.

The explanation for the elevation of the concentration of 20:3*n*-9 in EFA deficiency can be readily appreciated from Fig.2.4. During the course of evolution, animals lost the ability (retained by plants) to insert double bonds in positions 12 and 15. Each 'family' of unsaturated fatty acids (*n*-3, *n*-6, *n*-9) is biochemically distinct and its members cannot be interconverted in animal tissues. The 'parent' or 'precursor' fatty acid of the *n*-9 family, oleic acid, can originate from the diet or from biosynthesis in the body, whereas linoleic and α-linolenic acids, the precursors of the *n*-6 and *n*-3 families, respectively, must be obtained from the diet. They are thus essential nutrients. All three precursors can compete for desaturation by the same Δ6-desaturase enzyme as illustrated in Fig. 2.4. The affinity of the substrates for the Δ6-desaturase is in the order 18:3> 18:2> 18:1. In humans and many other mammals, quantitatively the most important metabolic pathway is the one in which linoleic acid is converted into arachidonic acid. Normally, the diet contains sufficient linoleic acid for this pathway to be able continually to supply the quantity of arachidonic acid needed by the membranes of body tissues. When the amount of linoleic acid in the diet is very low, however, the *n*-9 pathway takes precedence. Hence, there is a relative accumulation of the Mead acid and a relative depletion in arachidonic acid. In practice, although a relatively large proportional increase in 20:3*n*-9 may be observed early in the progression to EFA deficiency, noticeable depletion of arachidonic acid stores requires long-term consumption of diets containing little or no linoleic acid.

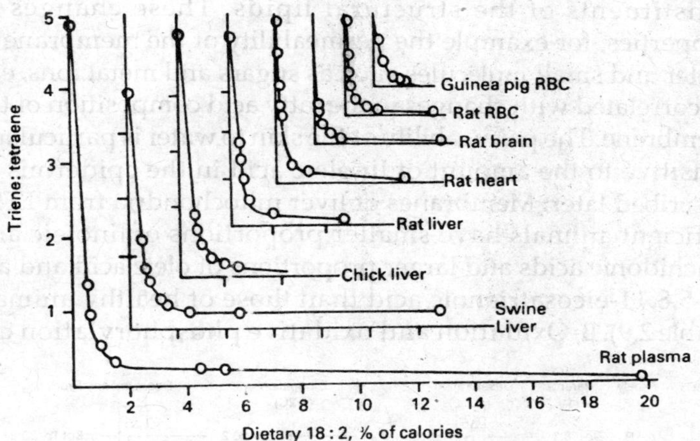

Fig. 2.3 Influence of dietary linoleic acid on the ratio of 20:3*n*-9 to 20:4*n*-6 (the triene/tetraene ratio) in different species. The relative scales are similar in all plots; the inset plots indicate the similarity in the relationship over a wide range of tissues and species.

It will be clear from Fig. 2.4. that there is also potential for competition for the Δ6-desaturase between 18:2*n*-6 and 18:3*n*-3 as well as the competition between the *n*-6 and *n*-9 substrates discussed above. Furthermore, there may also be competition at the level of the Δ5-desaturase, although this has been less well researched. Notwithstanding the high affinity of α-linolenic acid for the Δ6-desaturase, the flux through the *n*-3 pathway is generally low because of the small proportion of α-linolenic acid in most diets compared with linoleic acid.

FUNCTIONS OF ESSENTIAL FATTY ACIDS

Two distinct roles, to account for their essentiality, have been discussed for the EFA. These are as membrane components and as precursors for biologically active metabolites, the eicosanoids. Each will be discussed in turn, followed by a discussion of the links between the two roles.

Essential fatty acids in membranes

Among the results of EFA deficiency are changes in the properties of biological membranes of which PUFA are major constituents of the structural lipids. These changes in properties, for example the permeability of the membrane to water and small molecules such as sugars and metal ions, can be correlated with changes in the fatty acid composition of the membrane. The permeability of the skin to water is particularly sensitive to the amount of linoleic acid in the epidermis as described later. Membranes of liver mitochondria from EFA-deficient animals have smaller proportions of linoleic and arachidonic acids and larger proportions of oleic acid and all-*cis*-5,8,11-eicosatrienoic acid than those of healthy animals (Table 2.9). β-Oxidation and oxidative phosphorylation are

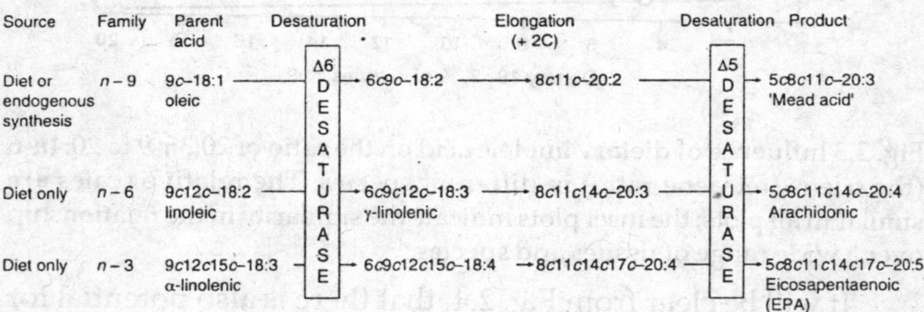

Fig. 2.4 Principal metabolic pathwas for desaturation and elongation of 'parent acids' to long-chain polyunsaturated fatty acids in the *n*-9, *n*-6 and *n*-3 families. A minor family, *n*-7 (whose 'parent' acid is palmitoleic acid, 9*c*-16:1) has been omitted for simplicity. As far as we know it has little nutritional significance. The figure aims mainly to illustrate the sequence

of alternate desaturations and elongations of the parent acids in each family and the competition between families for the Δ5 and Δ6-desaturases. For example, it should be readily apparent how the Mead acid accumulates in preference to arachidonic acid when there is a dietary deficiency of 18:2*n*-6 and an excess of 18:1*n*-9. An important continuation of the *n*-3 pathway via further elongation to C22 and further metabolism to form decosahexaenoic acid (DHA, 22:6*n*-3) has also been omitted.

Table 2.9 Changes in the fatty acid composition of the phospholipids of mitochondrial membranes in *n*-6 essential fatty acid deficiency (g per 100 g total fatty acids)

Fatty acid	*n*-6 EFA sufficiency	*n*-6 EFA deficiency
16:0	31	31
18:0	10	9
18:1*n*-9	13	32
18:2*n*-6	21	10
20:3*n*-9	0.6	3.7
20:3*n*-6	1.7	1.0
20:4*n*-6	14	7
20:3*n*-9/20:4*n*-6	0.04	0.53

less efficient. These changes at the molecular and cellular levels are reflected in the animal's poorer performance in converting food energy into metabolic energy for growth and in the maintenance of body function.

Organs and tissues performing functions such as storage (e.g. adipose tissue), chemical processing (e.g. liver), mechanical work (e.g. muscle) and excretion (e.g. kidney) tend to have membranes in which the *n*-6 fatty acids predominate, with arachidonic acid as the major component. In contrast, nervous tissue, reproductive organs and the retina of the eye have a greater proportion of the longer chain acids with 5 or 6 double bonds, predominantly of the *n*-3 family. It has been assumed that fatty acids of the *n*-3 family have specific roles in vision, reproduction and nerve function and there is some evidence in rats to show that when the ratio of *n*-3 to *n*-6 acids is modified experimentally, some changes do occur in the electroretinograms and in hehavioural responses. We still do not have sufficient evidence, however, to pinpoint the specific functions for these fatty acids.

Essential fatty acids as precursors for eicosanoids

We described the biosynthesis of eicosanoids from polyunsaturated fatty acids of the *n*-3 and *n*-6 families and indicated the balance of eicosanoids produced was important in, for example, maintaining normal vascular function. Several studies have demonstrated that altering the amounts and types of *n*-6 and *n*-3 fatty acids in the diet can influence the spectrum of eicosanoids produced. For example, substitution of fish oils, in which *n*-3 polyunsaturated fatty acids predominate, for diets in which linoleic acid (*n*-6) is the main polyunsaturated fatty acid (as typified by the UK and US diets) results in significant changes in plasma and platelet fatty acid profiles. Arachidonic (*n*-6) gives way to eicosapentaenoic (*n*-3) acid as the predominant polyunsaturated fatty acid and there is a reduction in the formation by platelets of thromboxane A_2, an eicosanoid that stimulates their aggregation (Table 2.10).

The physiological effects of eicosanoids are so potent in such minute quantities that they need to be generated locally and destroyed immediately after they have produced their effect by enzymes that convert them into inactive metabolites. These properties are typical of 'local chemical mediators' or autocoids', which distinguish them from hormones; the latter are produced in specific glands and act at more distant sites. The excretion of these metabolites in the urine has been used as a quantitative method for estimating the daily production of eicosanoids and assessing the quantities required by the body. The mechanism that ensures efficient local production is the release of the precursor polyunsaturated fatty acids from membrane phospholipids by specific phospholipases and transfer to the enzymes of eicosanoid biosynthesis, which are also located within the membrane. The membrane phospholipids can therefore be regarded as a vast body store of essential fatty acids that are immediately available for eicosanoid biosynthesis when required. As they are depleted, they must be replaced by new parent acids of dietary origin or LCPUFA biosynthesized by the pathways illustrated in Fig. 2.4. This in part explains the dynamic turnover of fatty acids in biological membranes, although eicosanoid formation represents a very small proportion of fatty acid turnover.

Several aspects of eicosanoid metabolism and function, however, need further research. These include: the mechanism by which the relative proportions of the different eicosanoids are regulated; the significance of changes of dietary fatty acid composition for whole-body eicosanoid production; the quantitative significance of the different pathways and sites of synthesis; and the quantitative relationships between the requirements for EFA, which are measured in grams and the daily production of eicosanoids, which is measured in micrograms.

Table 2.10 Influence of dietary fatty acid composition on eicosanoid production by cardiac tissue

Animal species	Eicosanoid	Dietary fat (pg mg^{-1} dry weight)		
		Sheep fat	Sunflower seed oil	Tuna fish oil
Rat (*n*=12)	PGI_2	122[a]	104[a]	61[a]
	TXA_2	36[a]	19[a]	9[a]
Marmoset (*n*=6)	PGI_2	167[a,b]	114[a]	88[b]
	TXA_2	69[a]	24[a]	12[a]

Animals were given the diets for 12 months (rats) or 24 months (marmosets). Minced heart tissue was incubated in buffer. Eicosanoids were determined by radioimmunoassays. Common superscripts within a row indicate a statistically significant difference at $P < 0.05$. PGI_2 is prostacyclin; $T \times A_2$ is thronborane A_2.

THE PRECURSOR FATTY ACIDS

Historically, most research attention has been devoted to linoleic acid. As far as we know, linoleic acid is essential for almost every animal species at a dietary level of about 1% of energy, although there is evidence that some insects and protozoa can produce this fatty acid.

Linoleic acid seems to have a specific role in the skin epidermis that cannot be substituted by arachidonic acid, suggesting that the classic skin symptoms of EFA are not due to lack of eicosanoids. This conclusion is strengthened by the fact that ingestion of aspirin, which would inhibit eicosanoid formation from linoleic acid, has no adverse effect on epidermal function. In EFA (linoleic acid) deficiency, the linoleate of several specific epidermal glycosphingolipids is replaced by oleate, a change that is associated with a loss of barrier function. Moreover, when EFA deficiency is cured by giving a mixture of linoleic and arachidonic acids in the diet, only linoleate appears in the epidermal sphingolipids. These results suggest a biochemical mechanism for some of the EFA defects in skin whereby linoleate *per se* is an essential part of the human diet.

There has always been less certainty about α–linolenic acid. Fish, whose lipid metabolism is geared to processing a high dietary intake of *n*-3 fatty acids, seem to have a definite and high requirement (3% of energy) for fatty acids of this family. The brain and retinal rods of most species, including the rat, are characterized by a high proportion of the *n*-3 fatty acids and we can infer that, since these acids cannot be synthesized in the body, there is some dietary requirement for them, however small. Capuchin monkeys maintained on a diet containing adequate linoleic acid, but little or no α-linolenic acid, suffered symptoms closely resembling those of classical EFA deficiency. These were cured by the addition of linseed oil (which contains appreciable amounts of α-linolenic acid) to the diet. In rodents, there is documented evidence that α-linolenic cannot substitute for linoleic acid in reproduction in male and female rats. It has not been possible, however, to demonstrate unambiguously a function for α-linolenic acid that could not be replaced by linoleic acid.

In humans, the essentiality of α-linolenic acid remained in doubt until Holman and his colleagues in 1982 describe a case of a young girl who displayed neurological symptoms 4-5 months after being on total parenteral nutrition. The fat component of the parenteral feed contained mainly linoleic acid and only a minute amount of α-linolenic acid. When safflower oil was replaced by soybean oil containing much more α-linolenic acid, the neurological symptoms disappeared. Later, a Norwegian group provided evidence for α-linolenic acid deficiency in elderly patients fed by gastric tube. It seems quite certain that only very small amounts of this nutrient are needed in human diets to prevent overt deficiency signs, although there is much debate about the amounts required to normalize functions that may not be manifested by classic deficiency signs.

Cats are unusual among mammals in that neither linoleic

nor α-linolenic acid alone is sufficient to protect against the effects of fatty acid deficiency: these animals require arachidonic acid in their diets because they lack the Δ6-desaturase as well as the Δ12-desaturase.

The desaturation/elongation products

In studies of physiological processes such as spermatogenesis and parturition, it is rarely clear whether the EFA are functioning mainly or solely through their contribution to membrane structure or as precursors of one of the eicosanoids. If the latter, then it is likely that elongation/desaturation of the precursor acid is a critical process, since most important eicosanoids are derived from long-chain polyunsaturated fatty acids rather than their precursors. Students who read further in this field of research may be confused to find that different authors use the term essential fatty acids' in apparently different ways. To nutritionists, there are only two essential fatty acids, linoleic and α-linolenic acids. They have to be supplied in the diet and are *essential nutrients*. The longer chain derivatives, it is argued, can be formed in the body. They are *'essential metabolites'* but not essential nutrients. (Cholesterol provides a similar example in the steroid field.) Others have argued that all these structures, whether precursors or long-cham products, are 'essential fatty acids' because they are essential to life and are precursors of eicosanoids. Both schools of thought may be right to some extent if there are limitations on the supply of the elongation/desaturation products as discussed in the next section.

The quantitative requirements for essential fatty acids in the diet?

It seems possible that there are different levels of requirements for EFA for their different functions. For example, there may be a relatively low level of requirement (as for most vitamins), which is necessary to prevent overt signs of EFA deficiency such as skin conditions and growth retardation. This may be around 1-3% of dietary energy as linoleic acid and about one-fifth of this amount of α–linolenic acid. The upper figures for 18:2*n*-6 have been suggested as the increased requirement in pregnancy (to supply the additional needs of the foetus) and in lactation (for milk fat). However, little is known about the ability of women to adapt their metabolism to 'spare' their essential fatty acids in times of high demand. Intakes of EFA may be low enough to induce deficiency signs when hospital patients are given poorly designed enteral or parenteral feeds, when babies are given inadequate artificial formulas, in cases of malabsorption (when the amount of the EFA in the diet itself may be adequate) and in general malnutrition.

Over and above the requirement to satisfy classic EFA deficiency, there may be a larger but quantitatively less easily definable quantity required for such functions as maintaining an optimal balance of membrane fatty acids, preserving a reservoir for eicosanoid formation and maintaining optimal concentrations of the different plasma lipoproteins. The absolute amounts required at this level are likely to be influenced by the amounts and types of other fatty acids in the diet because of the competition for metabolic pathways between EFA and non-EFA as well as competitions between the EFA families. Because primary EFA may be only slowly converted into their long-chain polyunsaturated fatty acid derivatives, there may sometimes be a dietary requirement for the latter. These may be termed 'conditionally essential fatty acids', which have their counterparts in the conditionally essential amino acids.

Long-chain polyunsaturated fatty acids downstream of linoleic and α-linolenic acids may become conditionally essential in the following circumstances.

Desaturase abnormalities

Low activity or complete absence of a specific desaturase may result from single or multiple gene defects. The American biochemist Ralph Holman, who was a pioneer in our understanding of the essential fatty acids, has used the technique of 'plasma fatty acid profiling. This involves comparison of the detailed fatty acid composition of normal plasma with the plasma profile of patients with a disease to pinpoint aberrations in the pathways of PUF A metabolism, which might be due to defects in desaturases. Such aberrations have been found in Sjörgren-Larsson, Reyes and Prader-Willi syndromes and in cystic fibrosis. Several desaturases have now been cloned and this will allow more rapid progress in pinpointing the nature of the metabolic abnormalities than was possible using these indirect techniques.

Δ6-Desaturase activity decreases with age and is also depressed by the so-called 'stress hormones', adrenaline and cortisol, by alcohol, by fatty acids of the *n*-3 family and in the condition of diabetes mellitus. One strategy to overcome the problem of low or absent Δ6-desaturase activity in the *n*-6 pathway is to provide γ-linolenic acid (*n*-6) in the diet, thereby by-passing the faulty Δ6-desaturase. The main dietary sources of γ-linolenic acid are evening primrose, borage and blackcurrant oils. The literature is replete with claims for the therapeutic benefits of 18:3*n*-6 and disorders for which its efficacy has been tested in controlled clinical trials include: atopic dermatitis; rheumatoid arthritis; diabetic neuropathy; multiple sclerosis; ulcerative colitis; hypertension; premenstrual syndrome; schizophrenia; hyperactivity; and several cancers. In almost all categories the results have been inconsistent. There are preliminary indications that evening primrose oil may be more effective in some of the physiological effects described above than the other oils in which 18:3*n*-6 occurs. One explanation that has been offered is that 18:3*n*-6 is present in evening primrose oil almost entirely as molecular species of triacylglycerol in which one 18:3*n*-6 is combined with two linoleic acid molecules: dilinoleoyl-monogammalinolenoyl-glycerol. Another possibility is that minor components of evening primrose oil, not 18:3*n*-6, are responsible for some of the effects.

Metabolic competition

The competition between the *n*-9 and *n*-6 pathways that gives rise to an accumulation of 20:3*n*-9 in linoleic acid deficiency has already been discussed. There may, however, be more subtle changes in concentrations of 20:3*n*-9 and other polyunsaturated fatty acids without concomitant overt EFA deficiency signs, when intakes of 18:2*n*-6 are on the low side of normal and intakes of 18:1 (which is normally the most abundant dietary fatty acid) are exceptionally high. One of the arguments for limiting the amounts of *trans*-unsaturated fatty acids in the diet has been that high levels may result in changes in fatty acid profiles indicative of 'covert' EFA deficiency even when intakes of 18:2*n*-6 are apparently adequate. However, despite much debate, there is as yet no consensus about the biological significance of such metabolic changes.

During the 1990s, the question of appropriate dietary intakes of the *n*-3 family has been hotly debated. It is fair to say that, hitherto, this fatty acid family had been somewhat neglected. In Western industrialized countries, intakes of *n*-3 PUFA are generally very low compared with intakes of the *n*-6 PUFA. The parent acid, α-linolenic is obtained from green vegetables, rapeseed and soybean oils and products manufactured from them. The only significant sources of the longer chain *n*-3 fatty acids are certain fish oils (see Table 2.1). Populations consuming large amounts of fatty fish (for example the Inuit and many Japanese) tend to have much higher intakes of *n*-3 PUFA than in the West.

Increases in daily consumption of fatty fish are quickly reflected in elevated concentrations of these fatty acids, mainly 20:5*n*-3 (EPA) and 22:6*n*-3 (DHA), in the plasma and membranes of red and white blood cells. A major effect of such membrane changes is to alter the composition of eicosanoids formed through the action of the enzymes cyclooxygenase and lipoxygenase. As well as increased production of weakly inflammatory prostanoids and leukotrienes from EPA, there is also inhibition of the strongly inflammatory eicosanoids from arachidonic acid (Fig. 2.5).

Some authors (for example, see the review by Okuyama *et al* in the Further Reading section) have argued a detailed case that when the ratio of dietary *n*-6 to *n*-3 fatty acids is too high, there is an increased risk of certain cancers, allergic diseases—including asthma—and thrombosis, all of which are prevalent in Western countries. Other authors have discussed evidence that certain depressive conditions may also result from chronic dietary intakes with a high *n*-6/*n*-3 ratio. They cite evidence for disruption of neurotransmitter systems, defective signal transduction pathways, reduced insulin receptor sensitivity, and enhanced calcium channel activity, all of which are a consequence of the relative depletion of *n*-3 in nervous tissue membranes.

Developmental immaturity

At certain critical periods in the life-cycle (for example, pregnancy/foetal life or lactation/neonatal life), the high demands of tissues for particular LCPUFA may outstrip their availability, even though the supply of precursors would normally be adequate and there are no specific enzyme abnormalities. The flux along the metabolic pathway may simply be inadequate for the task. Under these conditions, the LCPUFA may become conditionally essential and need to be supplied in the diet (or in foetal development, from the mother's blood) for a limited period of time.

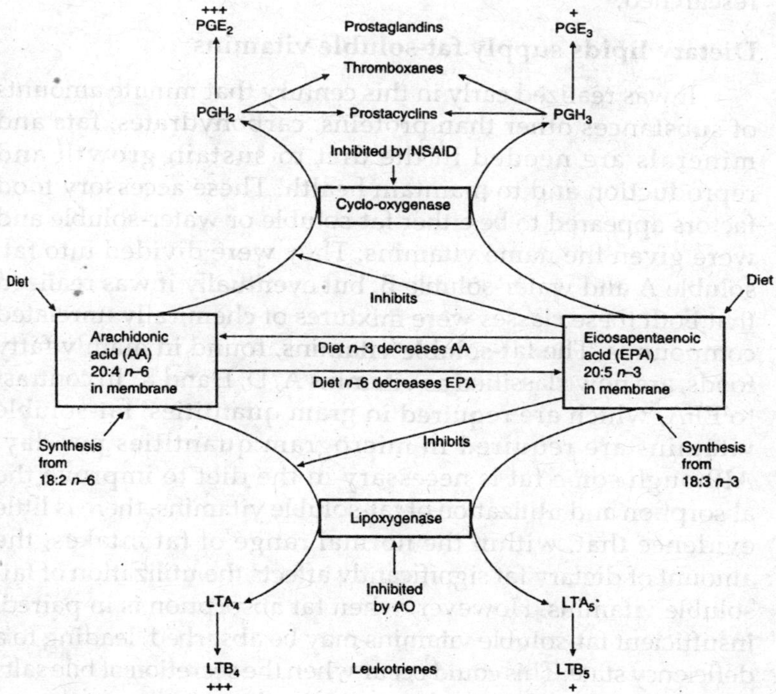

Fig. 2.5 Influence of dietary *n*-3/*n*-6 polyunsaturated fatty acids and drugs on inflammatory eicosanoid production. NSAID: non-steroidal anti-inflammatory drugs; AO: antioxidants; + : signifies weakly inflammatory; +++: signifies strongly inflammatory.

Finally, our discussion of 'conditional essentiality' has followed the generally made assumption lhat the body's requirements for saturated and monounsaturated fatty acids can always be met entirely by endogenous biosynthesis; in other words there is never a dietary requirement for them. There is evidence, albert limited, that fatty acid biosynthesis is a slow process in human tissues even when people are consuming diets with a low fat content. It is worth considering, therefore, whether there may be circumstances of high demand for saturated fatty acids (for example in brain or lung development), during which even saturated fatty acids become conditionally essential. Such a concept has not been actively researched.

Dietary lipids supply fat-soluble vitamins

It was realized early in this century that minute amounts of substances other than proteins, carbohydrates, fats and minerals are needed in the diet to sustain growth and reproduction and to maintain health. These accessory food factors appeared to be either fat-soluble or water-soluble and were given the name vitamins. They were divided into fat. soluble A and water-soluble B, but eventually it was realized that both these classes were mixtures of chemically unrelated compounds. The fat-soluble vitamins, found in mainly fatty foods, are now classified as vitamins A, D, E and K. In contrast to EFA, which are required in gram quantities, fat-soluble vitamins are required in microgram quantities per day. Although some fat is necessary in the diet to improve the absorption and utilization of fat-soluble vitamins, there is little evidence that, within the normal range of fat intakes, the amount of dietary fat significantly affects the utilization of fat-soluble vitamins. However, when fat absorption is in paired, insufficient fat-soluble vitamins may be absorbed, leading to a deficiency state. This could occur when the secretion of bile salts is restricted (as in biliary obstruction), when sections of the gut have been removed or damaged by surgery or in diseases, such as tropical sprue and cystic fibrosis, all of which are associated with poor intestinal absorption. When people have access to varied diets, there is little likelihood of dietary deficiencies of fat-soluble vitamins. However, some fat-soluble vitamins are present in a limited number of foods; inevitably, when food supplies are limited by economic conditions, deficiency diseases can then become a significant problem.

Vitamin A

The chemical name for vitamin A is all-*trans*-retinol (Fig.2.6 CA). Retinol itself is found only in animal fats. Plant materials, such as dark green leaves and vegetable oils, like palm oil, contain a precursor, β–carotene (provitamin A), which can serve as a source of retinol in the body. Dietary β-carotene (Fig. 2.6 CB) is split by an enzyme in the small intestinal mucosa, β-carotene 15,15' -oxygenase, yielding two molecules of all-*trans*-retinal (Fig. 2.6 reaction 1), which are then oxidized to all-*trans*-retinol (vitamin A) by an NADH-dependent reaction. After absorption, or conversion from carotene, retinol is esterified mainly with palmitic acid (irrespective of the composition of the dietary esters) in the enterocytes in a reaction catalysed by one of two microsomal enzymes. The first is phosphatidylcholine:retinol acyltransferase, which transfers palmitic acid from position *sn*-1 of phosphatidylcholine. The second type of reaction uses the CoA thiolester as a substrate and is catalysed by acyl-CoA:retinol acyltransferase.

Retinyl esters are the main storage forms of vitamin A in the body, mainly in the liver. They can be hydrolysed by a specific retinyl palmitate esterase, the activity of which can increase some 100-fold during vitamin A deficiency. Transport in the blood requires hydrolysis of the ester followed by binding of retinol to a specific transport protein, retinol binding protein. Retinol is delivered to target tissues (e.g. visual cells) by interaction of the retinol binding protein to cell-surface receptors.

Vitamin A, with its analogues and metabolites, has several distinct functions: in the visual process, in cellular differentiation, embryogenesis and in the immune response.

Vision

This is the best-defined function in molecular terms. The reactions involved in the visual process take place in the outer

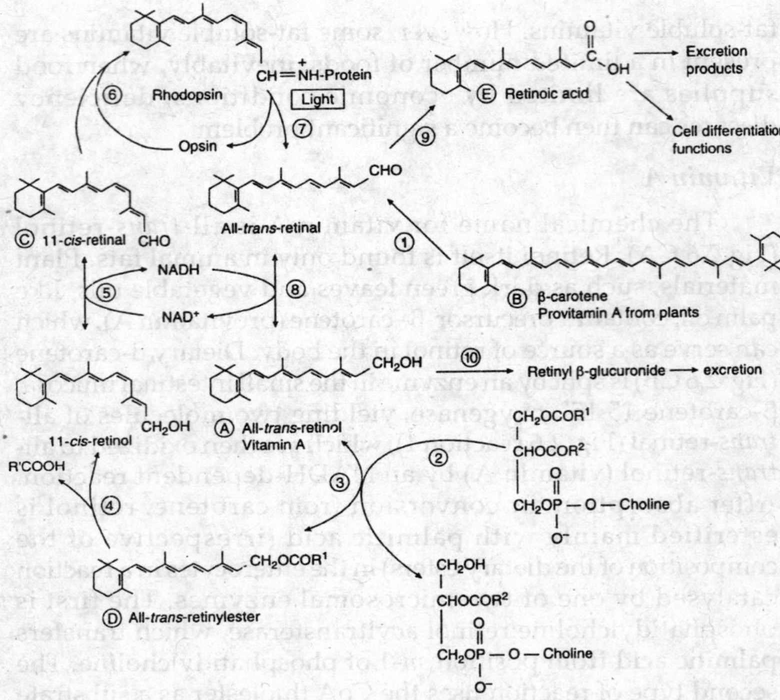

Fig. 2.6 Some important metabolic transformations of vitamin A. The parent compound of the vitamin A group is all-*trans*-retinol (A). It can be obtained directly from the diet or formed from the breakdown of dietary β-carotene (provitamin A) from plants (B). The active form of vitamin A in the visual process is 11-*cis*-retinal (C). Retinyl esters (mainly palmitate) provide a storage form of the vitamin in tissues, especially the visual pigment at the back of the eye (0). All-*trans*-retinal can be converted into retinoic acid (E), which has no visual function, but is involved in processes of cellular differentiation. Vitamin A can be excreted via the formation of giucuronides from retinoic acid or from vitamin A itself.

segment of rod cells and in the colour-sensing cone cells of the retina. All-*trans*-retinol, carried in the circulation on retinol binding protein, is discharged to the visual cells and converted into 11-*cis*-retinol. The reaction is complex, since it involves first the esterification of retinol to retinyl palmitate by transfer of an acyl group from phosphatidylcholine in the pigment epithelium at the back of the eye. Then retinyl palmitate breaks down in a concerted reaction that couples isomerization of the retinol and release of a free fatty acid. 11 *cis*-Retinol is oxidized to the aldehyde, 11-*cis*-retinal, which then reacts with a lysine residue in the protein, opsin, through a Schiff base linkage to form rhodopsin. This complex can absorb a photon of visible light (reaction 7), converting 11-*cis*-retinal into all-*trans*-retinal and triggering a conformational change in the protein. The resulting change in membrane potential is transmitted to the brain and perceived as sight. Following the conformational change, all*trans*-retinol is discharged from the protein and the active 11-*cis* form may be regenerated by the reaction sequence 8, 3, 4, 5 shown in Fig. 2.6.

Cellular differentiation

Just as 11-*cis*-retinal is the analogue of vitamin A involved in vision, so 9-*cis*-retinoic acid is the analogue mainly involved in differentiation (Fig. 2.6). The cell nucleus contains two principal types of high-affinity receptor proteins for retinoic acid, designated RAR and RXR. Each may be present in α, β or γ isoforms. Furthermore, each set of receptors has six different domains that are involved in the transcription of genes. The receptors are coded by three different genes expressed at different times and places during differentiation. RAR bind either all-*trans*- or 9-*cis*-retinoic acids, whereas RXR bind only the 9-*cis* isomer. Retinoic acid isomers, together with their nuclear (or 'intracellular') receptors can both stimulate and inhibit gene expression, depending on the types of interactions described above. RXR appears to interact with a number of other intracellular receptors such as the vitamin D receptor and the PP AR. This raises the possibility of interesting interactions between vitamin A and other chemical messengers.

Retinoic acid can also form covalent bonds with certain proteins. These retinoylated proteins are similar in size to the nuclear retinoic acid receptors and may play roles similar to those of lipid-anchored proteins.

Immune response

Risk of infection is increased markedly in vitamin A deficiency and both humoral and cell-mediated immune responses are impaired. A major site of vitamin A action in the immune response is thought to be the T -helper cell and there seems to be a requirement for a specific form of vitamin A - 14-hydroxyretroretinoic acid-rather than retinoic acid. Understanding of the mechanisms by which vitamin A is involved in immunity at the molecular\ level, as well as in embryogenesis, spermatogenesis and taste perception is still in its infancy.

Nutritional aspects

Major sources of vitamin A are green vegetables and carrots (provitamin A), liver (especially fish), milk and some fat spreads. No more than 750 μg vitamin A is required by the average person daily. Compare this with the gram quantities required for EFA intake.

The most tragic manifestation of vitamin A deficiency is blindness in young children. The first effects are seen as severe eye lesions, a condition known as xerophthalmia, which is eventually followed by keratomalacia with dense scarring of the cornea and complete blindness. Xerophthalmia is considered to be one of the four commonest preventable diseases in the world. Although there are large-scale programmes for the supplementation of children's diets with vitamin A, these are difficult to implement successfully and the World Health Organization considers that if the consumption of green leafy vegetables and suitable fruits by young children could be substantially increased, there is every reason to believe that the problem would be solved. There might be a case for trying to increase the overall fat content of the diet, too. Here, we are more concerned with solving problems of economics and distribution and with changing local eating habits than with biochemistry and nutrition, where the knowledge is al1ready to hand.

Epidemiological evidence shows that people with above-average blood retinol concentrations or above-average β-carotene intakes have a lower than average risk of developing some types of cancer.

In contrast to most water-soluble vitamins, excessive intakes of fat-soluble vitamins can be harmful, since in general they accumulate in tissues rather than being excreted. Thus, vitamin A, if taken in excess, accumulates in the liver. Chronic over-consumption may then cause not only liver necrosis, but also permanent damage to bone, vision, muscles and joints. Vitamin A can be excreted via its conversion into glucuronides formed from retinoic acid or from all-*trans-retinol* itself (Fig. 2.6).

Vitamin D

Vitamin D is the generic name for two sterols with the property of preventing the disease rickets. Ergocalciferol (vitamin D_2) is formed by irradiation of the plant sterol, ergosterol, and is the main dietary source of vitamin D. Cholecalciferol (vitamin D_3) is produced in the skin by ultraviolet irradiation of 7-dehydrocholesterol present in the skin surface lipids. It is the main source of the vitamin for most human beings. Dietary sources are fish liver oils (e.g. cod), eggs, liver and some fat spreads.

The parent forms of vitamin D are biologically inactive. The active metabolites are formed in the liver and kidneys. After absorption from the diet or formation in the skin, vitamin D is carried in the circulation bound to a specific transport protein of the Cα-globulin class. In the liver, cholecalciferol is first hydroxylated to 25-hydroxycholecalciferol (Fig. 2.5). The 25-hydroxylase requires NADPH, molecular oxygen and Mg^{2+} ions. It is the main transport form of vitamin D and its concentration in plasma is used as an indicator of vitamin D status. Although this metabolite has modest biological activity, it is carried to the kidney, also attached to the transport protein, where it is further hydroxylated either to 1,25-dihydroxy-cholecalciferol (sometimes called calcitriol) or 24,25-dihydroxy derivative (Fig. 2.7) by enzymes similar to the liver 1-α-hydroxylase. 1,25-Dihydroxycholecalciferol is the most biologically active vitamin D metabolite, some 100 times more active than the 25-hydroxy metabolite. Its circulating concentration is tightly controlled by the plasma level of parathyroid hormone and the body's overall vitamin D status.

Thus, in relative vitamin D deficiency, the circulating concentration of 1,25-dihydroxycholecalciferol is high as a result of the stimulation of the liver 1-α-hydroxylase by parathyroid hormone, whereas that of the 24,25-dihydroxy metabolite is low. Calcitriol functions as a hormone in several distinct ways.

Calcium homeostasis

The best-studied function of calcitriol is in calcium homeostasis. Several mechanisms seem to be involved. The first

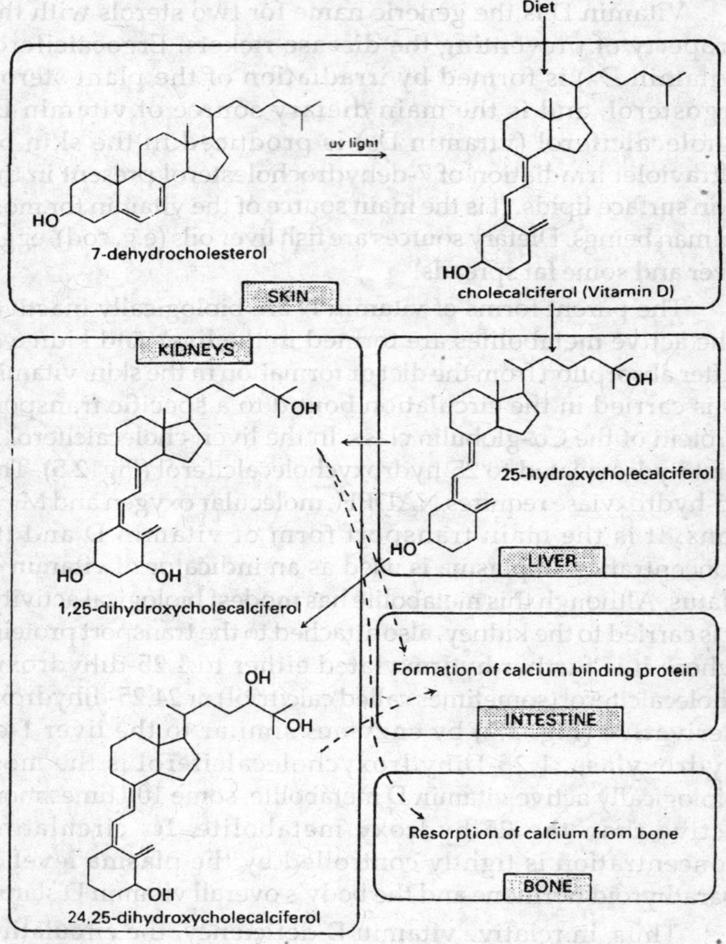

Fig. 2.7 Metabolism of vitamin D.

is to increase. the absorption efficiency of calcium by increasing its transport across the enterocyte brush border membrane and its subsequent export from the enterocyte into the blood. Calcitriol, a lipid-soluble compound, easily passes through the enterocyte cell membrane, binds to a specific receptor and is translocated to the nucleus where it binds to the DNA of specific response genes. Loops in the receptor, known as 'zinc fingers' enable the receptor to interact with DNA to induce the transcription of mRNA for a specific Ca-binding protein that aids active transport of Ca from the gut to maintain blood concentrations within the narrow range of 2.1-2.6mM. A second role for calcitriol, in the presence of parathyroid hormone, is to activate bone osteoclasts to resorb calcium from bone, again to maintain an appropriate plasma calcium concentration. However, the role of calcitriol in bone is more complex than implied here, since it can also stimulate bone formation by osteoblasts, thus participating in the regulation of the overall process of bone turnover.

Other functions

The discovery of receptors for 1,25-dihydroxycholecalciferol in many tissues other than those involved in calcium and bone metabolism (e.g. pancreatic islet cells, skin keratinocytes, mammary epithelium and some neurones) suggests a wider involvement in aspects of tissue development. Indeed, calcitriol seems to be a general developmental hormone, inhibiting proliferation and promoting differentiation in many tissues. Present knowledge suggests functions that include regulation of gene products associated with mineral metabolism, differentiation of cells in skin and in the immune system and regulation of DNA replication and cell proliferation. A recent finding is that when 1,25-dihydroxycholecalciferol binds to a response element in the promoter region of its target gene, it requires an accessory protein. It is now thought that this protein may be identical to the retinoic acid X receptor discussed in the previous section, suggesting an interesting convergence of vitamin A and D functions that is yet to be properly elucidated.

Nutritional aspect

It is difficult to decide upon a precise dietary requirement for vitamin D because much is derived from the skin lipids rather than the diet. Thus the UK committee that advises on dietary requirements set no dietary reference value for vitamin D for people between the ages of 4 and 50 years. Two groups of people, however, may have a special need to obtain vitamin D from the diet. In the first group are children and pregnant and lactating women whose requirements are particularly high. In the second group are people who are little exposed to sunlight, such as the housebound elderly and people in far northern latitudes or those who wear enveloping clothes. Dark-skinned immigrants to Northern Europe are especially vulnerable. Infants and children who obtain too little vitamin D develop rickets, with deformed bones that are too weak to support their weight. The reason why, in the UK and some other countries, vitamin D preparations are provided for children and pregnant women, and margarine is fortified with it, is because these degenerative changes soon become permanent if supplementation is not started early enough.

Like vitamin A, vitamin D is toxic in high doses. Even amounts that are no more than five times the normal intake can be toxic. Too high an intake causes more calcium to be absorbed than can be excreted, resulting in excessive deposition in, and damage to, the kidneys.

Vitamin E

Vitamin E activity is possessed by eight tocopherols and tocotrienols (Fig. 2.8). α-Tocopherol is the most potent of these, the other compounds having between 10% and 50% of its activity. α-Tocopherol is also the most abundant form of vitamin E in animal tissues, representing 90% of the mixture. The form used in commercial preparations is synthetic racemic α-tocopherol. (the natural compounds have optical activity), often in the acetylated form as a protection from oxidation.

In 1922, the Americans, Evans and Bishop, discovered that vitamin E prevented sterility in rats reared on fat-deficient diets fortified with vitamins A and D. Because vitamin E is so widespread in foods and like all other fat-soluble vitamins is stored in the body, deficiency states are rarely if ever seen, possible exceptions being in premature infants with low fat stores and in people with severe malabsorption. The richest sources are vegetable oils, cereal products and eggs.

Vitamin E is carried in the blood by the plasma lipoproteins.

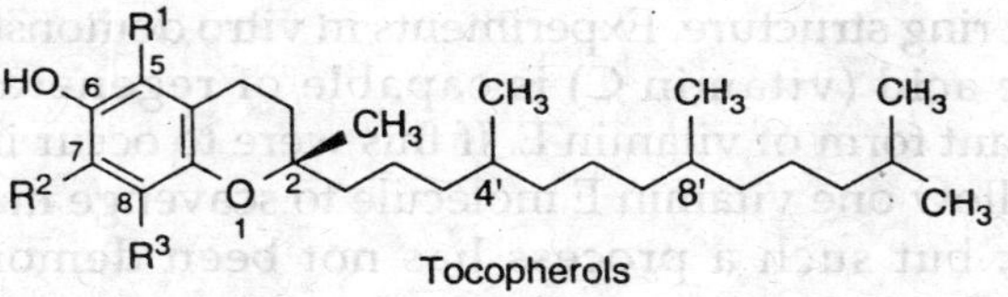

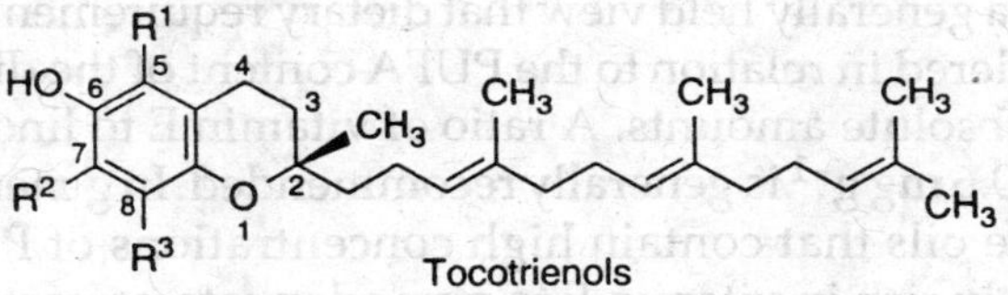

Tocopherols

Tocotrienols

Fig. 2.8 Structures of compounds with vitamins E activity. All molecules comprise a chromanol ring system and a 16 C phytyl side chain. Tocopherols have a saturated side chain; tocotrienols contain three unsaturated bonds in the side chain. Analogues are named α, β, γ and δ-tocopherols or tocotrinols as follows:

α : $R^1 = CH_3$; $R^3 = CH_3$

β: $R^1 = CH_3$;$R^2 = H$; $R^3 = CH_3$

γ : $R^1 = H$; $R^2 = CH_3$; $R^3 = CH_3$

δ: $R^1 = H$; $R^2 = H$; $R^3 = CH_3$

There is evidence for specific tissue vitamin E binding proteins, but research into this topic is still in its infancy.

Vitamin E as an antioxidant

Until the last decade of the twentieth century, almost all interest in vitamin E was concerned with its putative role as a natural antioxidant. α–Tocopherol is present in the lipid bilayers of biological membranes and may play a structural role there. However, its main function is thought to be the prevention of the oxidation of the unsaturated lipids present in the membrane. The products of lipid peroxidation can cause damage to cells if the oxidative process is not kept in check. Such damage appears to be exacerbated in animals fed diets deficient in vitamin E. α–Tocopherol primarily acts as a terminator of the lipid peroxidation chain reactions by donating a hydrogen atom to a lipid radical. The resulting tocopheryl radical is relatively unreactive and unable to attack adjacent unsaturated fatty acids because the unpaired electron becomes delocalized in the aromatic ring structure. Experiments in vitro demonstrate that ascorbic acid (vitamin C) is capable of regenerating the antioxidant form of vitamin E. If this were to occur in vivo, it would allow one vitamin E molecule to scavenge many lipid radicals, but such a process has not been demonstrated conclusively in vivo.

It is a generally held view that dietary requirements should be considered in relation to the PUFA content of the diet rather than in absolute amounts. A ratio of vitamin E to linoleic acid of about 0.6 mg g^{-1} is generally recommended. In general, those vegetable oils that contain high concentrations of PUFA are sufficiently rich in vitamin E to give adequate protection.

Other functions

There is now accumulating evidence that vitamin E may also play other biological roles that do not necessarily involve its antioxidant function. A possible structural role in the maintenance of cell membrane integrity has already been mentioned. Some observations imply diverse roles in immune function. Vitamin E has anti-inflammatory effects, possibly by interacting with the prostaglandin synthetase that produces inflammatory eicosanoids and also appears to stimulate the immune response. α-Tocopherol seems to be involved in the regulation of intercellular signalling and cell proliferation through the modulation of protein kinase C activity. Finally, it may have a role in DNA biosynthesis. The details of each of these 'other functions' are still far from clear and this should provide a fascinating area of future research.

Vitamin K

Vitamin K is the generic name given to a group of

compounds, having in common a naphthoquinone ring system (menadione) with different side chains (Fig. 2.9). Plants synthesize phylloquinone (vitamin K_1) with a phytyl side chain identical to that in chlorophyll. Bacteria synthesize menaquinones (vitamin K_2) the side chains of which comprise 4-13 isoprenyl units. Major dietary sources of vitamin K_1 are fresh green leafy vegetables, green beans and some seed oils. Cereals are poor sources and, of animal foods, only beef liver is a significant source. Vitamin K_2 is found in fermented foods, such as cheeses and yoghurt and in ruminant liver. Bacteria in the human gut synthesize menaquinones, but there is uncertainty about their bioavailability and hence their significance as a source of vitamin K.

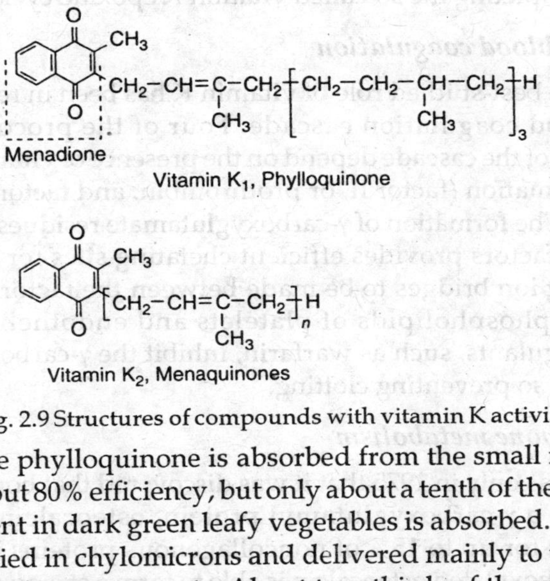

Menadione

$CH_2-CH=C(CH_3)-CH_2[CH_2-CH_2-CH(CH_3)-CH_2]_3H$

Vitamin K_1, Phylloquinone

$[CH_2-CH=C(CH_3)-CH_2]_nH$

Vitamin K_2, Menaquinones

Fig. 2.9 Structures of compounds with vitamin K activity.

Free phylloquinone is absorbed from the small intestine with about 80% efficiency, but only about a tenth of the vitamin K_1 present in dark green leafy vegetables is absorbed. Vitamin K is carried in chylomicrons and delivered mainly to the liver, in chylomicron remnants. About two-thirds of the vitamin K intake is soon excreted from the body (into the faeces via the bile) and as there is no evidence of an entero-hepatic circulation, it is presumed that a constant intake is essential.

Function in enzymic carboxylation reactions

Like many water-soluble vitamins, but in contrast to other

fat-soluble vitamins, vitamin K has a specific function as an enzyme cofactor. The enzyme, present in the endoplasmic reticulum of liver cells and in some other tissues, is γ-glutamyl- (or vitamin K-dependent-) carboxylase. It requires both molecular oxygen and carbon dioxide and inserts a carboxyl group into the glutamyl residue of several specific calcium-binding proteins. The reaction is a post-translational modification of the protein. The active form of vitamin K in this reaction is the reduced quinol form, which donates hydrogen to glutamic acid, being transformed in the process to an epoxide. The 2,3-epoxide is then transformed to the quinone form by vitamin K epoxide reductase in a dithiol-dependent reaction and the quinone can be further reduced to the quinol, thus completing the so-called vitamin K epoxide cycle.

Role in blood coagulation

The best-studied role of vitamin K has been in relation to the blood coagulation cascade. Four of the procoagulant proteins of the cascade depend on the presence of vitamin K for their formation (factor II, or prothrombin, and factors VII, IX and X). The formation of γ-carboxyglutamate residues in these protein factors provides efficient chelating sites for Ca ions, enabling ion bridges to be made between the factor and the surface phospholipids of platelets and endothelial cells. Anticoagulants, such as warfarin, inhibit the γ-carboxylation reaction, so preventing clotting.

Role in bone metabolism

It was only in 1975 that it was discovered that bone tissue contains a γ-carboxyglutamyl protein, osteocalcin, which accounts for up to 15% of noncollagenous proteins in bone. Fully carboxylated osteocalcin is able to form a strong complex with hydroxyapatite, the calcium phosphate mineral component of bone. Much circumstantial evidence suggests a key role for vitamin K in bone health. For example, treatment of pregnant mothers with vitamin K anticoagulants, such as warfarin, can lead to bone defects in their infants. Measurement of 'undercarboxylated' species of osteocalcin has been explored as an index of vitamin K deficiency. Yet in practice, there is no direct evidence that vitamin K deficiency has long-term adverse effects on bone health. This apparent paradox is probably due to insufficient detailed information and this area promises to be an important growth point in bone research.

Nutritional aspects

In vitamin K deficiency, the time taken for blood to clot is prolonged and the activities of factors II, VII, IX and X are reduced. It is rare to see vitamin K deficiency in an adult except when fat absorption is impaired. Deficiency signs occur more frequently in infants. This is because they have an immature liver that synthesizes prothrombin only slowly. They also have an undeveloped intestinal flora, that might otherwise provide some menaquinones and, finally, human milk contains only small amounts of vitamin K. Dietary amounts needed to satisfy most individuals' requirements are thought to be about 10 μg day^{-1} for infants and between 40 and 80 μg day^{-1} for adults. In some new-born infants, insufficient intakes of vitamin K give rise to bleeding into tissues, known as haemorrhagic disease of the new-born. In some countries (e.g. the UK) vitamin K is given routinely at birth, but this is not universally the practice. There is little evidence of harm from high intakes of natural vitamin K, although high intakes of menadione can lead to haemolysis and liver damage in infants.

The palatability of foods

Whatever its nutrient composition might be, food has no nutritive value if it is not eaten. In countries where choice of food is abundant, palatability is a major factor in determining food choice and, therefore, nutrient intakes.

Fats contribute to palatability mainly in two ways. Firstly by responses to their texture in the mouth (sometimes called mouthfeel) and secondly by olfactory responses, namely taste in the mouth and aroma or odour in the nose; together, these are called flavour. Fat is, on the one hand, a source of taste and aroma compounds and, on the other, a medium that regulates the distribution of these compounds between water, fat and vapour phases, which influence their perception by the sense organs. Some flavour compounds result from the decomposition of lipids by lipolysis, oxidation, and microbial or thermal degradation. These processes may produce free fatty acids, aldehydes, ketones, lactones and other volatile compounds (Fig. 2.10). PUFA play a dominant role because of their susceptibility to oxidation. Figure 2.10 illustrates the generation of a specific odour compound, whereas Fig. 2.11 summarizes the different aspects of food quality that may be influenced by oxidative metabolites of PUFA.

Odour

When low molecular weight volatile compounds interact with receptors in the nose, they give rise to the sensation of odour or aroma. This sensation depends on features of the molecular structure of the odour compounds and on their partitioning between the fat and the vapour phases. Thus, shortchain length fatty acids (e.g. in butter) have a more intense odour than longer chain acids; those with chain lengths greater than about ten carbon atoms have little odour because they are not sufficiently volatile. The fat content of the food has an important influence on aroma. A high fat content favours partition into the fat phase and may slow down the loss of volatiles, which are generally hydrophobic.

Sometimes an advantage of a high fat content may be in lowering the intensity of the sensation of an unpleasant odour. Frequently the perception of a particular odour is quite different (and often more unpleasant) at high compared to low concentrations and the effective concentrations will be much influenced by the fat content of the foods. If food products do not contain much lipid, then there will be an early release of aroma compounds as the food is eaten. The nature and time-scale of the perceived sensation will be quite different, and possibly less pleasant, than when a fatty food is eaten.

Taste

As food is eaten, taste compounds, which may elicit acid, sweet, salty or bitter tastes, must pass into the aqueous phase before they are sensed by the taste receptors in the mouth. The aqueous phase is composed of the water from the food mixed with saliva in the mouth. Thus, factors such as water solubility, pH and salt formation, which influence the partitioning of taste compounds between aqueous and fat phases, influence flavour perception. For example, the presence of fat may retard their passage into saliva and thus limit the perception of taste. The way in which fat and water are distributed in a food, and the

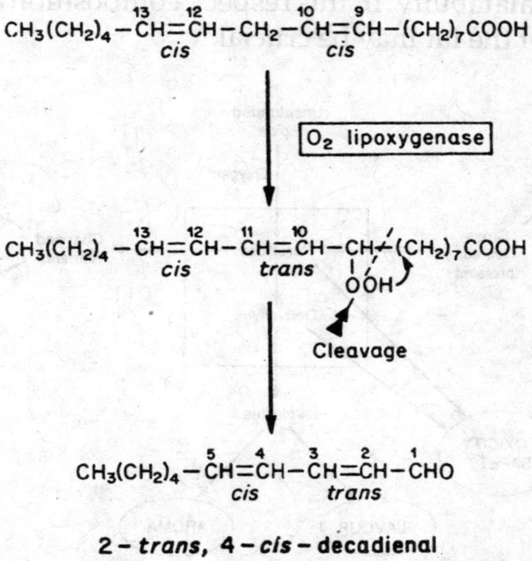

$$\underset{}{CH_3(CH_2)_4}-\underset{cis}{\overset{13}{C}H=\overset{12}{C}H}-CH_2-\underset{cis}{\overset{10}{C}H=\overset{9}{C}H}-(CH_2)_7COOH$$

↓ O_2 lipoxygenase

$$CH_3(CH_2)_4-\underset{cis}{\overset{13}{C}H=\overset{12}{C}H}-\underset{trans}{\overset{11}{C}H=\overset{10}{C}H}-\underset{OOH}{CH}-(CH_2)_7COOH$$

↓ Cleavage

$$CH_3(CH_2)_4-\underset{cis}{\overset{5}{C}H=\overset{4}{C}H}-\underset{trans}{\overset{3}{C}H=\overset{2}{C}H}-\overset{1}{C}HO$$

2 - *trans*, 4 - *cis* - decadienal

Fig. 2.10 The formation of an odour compound. 2-*trans*, 4-*cis* Decadienal is responsible for some of the flavour of oxidized soybeans.

proportions of each, influence flavour perception. Many foods have a poor flavour if they contain too little fat (cheese is a good example). The effect, while very obvious to the consumer, is not fully understood scientifically but may be related to the state of dispersion of the fat.

Texture

Fat may also have an important effect on the texture of food as perceived during eating. The physical state of the fat is important. Pure oil is unpleasant to swallow for most people, while an emulsion may be perceived as pleasantly creamy. An oil-in-water emulsion gives quite a different sensation from a water-in-oil emulsion of the same chemical composition. Thus, full fat milk, cream and butter each has its individual sensory characteristics. A high content of solid fat (e.g. in chocolate) gives a cooling effect on the tongue during eating because of the heat needed for melting. If the fat does not melt completely during eating, however, an impression of stickiness results. Firmness and graininess are other attributes affected by solid fat content. People vary greatly in their appreciation of fat in relation to palatability. In this respect, composition and state of dispersion of the fat may be crucial.

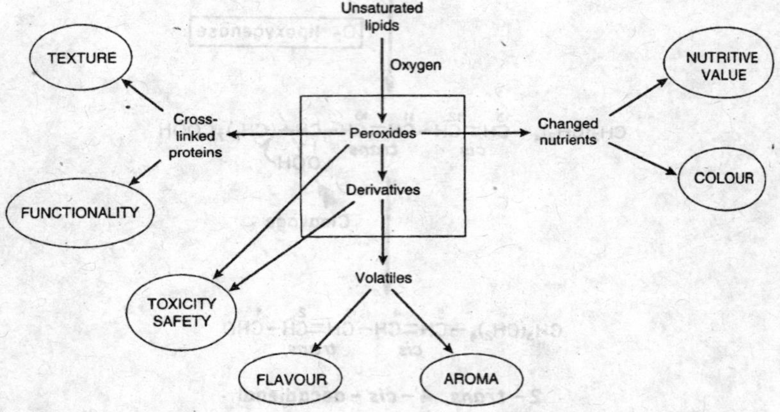

Fig. 2.11 Reactions of polyunsaturated fatty acids leading to changes in food quality.

Lipids influence immunity in several ways. Some glycolipids are components of cell-surface antigens. Dietary fat enhances immunity in malnourished people through its energy contribution. Individual lipids have more specific roles through effects on membrane structure, eicosanoid production, peroxidation and gene expression.

Components of the immune system and their functional assessment

It is well established that diet has an important influence on immunity. Energy malnutrition and deficiencies in specific nutrients result in increased susceptibility to infection and increased mortality, especially in children.

Lipids may be involved in immunity in several ways. The role of some glycosphingolipids in cell surface antigens is discussed lotcs. The following deals only with the contributions of dietary lipids. Because of their importance as a source of dietary energy, triacylglycerols may contribute to the enhancement of immune function in malnourished individuals. Individual lipids, including some fat-soluble vitamins and different types of fatty acids, play more specific roles through their contribution to membrane structure and integrity, their metabolism to eicosanoids, their participation in processes of lipid peroxidation, or their influence on gene expression.

The immune system has evolved to protect the body from constant attack by infectious organisms and other non-self molecules that could have detrimental effects on the host. It involves complex interactions between different types of cells, which produce substances that are toxic to invading pathogens. Lymphocytes are key players in the immune system. B-lymphocytes produce anti-bodies against specific antigens located on the surface of invading organisms (humoral immunity). T-lymphocytes do not produce antibodies. They recognize peptide antigens attached to major histocompatibility complex proteins on the surfaces of so-called antigen-presenting cells. In response to antigenic stimulation, they secrete cytokines (e.g. interleukins, interferon, transforming growth factor, etc.), whose function is to promote the proliferation and differentiation of T-lymphocytes as well as other cells of the immune system. These include B-lymphocytes, monocytes, neutrophils, natural killer cells and macrophages. Whereas life would hardly be possible without this well-integrated system, failure adequately to regulate these processes can lead to damage to the body's own tissues as seen in severe inflammation and in autoimmune diseases.

In the 1970s it was discovered that diets rich in PUFA could prolong the survival of skin allografts in mice and such diets were subsequently employed as adjuncts to conventional immunosuppressive therapy to reduce rejection of human kidney grafts. Such diets also appeared beneficial in treating patients with multiple sclerosis, an autoimmune disease in which an inappropriate immune response to one of the body's own proteins causes damage to the myelin membrane. These findings, and others over the intervening years, strongly suggested that PUFA were acting to suppress the immune system. Recent research has been directed to better understanding the mechanisms of the action of PUFA at the cellular and molecular levels.

Much of the research into the effects of dietary fatty acids on aspects of immune function has been done with lymphocytes 'ex vivo'. The cells are isolated from the blood of animals (including human subjects) given diets differing in fatty acid composition. Commonly used tests are (a) rates of cell division (measured by incorporation of tritiated thymidine into DNA) of lymphocytes stimulated with specific antigens or non-specific mitogens; (b) macrophage migration in an electric field; (c) cytokine production by different cell types; (d) expression of lymphocyte cell-surface molecules; (e) lymphocyte-mediated cytolysis; (f) cytotoxic T-lymphocyte activity; (g) natural killer cell activity. Such ex vivo tests have been supplemented by assessment of immune function in vivo by measuring (a) delayed-type hypersensitivity reactions; (b) graft versus host responses; (c) organ transplantation and (d) antibody production. It is not possible to describe these methods here and interested students should consult the reference lists in appropriate publications listed under Further Reading at the end of this chapter.

Interpretation of the results of all these studies is difficult for several reasons. Each of the aforementioned tests measures a different aspect of immune function and dietary lipids may not affect all functions in the same way. Results of tests involving lymphocytes in culture are often critically dependent on the type of serum used in the culture medium. When animal models are used, results may differ between species and may not always be relevant to man. Very importantly, dietary experiments have normally used natural fats and oils to represent different classes of fatty acids. For example beef tallow, olive oil, sunflower oil and fish oils have been given to provide dietary fats dominated by saturated, monounsaturated, *n*-6 polyunsaturated and *n*-3 PUFA, respectively. Of course all these fats contain mixtures of fatty acids, as well as many minor components, such as fatsoluble vitamins and sterols. Consequently, it can never be certain that the effects observed were due to the fatty acid of interest, even though it may be the major component of the mixture. Some authors have tried to overcome this problem by giving synthetic diets containing the ethyl esters of specific fatty acids. However, with this approach, potential problems of nutrient balance need to be addressed.

Summary of lipid effects on different components of immunity

Bearing in mind all the reservations expressed above, the main findings may be summarized in the following sections.

Influence on target cell composition

(1) Giving diets enriched with specific fatty acids generally leads to a corresponding enrichment of that fatty acid in the different cells of the immune system.

(2) It is frequently assumed that the observed changes occur in membrane lipids. However not all studies specifically isolated membrane phospholipid fatty acids. Some of the cell types of interest contain small stores of triacylglycerols, which are more susceptible to modification by diet.

(3) Membrane fatty acid changes do occur, however, especially after giving dietary *n*-3 PUPA, which tend to replace membrane arachidonic acid *(n*-6*)*.

Influence on lymphocyte functions ex vivo

(1) Animal studies indicate that high fat diets in general suppress lymphocyte proliferation compared with low fat diets.

(2) Among high fat diets the order of potency is fats rich in saturated fatty acids < oils rich in *n*-6 PUF A < olive oil < linseed oil < fish oil.

(3) Some, but by no means all, human studies replicate these results.

(4) Cytotoxic T -lymphocytes and natural killer cell activities also tend to be suppressed by high fat diets in a manner similar to the lymphocyte proliferation.

(5) Fewer studies have described effects on cytokine production or surface-molecule expression and these have mainly focused on *n*-3 PUFA. For example, when diets of healthy human subjects have been supplemented with fish oils, neutrophil leukotriene B_4 (L TB_4) production was suppressed, concomitantly with an increase in neutrophil EP A and a decrease in arachidonic acid content.

(6) At the same time neutrophil chemotactic responses to L TB_4 were also suppressed, as well as the expression of some surface molecules.

(7) Similar changes were found in subjects with rheumatoid arthritis.

Influence on antibody production

(1) Animal studies have found suppression of IgG and IgM antibody production following an antigenic challenge, when diets rich in *n*-6 PUFA were compared with a low fat diet or diets rich in SFA.

(2) Enhanced IgE production to ovalbumin has been reported when rats were given diets enriched with fish oil compared with SFA.

(3) The few human studies that have been reported found minimal effects of dietary fat on antibody production.

Influence on delayed-type hypersensitivity

(1) Delayed-type hypersensitivity is the result of a cell-mediated response to challenge with an antigen to which the individual has already been primed.

(2) In general, animal studies indicate that high fat diets reduce this type of response compared with low fat diets.

(3) Among high fat diets, the order of potency is SFA-rich fats < n-6 PUFA-rich oils < fish oils.

(4) Again there are few studies with human subjects and these reveal minimal effects of dietary fats.

Graft versus host and host versus graft reactions and organ transplants

(1) A limited number of animal studies suggests that high fat diets reduce graft versus host and host versus graft responses compared with low fat diets.

(2) Among high fat diets the order of potency is similar to that seen with delayed hypersensitivity.

(3) Diets rich in unsaturated lipids tend to prolong transplant survival, the effect being greatest with fish oils.

Survival after infection

Given the general conclusion that unsaturated fatty acids suppress immune responsiveness, an important question arises. Are animals that habitually consume diets rich in unsaturated lipids less able to combat infection? A small number of animal studies suggests that this is so but there have been no reports of compromised ability to fight infection in human subjects supplementing their diets with different kinds of unsaturated fatty acids.

Influence on autoimmune and inflammatory disease processes

(1) In general, studies have demonstrated a proinflammatory effect when diets were rich in n-6 PUFA.

(2) Meta-analyses of studies on the effect of supplementing the diets of rheumatoid arthritis patients with fish oils for several months confirmed statistically significant improvements in tender joint and morning stiffness symptoms.

(3) There is accumulating evidence for amelioration by fish oils of other inflammatory disease processes

including psoriasis, asthma, Crooo's disease and ulcerative colitis.

(4) The substantial weight loss that occurs in patients with pancreatic cancer has been alleviated by daily supplements of fish oils.

Mechanisms

Membrane properties

The observed enrichment of specific dietary unsaturated fatty acids in the membranes of cells of the immune system immediately suggests the possibility that membrane physical properties could be affected. Changes in so-called membrane fluidity may modify activities of membrane-bound enzymes. For example, G-protein activity changes can result in changes in adenylate cyclase, phospholipase A_2 and phospholipase C activities. Alterations in membrane phospholipids can influence the production of lipid mediators such as diacylglycerols.

Availability of eicosanoid precursors

The pattern of production of eicosanoids can be modified not only by changes in the nature of cyclooxygenase substrates (e.g. arachidonic acid replaced by EPA), but also in their availability via changes in the activity of phospholipase A_2, which releases these cyclooxygenase substrates from membrane phospholipids. A general observation is that eicosanoids formed from arachidonic acid are pro-inflammatory, whereas those formed from PUFA of the *n*-3 family reduce inflammation.

Availability of vitamin E

PUFA are highly susceptible to peroxidation unless adequate vitamin E is available as an antioxidant. When diets that contain high levels of *n*-3 PUFA from fish oils are given, some diminution in the concentration of plasma vitamin E may be observed. This is because such diets contain relatively low levels of vitamin E. Concomitantly, some increase in the concentration of the products of lipid peroxidation may be measured in the plasma. Such changes do not occur when the sources of *n*-3 PUFA are plant oils, because they are richer in vitamin E than fish oils. There is limited evidence that the marked suppression by *n*-3 PUFA of lymphocyte proliferation in response to antigens or non-specific mitogens, does not occur when the diet is supplemented with vitamin E. Functions such as neutrophil phagocytosis are suppressed by vitamin E deficiency and enhanced by vitamin E supplementation. However, vitamin E supplementation decreases bactericidal activity, probably because such activity is promoted by reactive oxygen species, which are suppressed by vitamin E. Many authors have routinely supplemented diets with vitamin E, whether they contained high levels of *n*-3 PUFA or not, whereas in other studies there has been no supplementation. The evidence cited above suggests that functions such as lymphocyte proliferation are inhibited by peroxidation damage when vitamin E is limiting. Therefore, many of the results of studies on the effects of PUFA on immune function need to be interpreted with caution until the contribution of reactive oxygen species has been fully elucidated.

Gene expression

PUFA of the *n*-3 family appear to be able to modify gene expression directly or indirectly, depending on the gene in question. Direct effects are most probably mediated by their ability to bind to positive and/or negative regulatory transcription factors, whereas indirect effects seem to be mediated through alterations in the generation of intracellular lipid second messengers such as diacylglycerols or ceramides.

Implications for dietary advice

In many industrial countries large increases in intakes of linoleic acid *(n-6)* have occurred. mainly as a result of advice that such dietary changes are likely to result in lowering of plasma LDL-cholesterol concentration and a general reduction in coronary heart disease incidence and mortality. At the same time, intakes of *n*-3 PUF A have not increased, and may even have decreased, resulting in significant reductions in the ratio of dietary *n*-3/*n*-6 PUFA.

Many people have pointed out the correlation between the fall in the *n*-3/*n*-6 PUFA ratio and the concomitant rise in the incidence of inflammatory diseases such as asthma. Whereas such statistical correlations do not show that one is the result of the other, the foregoing section has illustrated how there may be some biological basis for such an association. Many nutritionists are now recommending increases in consumption of *n*-3 PUFA and advising against further increases in the *n*-6 family. Some of them are coupling their advice on *n*-3 PUFA supplementation with advice on vitamin E intakes, but others are not. It is clear that immune suppression may be advantageous in some conditions and hazardous in others. Moreover, different dietary components have different effects on immunity depending on the immune function under consideration and their relationship with other components of the diet. The current scientific debates about the appropriate inclusion of different fatty acids in the diet in relation to optimum health are important in reminding us that it is unwise to view the effects of a single nutrient in isolation. Interactions between a whole host of nutrients have to be taken into account.

LIPIDS AND CANCER

Lipids may influence processes leading to cancer development by affecting membrane properties, cell signalling pathways, gene expression and immune function. Some dietary lipids may exacerbate carcinogenesis, others retard or prevent it. Liposomes may be used to target drugs in cancer therapy.

Cancer represents the uncontrolled proliferation of cells. It arises from mutations in genes that normally control cell division or promote programmed cell death (apoptosis). The development of cancer involves a number of stages. Initiation is the initial process of DNA mutation; promotion is a stage at which a potentially cancerous cell, through accumulation of further mutations, becomes a fully cancerous cell. Then there is a stage of progression, or development by cell division, following which the cancerous cells may spread and invade other tissues (metastasis).

Lipids are involved in cancer in many ways. Dietary lipids, including micronutrients such as fat-soluble antioxidants, may play a role in predisposing to, or protecting from cancer. The development of cancer involves cellular changes that include alterations to lipid components of the cell (and this may be relevant to the design of new treatments). Finally, there may be a role for certain dietary lipids in the control or treatment of certain cancers or aspects of cancer.

Dietary lipids and cancer

There is a large body of evidence on the role of dietary lipids in cancer. It is based on epidemiological studies of various sorts in humans, and on feeding studies in animals. It should be stressed that the evidence is in all cases somewhat conflicting. In 1997-1998, two major reports on diet and cancer were published in which all existing evidence was reviewed (see Further Reading). The expert groups who compiled these reports assessed the literature and graded the evidence for associations between dietary components and specific cancers on a scale from 'convincing' to 'insufficient'. For dietary fat, no association was found to be convincing, and a few 'possible'. There was considered to be strong evidence, however, for a relationship between obesity (particularly central fat deposition) and risk of both endometrial cancer and breast cancer in post-menopausal women.

Ken Carroll, a Canadian biochemist who studied this subject extensively, wrote two thorough review articles shortly before his death in 1998, ironically from stomach cancer. In those reviews, Carroll distinguished indirect from direct effects. Indirect effects are mainly related to dietary fat as a form of energy, leading to overweight or obesity.

As reviewed above, the evidence is strong for a link between adiposity and breast cancer and for endometrial cancer, both in post-menopausal women. The most consistent theory implicates the production of oestrogens by the excess adipose tissue. Adipose tissue is an important site for the conversion of androgens (particularly androstenedione, secreted from the adrenal cortex) to oestrogens, oestradiol and oestrone. The development of some forms of breast and endometrial cancer is stimulated by oestrogens, and hence the association with dietary fat or obesity. The intake of some antioxidants, including tocopherols and lycopene (the red carotenoid pigment of tomatoes), has been linked to protection against certain cancers, including prostate. Even if these findings should be confirmed, it should be emphasized that cancer prevention may not be as simple as supplementation with the antioxidant vitamins at high doses. A number of supplementation studies has been carried out, none so far with positive results, and in one study of β-carotene supplementation, there was an excess of mortality from lung cancer (in those already at high risk).

Cancer of the colon is common, but its incidence varies widely between countries. In comparisons between countries, colon cancer incidence is correlated with meat and fat consumption. However, the interpretation of such studies needs caution as many other factors may be different between countries. The evidence from other types of study, such as case-control studies (where each 'case', or person suffering from the disease, is matched with a similar non-sufferer), for the link between meat consumption and colon cancer is very conflicting, with some studies showing clear positive links and others the opposite. There is moderately consistent evidence, however, for a protective effect of dietary fibre intake against colon cancer. Whilst dietary fibre (or non-starch polysaccharides) is not a dietary lipid, this association is relevant to this book. The mechanism responsible is believed by most to be that fermentation of non-starch polysaccharides by bacteria in the colon produces the short-chain fatty acids, acetic, propionic and butyric. Butyric acid is an important and preferred fuel for the colonic mucosa, but there is now also strong evidence that butyric acid exerts anti-proliferative effects on the mucosal cells.

Evidence for direct effects of dietary fat on cancer is based mainly on animal studies. Ken Carroll and other workers have consistently found that *n*-6 PUFA in the diet act as mammary tumour promoters when tumours are induced by a carcinogen such as 7,12-dimethylbenz(α)anthracene. Saturated fatty acids in the diet had no such effect, and *n*-3 PUPA have been found to inhibit tumorigenesis. However, the relevance to human cancer remains unclear. Likewise, the therapeutic usefulness of CLA for human cancer patients remains to be demonstrated.

Cellular lipid changes in cancer

Cells express a number of carbohydrate and lipid derivatives on their surfaces. Amongst these are glycosphingolipids, which are involved in cellular recognition, cell adhesion and regulation of cell growth. In 1968 the Japanese biochemist Hakomori observed that transformed tissues (i.e. cancer cells) expressed variant forms of glycosphin-golipids, often in larger amounts than normal. These glycosphingolipids may play a role in the uncontrolled cell growth of cancer. There are characteristic changes in glycosphingolipids in different cancers. In general, there is a reduction in the more complex glycosphingolipidsand an over-expressionof some of the more unusual glycosphingolipids that may normally only be present in trace amounts.

The more complex glycosphingolipids act as cell surface antigens; that is, they may be recognized by specific antibodies. The blood group antigens, which cause an immune reaction if blood of one group is infused into someone of another blood group, are of this type. These cell-surface antigens are released to some extent into the circulation and may therefore provide a means of diagnosis of particular cancers. Equally exciting is the idea that specific antibodies against these antigens might be used to target drugs to the cancer cells. Either drugs could be coupled to the antibodies, or the presence of the antigens might be used as the basis for developmentof a specific vaccine. These ideas have been promoted consistently by Hakomori, who is still very active in the field. Amongst Hakomori's suggestions is the idea that the glycosphingolipids cf cancer cells are involved in tumour cell adhesion and in signal transduction, and that it may be possible to develop therapeutic agents that disrupt adhesion ('anti-adhesion therapy') or normalize signalling pathways ('ortho-signalling therapy')or normalize signalling pathways ('ortho-signalling therapy'). Although trials in cell culture have demonstrated the possibility of antibody-directed tumour killing, these ideas have yet to be developed to a clinically useful stage.

Lipids and the treatment of cancer

The expression of variant forms of glycolipids on cell surfaces may offer a means of directing specific drugs to kill the tumour cells, as discussed above. In addition, several potential new treatments based on phospholipid analogues have been developed and in some cases tested in clinical trials. These compounds are intended to incorporate into target membranes, either the cell membrane or intracellular membranes such as the endoplasmic reticulum. Thus, their mode of action differs from conventional anti-cancer drugs that target DNA replication. Several types of compounds have been tested, although most are analogues of phosphatidylcholine or of lysophosphatidy1choline, e.g. with a sugar molecule ether-linked at the *sn*-2 position. They are believed to act in diverse ways, all of which could be seen as targeting cells with a high rate of cell division. They may act by disrupting signal transduction and signalling pathways, e.g. inhibition of the phosphoinositide-specific phospholipase C with suppression of the diacylglycerolprotein kinase C pathway, or inhibition of phosphocholine cytidylyltransferase leading to inhibition of phosphatidylcholine biosynthesis *de novo*. These compounds have shown great promise in cellular systems, but their clinical application has been limited by problems of toxicity to the intestinal tract (whose cells are also characterized by a high rate of division). They have found some application as topical therapies (direct application, e.g. to skin lesions).

Cancer causes death through a number of mechanisms including direct invasion of critical organs. However, a common and life-threatening feature of many cancers is a marked loss of body fat, cachexia. There have been many investigations of the cause of cancer cachexia. One theory is that a circulating factor, known as cachectin, leads to inhibition of adipose tissue lipoprotein lipase and consequent failure of fat storage (so there is continuous net fat loss). Cachectin is now known to be identical to tumour necrosis factor-α, a cytokine produced by many cell types including macrophages and adipocytes. The evidence for this mechanism has largely been based on animal studies. In humans, measurements of energy intake have shown that the loss of body fat is mainly a problem of energy balance. Patients lose appetite and in many cases involving cancers of the gastrointestinal tract have difficulty eating. There have been some interesting developments in the treatment of cancer cachexia by the oral administration of relatively large amounts of *n*-3 PUFA (1-2g day^{-1} of eicosapentaenoic acid, 20:5*n*-3), together with a conventional nutritional supplement. Patients who were previously losing weight have shown an increase in body weight during this treatment. It is likely that the mechanism relates to a general antiinflammatory effect of large doses of the *n*-3 PUFA, mediated in part by their role as precursors of eicosanoids of the 3-series rather than the 2-series. The *n*-3 PUFA also tend to reduce the production of pro-inflammatory cytokines including tumour necrosis factor-α and interleukin-6.

SUMMARY

Lipids are important constituents of the diet, although the proportion of energy provided by dietary lipids varies widely between species, between different human societies and between individuals. Quantitatively, the most important lipids in the human diet are the triacylglycerols that are provided mainly by the adipose tissue of farm animals, the oils of fish and the seed oils of plants. Their fatty acids may be predominantly saturated and monounsaturated, as in ruminant fats, coconut and palm oils, di- and triunsaturated as in many seed oils or highly polyunsaturated as in many marine oils. Their physical form and chemical composition is usually modified during food manufacture and processing, for example in the hydrogenation employed in the manufacture of margarine to reduce the degree of unsaturation.

Lipids confer palatability to the diet through the textural

properties they give to foods and through low molecular weight flavour and odour compounds produced by enzymic oxidation, heating and fermentation. Modem industrial techniques now allow the production of dietary fats with reduced or even zero energy value and of different textures. Triacylglycerols provide a concentrated form of dietary energy and act as vehicles for the fat-soluble vitamins. They also provide the fatty acids required for cellular structures such as biological membranes. Whereas the human body has the capacity to synthesize for itself saturated and monounsaturated fatty acids and cholesterol, it is unable to synthesize linoleic and α-linolenic acids (essential fatty acids) and vitamins A, E and K. The amounts required daily in the diet can be assessed, albeit rather imprecisely. Requirements vary between individuals and depend on the rate of growth, age and physiological states such as puberty, pregnancy and lactation.

Linoleic and α-linolenic acids are converted in many tissues into two series of longer chain, more highly polyunsaturated fatty acids, classified as the *n*-6 and *n*-3 families, respectively. These are important in contributing to the structural integrity of membranes and in acting as precursors for a range of oxygenated fatty acids, the eicosanoids. The latter are produced locally, have short half-lives and exert powerful physiological activities such as muscle contraction and platelet aggregation at extremely low concentrations. Although many human tissues contain the enzymes for converting the parent fatty acids of the *n*-6 and *n*-3 families (linoleic and α-linolenic) into longer chain, more highly polyunsaturated fatty acids by a series of desaturations and elongations, the measured activities of these enzyme systems are often very weak. There is uncertainty, therefore, as to whether essential fatty acid requirements can always be satisfied by dietary consumption of the parent fatty acids alone, or whether there may sometimes be a specific requirement for the long-chain polyunsaturates. This additional requirement may be particularly acute in times of high demand, for example during brain development, especially in premature infants. There is continuing debate about the potential requirements for long-chain polyunsaturates which can only be resolved by further research.

The metabolism of essential fatty acids is influenced by the amounts and types of other fatty acids in the diet as a result of competition for the elongases and desaturases. Consequently, the ratio between the different families of unsaturated fatty acids in the dietary lipids may be as important in nutrition as the absolute amounts of dietary linoleic and ex-linolenic acids.

Lipids influence immunity in several ways. Some glycolipids are components of cell-surface antigens. Dietary fat enhances immunity in malnourished people through its energy contribution. Individual lipids have more specific roles through effects on membrane structure, eicosanoid production, peroxidation and gene expression.

Lipids may influence processes leading to cancer development by affecting membrane properties, cell signalling pathways, gene expression and immune function. Some dietary lipids may exacerbate carcinogenesis, others retard or prevent it. Liposomes may be used to target drugs in cancer therapy.

Chapter 3

Lipid Transport

DIGESTION AND ABSORPTION

To be usable by the body, dietary fats must be digested in the lumen of the small intestine. The digestion products pass through the gut wall and are resynthesized in the intestinal epithelial cells and packaged for transport in the bloodstream.

Intestinal digestion of dietary fats involves breakdown into their component parts by a variety of digestive enzymes

Typically 90-95% of fat in the human diet is provided by triacylglycerols, with smaller contributions from phospholipids and cholesterol. The dietary triacylglycerols must be partially hydrolysed before the body can assimilate them.

At birth, the new-born animal has to adapt to the relatively high fat content of breast milk after relying mainly on glucose as an energy substrate in foetal life. It is presented with two major problems in fat digestion: the pancreatic secretion of lipase is rather low and the immature liver is unable to provide sufficient bile salts to solubilize the digested lipids. These problems are even more acute in the premature infant. Yet the new-born baby can digest fat, albeit less efficiently than the older child or adult. There are two mechanisms that allow this to occur. Breast milk contains a lipase that is stimulated by bile salts, known as bile-salt stimulated lipase (BSSL). Milk also contains lipoprotein lipase, although this is not thought to play a role in fat digestion. BSSL may assist the neonate in digesting the milk triacylglycerols. In addition, the neonate secretes a triacylglycerol lipase from glands in the stomach. This is an acid lipase, with a pH optimum of around 4-6, although it is still active even at a pH of 1. In other mammals a homologous lipase may be secreted higher up the gastrointestinal tract–from the serous glands of the tongue in rodents (lingual lipase) and from the pharyngeal region in ruminants. The realization that humans (neonatal and adult) secrete a gastric lipase is relatively recent, and there are still references to human lingual lipase in the literature.

The products of these lipases are mainly 2-monoacylglycerols, diacylglycerols and non-esterified fatty acids, the latter being relatively richer in medium-chain length fatty acids than the original acylglycerols. The milk fat of most mammals is relatively rich in medium-chain length fatty acids (8C-12C) rather than the usual 16C and 18C fatty acids. The relative ease with which lipids containing medium-chain fatty acids can be absorbed certainly helps lipid uptake in babies. Human milk triacylglycerols (along with those of some other mammals including pigs) have a predominance of palmitic acid esterified at position 2. This will lead to the production of 2-palmitoylglycerol, which is readily absorbed (see below). Since the absorption of saturated (non-esterified) fatty acids (which are solid at body temperature unless emulsified) may be less efficient that that of unsaturated fatty acids, the structure of human milk triacylglycerols seems to be a way of optimizing absorption of palmitic acid.

As the baby is weaned onto solid food, the major site of fat digestion shifts from the stomach to the duodenum. Gastric lipase continues to play a role, however, even into adulthood. It has been estimated that gastric lipase is responsible for 25% of the partial triacylglycerol hydrolysis necessary for absorption to occur. In addition, the action of gastric lipase seems to produce lipid droplets that are better substrates for the later action of pancreatic lipase. The stomach also plays a role with its churning action, creating a coarse oil-in-water emulsion, stabilized by phospholipids. Also, proteolytic digestion in the stomach serves to release lipids from the food particles where they are generally associated with proteins. The acidic fat emulsion that enters the duodenum from the stomach is neutralized and modified by mixing with bile and pancreatic juice. Bile supplies bile salts that in humans are mainly the glycine and taurine conjugates of tri- and di-hydroxycholanic acids (usually known as cholic and chenodeoxycholic acid), formed from cholesterol in the liver and phospholipids. Much of the intestinal phospholipid in humans is of biliary origin and is estimated at between 7 and 22 g day^{-1} compared with a dietary contribution of 4-8 gday^{-1}. Biliary secretion is enhanced by the enzyme cholecystokinin, secreted from the duodenal mucosa, in response to entry of the acidic mixture from the stomach (chyme) into the duodenum. Pancreatic juice, whose secretion is also stimulated by cholecystokinin, supplies bicarbonate to neutralize the acidic chyme, and enzymes that catalyse the hydrolysis of fatty acids from triacylglycerols, phospholipids and cholesterol esters (Fig. 3.1).

In most adults the process of fat digestion is very efficient and the hydrolysis of triacylglycerols is mainly accomplished in the small intestine by a triacylglycerol lipase secreted from the pancreas, pancreatic lipase. This enzyme is related to lipoprotein lipase and hepatic lipase. Pancreatic lipase attacks triacylglycerol molecules at the surface of the large emulsion particles (the oil-water interface), but before lipolysis can occur, the surface and the enzyme must be modified to allow interaction to take place. Firstly, bile salt molecules accumulate on the surface of the lipid droplet, displacing other surface-active constituents. As amphipathic molecules they are adapted for this task since one side of the rigid planar structure of the steroid nucleus is hydrophobic and can essentially dissolve in the oil surface. The other face contains hydrophilic groups that interact with the aqueous phase (Fig. 3.2).

The presence of the bile salts donates a negative charge to

the oil droplets, which attracts a protein to the surface. This protein has a molecular mass of 10 kDa and is called colipase. Colipase, also secreted from the pancreas, is essential for the activity of pancreatic lipase, which is otherwise strongly inhibited by bile salts. Thus, bile salts, colipase and pancreatic lipase interact in a ternary complex, which also contains calcium ions that are necessary for the full lipolytic activity.

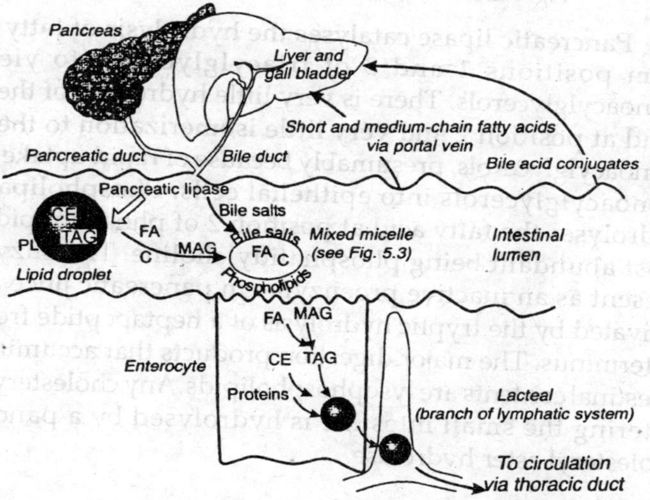

Fig. 3.1 The digestion and absorption of dietary fat in the small intestine. Lipid droplets entering the small intestine from the stomach are subjected to the action of pancreatic lipase, phospholipase A_2 and cholesterol esterase, which hydrolyse triacylglycerol (TAG) to produce monoacylglycerols (MAG) and fatty acids (FA), phospholipids (PL) to produce lysophospholipids and fatty acids; and cholesterol esters (CE) to liberate cholesterol (C) and fatty acids. These are emulsified with bile salts (from the gall bladder) to produce a micellar suspension (the mixed micelles) from which components are absorbed across the epithelial cell (enterocyte) membranes. Short- and medium-chain fatty acids pass through into the circulation (into the hepatic portal vein), and bile salts are reabsorbed, along with further cholesterol, in the lower part of the small intestine. Within the enterocyte, the components are reassembled, and packaged into chylomicrons, the largest of the lipoprotein particles (see Fig. for further details of this process). The chylomicrons are secreted into small branches of the lymphatic system, the lacteals.

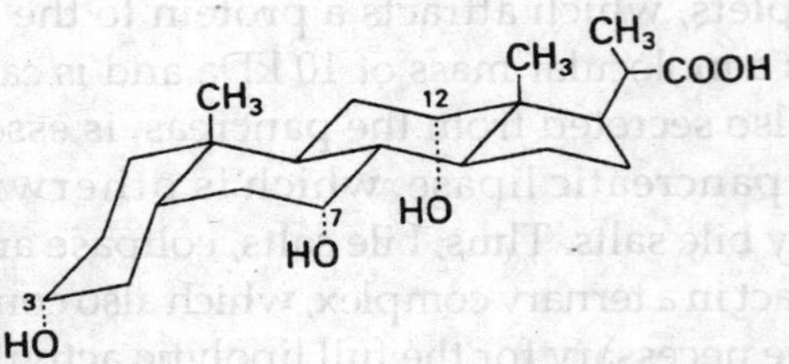

Fig. 3.2 The structure of a bile acid (cholic acid).

Pancreatic lipase catalyses the hydrolysis of fatty acids from positions 1 and 3 of triacylglycerols to yield 2-monoacylglycerols. There is very little hydrolysis of the ester bond at position 2 and very little isomerization to the 1(3)-monoacylglycerols, presumably because of rapid uptake of the monoacylglycerols into epithelial cells. Phospholipase A_2 hydrolyses the fatty acid at position 2 of phospholipids, the most abundant being phosphatidylcholine. The enzyme is present as an inactive proenzyme in pancreatic juice and is activated by the tryptic hydrolysis of a heptapeptide from the N-terminus. The major digestion products that accumulate in intestinal contents are lysophospholipids. Any cholesteryl ester entering the small intestine is hydrolysed by a pancreatic cholesteryl ester hydrolase.

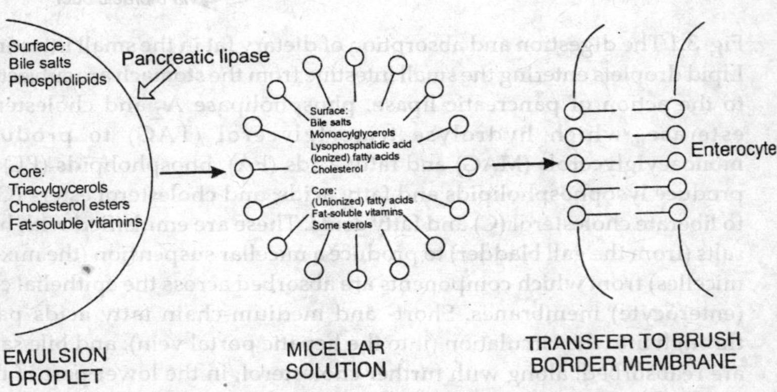

Fig. 3.3 Role of the mixed micelle in fat absorption.

As these enzymes act upon the contents of the large emulsion particles, which may be around l) μm in diameter, the products of their hydrolytic action disperse and form multimolecular aggregates called mixed micelles (Fig. 3.3), typically 4-6 nm in diameter. Lysophospholipids and monoacylglycerols are highly amphipathic substances and, with bile salts, stabilize these aggregates. In the more neutral environment of the small intestine, fatty acids are found largely in the ionized form and are therefore also amphipathic. The presence of these amphiphiles helps to incorporate insoluble non-polar molecules like cholesterol and the fat-soluble vitamins into the micelles and aid their absorption.

In ruminant animals, the complex population of micro-organisms contains lipases that split triacylglycerols completely to glycerol and nonesterified fatty acids. Some of these are fermented to acetic and propionic acids, which are absorbed directly from the rumen and carried to the liver, where they are substrates for gluconeogenesis. A proportion of the remaining long-chain unsaturated fatty acids undergoes several metabolic transformations catalysed by enzymes in the rumen micro-organisms before passing into the small intestine where they are absorbed. Principal among these is hydrogenation, in which double bonds are reduced by a process that is strictly anaerobic. During hydrogenation, some double bonds are isomerized from the *cis* to the *trans* geometrical configuration. Positional isomerization also occurs, in which the double bonds, both *cis* and *trans,* migrate along the carbon chain. The result is a complex mixture of fatty acids, generally less unsaturated than the fatty acids in the ruminant's diet and containing a wide spectrum of positional and geometrical isomers. Because there is very little monoacyl glycerol present in the digestion mixture, the mixed micelles in ruminants are composed largely of non-esterified fatty acids, lysophos-pholipids and bile salts.

The intraluminal phase of fat absorption

Lipid absorption in humans begins in the distal duodenum and is completed in the jejunum. The principal molecular species passing across the brush-border membrane of the enterocyte are the monoacylglycerols and non-esterified long-chain fatty acids. The bile salts themselves are not absorbed in the proximal small intestine, but pass on to the ileum where they are absorbed and recirculated in the portal blood to the liver, from whence they are re-secreted in the bile. This is also true for some cholesterol that is a constituent of bile. The recirculation of bile constituents in this way is referred to as the entero-hepatic circulation. It may be interrupted by resins (given by mouth) that bind cholesterol and bile salts and prevent their reabsorption, leading the liver to synthesize new bile salts from cholesterol. This is therefore a mechanism for depleting the body of cholesterol and may be useful in the treatment of high cholesterol levels.

The digestion products encounter two main barriers to their absorption. At the surface of the microvillus membrane is a region known as the unstirred water layer. The mixed micelles are small enough to diffuse readily into this layer. It is a few hundred micrometres thick and is retained by mucopolysaccharides secreted by the epithelial cells (mucosa). Its pH is relatively acidic. This will promote the protonation of non-esterified fatty acids (i.e. they become uncharged) so that they can more easily leave the micelles and move into the epithelial cell membrane. This membrane is highly convoluted (hence the description brush-border) to increase its absorptive capacity. There is a suggestion that pancreatic lipase may be loosely bound to the brush-border membrane, which would bring the products of its action even more closely into contact with the absorptive surface.

The second barrier is the brush-border membrane itself. It has been speculated for many years that fatty acids in their protonated form might diffuse across this membrane (and other cell membranes). However, it now looks increasingly likely that there is a specific transport protein to facilitate their movement into the cell. Two candidate proteins have been identified in enterocytes: plasma membrane fatty acid binding protein (FABPpm) and fatty acid translocase (FAT). The former, despite its name, is not related structurally to the family of intracellular fatty acid binding proteins.

FAT has been identified with a cell surface receptor CD36, which is a member of the family of scavenger receptors. The relative importance of these two proteins in fatty acid uptake by enterocytes is unknown, if indeed they play any role. This is an area of intense interest at present.

The intracellular phase of fat absorption

For efficient absorption into the enterocytes to occur, it is essential that an inward diffusion gradient of lipolysis products is maintained. Two cellular events ensure that this occurs. First, the long-chain fatty acids entering the cells bind to a cytosolic fatty acid binding protein (FABPc). There are two FABPc expressed in enterocytes. One is found only in the intestine and is called I-FABPc (also known as FABP2); the other is known as liver FABP (L-FABPc) and is also expressed in liver and kidney. They are smail proteins (molecular mass 14-15kDa) and bind one or two fatty acids, respectively. The FABPc are believed to playa role in targeting fatty acids within the cell, and also in protecting the cell from the potentially cytotoxic effects of high fatty acid concentrations. They also provide the concentration gradient that ensures efficient fatty acid uptake from the intestinal lumen. It is not clear whether the two FABPc isoforms have different roles, although it has been suggested that I-FABPc is involved in the uptake of dietary fatty acids for triacylglycerol synthesis, whilst L-FABPc in enterocytes is involved in uptake of fatty acids delivered in the blood, which may be substrates for oxidation and for phospholipid synthesis. Both these FABPc bind long-chain unsaturated fatty acids with higher affinity than saturated fatty acids, and this may explain the more rapid absorption of oleic than stearic acid.

Up to this stage, the absorption process has not been dependent on a source of energy. The next phase, which removes fatty acids, thereby maintaining a gradient, is the energy-dependent reesterification of the absorbed fatty acids into triacylglycerols and phospholipids. The first step in re-esterification is the ATP-dependent 'activation' of fatty acids tc their acyl-CoA thioesters. Again, the preferred substrates are the long-chain fatty acids. There are several isoforms of acyl CoA synthetase (ACS), one of which, ACS5, is prominent in the intestine. The ACSs are all membrane-associated and are situated on the endoplasmic reticulum as well as on mitochondrial and peroxisomal membranes. The acyl-CoA thioesters are themselves bound by a specific, 10 kDa, cytosolic acyl-CoA binding protein which, like FABPc, may direct the acyl-CoA within the cell.

In humans and other simple-stomached animals, the major acceptors for esterification of acyl-CoA are the 2-monoacylglycerols that together with the non-esterified fatty acids are the major forms of absorbed lipids. Resynthesis of triacylglycerols, therefore, occurs mainly via the monoacylglycerol pathway. In ruminant animals, the major absorbed products of lipid digestion are glycerol and non-esterified fatty acids and resynthesis occurs via the glycerol phosphate pathway after phosphorylation of glycerol catalysed by glycerol kinase. It should be noted, however, that lipids usually form a minor part of ruminant diets, so that glucose is probably the major precursor of glycerol phosphate in the ruminant enterocyte.

The main absorbed product of phospholipid digestion is monoacylphosphatidylcholine (lysophosphatidylcholine). A fatty acid is re-esterified to position 1 to form phosphatidylcholine by an acyl transferase located in the villus tips of the intestinal brush-border. The function of this phospholipid will be to stabilize the triacylglycerol-rich particles (chylomicrons) exported from the cell as described later. It is thought that the phosphatidylcholine used for the synthesis and repair of membranes in the enterocytes (cells with a rapid turnover) is synthesized by the classical CDP-choline pathway in cells at the villus crypts.

The absorption of cholesterol is slower and less complete than that of the other lipids, about half of the absorbed sterol being lost during desquamation of cells. Most of the absorbed portion is esterified either by reversal of cholesteryl esterase or via acyl-CoA:cholesterol acyl transferase. The latter enzyme is induced by high concentrations of dietary cholesterol.

During fat absorption, the biosynthetic activity of the enterocyte is geared to packaging the resynthesized absorbed lipids in a form that is stabilized for transport in the aqueous environment of the blood. Within minutes of absorption products entering the enterocyte, fat droplets can be seen within the cysternae of the smooth endoplasmic reticulum, where the enzymes of the monoacylglycerol pathway are located. The rough endoplasmic reticulum is the site of the synthesis of phospholipids and apolipoproteins, which provide the coat that stabilizes the lipid droplets. One apolipoprotein in particular, apolipoprotein B48 (apoB48), provides a 'skeleton' which associates with lipid during its synthesis, forming the immature chylomicron particle. These immature particles gain further triacylglycerol in a process that involves the microsomal triacylglycerol transfer protein (MTP). They migrate through the Golgi apparatus where carbohydrate moieties are added to the apolipoproteins, and the fully formed chylomicrons are exported in secretory vesicles that move to the basolateral surface of the enterocyte. The final phase of transport from the cells involves fusion with the membrane and secretion into the intercellular space by exocytosis (Fig. 3.1). Chylomicron assembly in the enterocyte is similar to the assembly of very low density lipoprotein (VLDL) particles in hepatocytes.

The chylomicrons do not enter the plasma directly. Instead they are secreted into the tiny lymph vessels that are found inside each of the intestinal villi, called lacteals because of their milky appearance after the ingestion of fat. From here the chylomicrons pass via the thoracic duct (a main branch of the lymphatic system) and enter the circulation in the subclavian vein, from where they reach the heart for distribution around the body. Thus, dietary lipids, contained in the chylomicrons, are unique amongst the products of intestinal digestion and absorption in that they do not enter the hepatic portal vein and traverse the liver before entering the systemic circulation.

Chylomicrons are the main route for the transport of dietary long-chain fatty acids. Those with chain lengths of less than twelve carbon atoms are absorbed in the non-esterified form, passing directly into the portal blood and are metabolized directly by β-oxidation in the liver. There are several reasons for this partition. Firstly, short- and medium-chain fatty acids are more readily hydrolysed from triacylglycerols and since they occupy mainly position 3, are not retained in the 2-

monoacylglycerols. Secondly, they are more likely to diffuse into the aqueous phase rather than the mixed micelles and for this reason are more rapidly absorbed. Around 4 g day^{-1} short- and medium-chain fatty acids enter the diet from dairy products or foods that incorporate coconut or palm kernel oils. Short-chain fatty acids derived by microbial fermentation of non-starch polysaccharides (dietary fibre) in the colon are also absorbed and contribute to lipid metabolism.

Malassimilation of lipids, through failure to digest or absorb lipids properly, can arise from defects in the gut or other tissues but may also be induced deliberately

Failure to assimilate lipids of dietary origin into the body may arise from defects in digestion (maldigestion) or absorption (malabsorption).

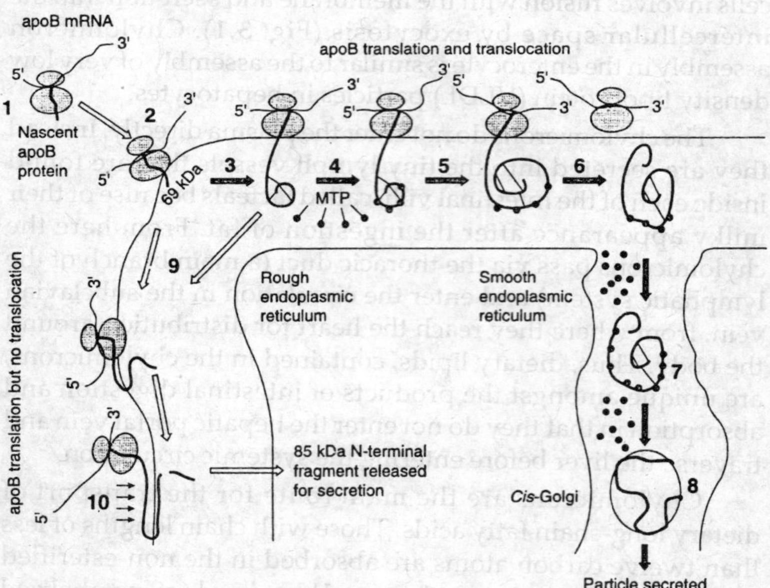

Fig. 3.4 Assembly of triacylglycerol-rich lipoprotein particles, chylomicrons in the enterocyte, and VLDL in the hepatocyte. Apolipoprotein B48 (apoB100 in hepatocytes) is synthesized on the rough endoplasmic reticulum (ER) (upper left). At step 4, the N-terminal region acquires a small amount of 'core lipid' (triacylglycerols and cholesteryl

esters) in a process mediated by the microsomal triacylglycerol transfer protein (MTP). After further protein chain synthesis (step 5), a lipid-poor 'primordial' particle is produced. There is debate about whether this is released into the ER lumen as shown at step 6, or whether the next stage, of bulk lipid addition (step 7), occurs whilst the primordial particle is still attached to the ER membrane. There is also debate about whether MTP is involved in this stage also (consensus is that it probably is). The left side of the diagram indicates that if lipid availability is low and lipid addition fails to occur, then the nascent apoB is directed into a degradative pathway (steps 9 and 10).

Maldigestion can occur because of incomplete lipolysis. Thus, pancreatic insufficiency, which may result from pancreatitis, pancreatic tumour, diseases of malnutrition such as kwashiorkor, or rarely from an inherited mutation in pancreatic lipase, can lead to a failure to secrete enough lipase or the production of an enzyme with reduced or no activity. Alternatively, the lipase may be fully functional but a failure to produce bile (generally arising from hepatic insufficiency) may result in an inability to effect the micellar solubilization of lipolysis products. This, in turn, feeds back to cause an inhibition of lipolysis. Gastric disturbances that result in abnormal acid secretion also inhibit pancreatic lipase and, furthermore, gastric problems may cause poor initial emulsification of the lipid in the stomach, further reducing the efficiency of digestion. Thus, maldigestion can arise from defects in a variety of organs contributing to different aspects of the digestive process.

Malabsorption may occur, even when digestion is functioning normally, due to defects in the small intestine affecting the absorptive surfaces. There may be a variety of causes, some common ones being bacterial invasion of the gut or sensitization of the gut to dietary components such as gluten, as in coeliac disease. Malabsorption syndromes (often called sprue) are characterized by dramatic changes in the morphology of the intestinal mucosa. The epithelium is flattened and irregular and atrophy of the villi reduces the absorbing surface. Tropical sprue is a prevalent disease in many countries of Africa and Asia. Malabsorption of fat can also occur in certain inherited disorders in which the formation of chylomicrons is impaired, so that fat cannot be transported out of the cell. Triacylglycerols then accumulate in the enterocyte. One such condition is known as chylomicron retention disease or Anderson's disease. The molecular defect is not known.

Digestion and absorption may also be interfered with deliberately to reduce fat absorption in an attempt to control body weight. Synthetic fat substitutes, in particular sucrose-polyesters (sucrose with long-chain fatty acids esterified to its hydroxyl groups), can be used in cooking, and are completely resistant to the action of pancreatic lipase. They are used at present in some snack foods that might otherwise contribute significantly to energy intake. Similarly, a bacterial metabolite known as tetrahydrolipstatin (or orlistat) is licensed for prescription as a drug to treat obesity. Tetrahydrolipstatin is a potent irreversible inhibitor of pancreatic lipase and so reduces the absorption of dietary fat. Certain plant sterols (phytosterols) can interfere with the absorption of cholesterol of both dietary and biliary origin, and are being marketed as components of spreads that can help to lower cholesterol levels in the blood.

A common feature of all fat malassimilation syndromes is increased excretion of fat in the faeces (steatorrhoea), which arises not only from unabsorbed dietary material but also from the bacterial population that usually proliferates in the gut and the breakdown of cells. The bacteria undoubtedly affect the composition of the excreted fat. For example, a major component (absent from the diet) of faecal fat, 10-hydroxy-stearic acid, was shown by tracer studies to be formed by bacteria from stearic acid. This is a normal component but found in particularly high concentration in patients with steatorrhoea.

Patients with poor fat absorption are very much at risk from deficiencies of energy, fat-soluble vitamins and of essential fatty acids. The clinical management of fat malassimilation is facilitated, according to the nature of the defect, by giving bile salts in tablet form or a pancreatic extract containing pancreatic lipase and other enzymes, or by replacing normal dietary fats by medium-chain triacylglycerols (MCT). This product is composed largely of triacylglycerols with C8 and C10 saturated fatty acids refined from coconut oil, and can be purchased as a cooking oil or as a fat spread. The medium-chain fatty acids are rapidly hydrolysed and efficiently absorbed into the portal blood, thereby bypassing the normal absorptive route of long-chain fatty acids and chylomicron formation.

Some people are unable to take any nutrition by mouth, usually because of inflammatory conditions of the bowel or infection following surgical intervention. In early attempts at complete nutrition of such people using the intravenous route, their energy requirements were supplied in the form of glucose and amino acids. Several such patients developed symptoms of essential fatty acid deficiency. In addition, the large carbohydrate loads caused metabolic problems, not least because such patients often have a reduced ability to handle glucose, associated with their inflammatory condition. In the early 1960s Arvid Wretlind in Sweden developed the first synthetic lipid emulsion that could be infused intravenously. It was based on an emulsion of soybean oil stabilized with lecithin, and was soon available commercially. Now there is a number of such intravenous lipid preparations with different combinations of fatty acids, including some enriched with the very long-chain polyenoic fatty acids 20:5*n*-3 and 22:6*n*-3.

TRANSPORT OF LIPIDS

Lipids, being insoluble in water, need to be combined with proteins for transport in the blood. These plasma lipoproteins represent a continuum of lipid-protein complexes in which the ratio of lipid to protein, and hence their density, varies.

Types of Lipoproteins

The biological problem of how to transport water-immiscible lipids in the predominantly aqueous environment of the blood has been solved by emulsification. Aggregates of hydrophobic molecules (particularly triacylglycerol and cholesteryl esters) are stabilized with a coat of amphipathic compounds: phospholipids, unesterified cholesterol and proteins. The resulting particles are the lipoproteins. The protein moieties are known as apolipo-proteins and, as will be discussed later, have more than a stabilizing role. They also confer specificity on the particles, allowing them to be recognized by specific receptors on cell surfaces, and they regulate the activity of some enzymes involved in lipoprotein metabolism.

The lipoprotein system developed quite early in evolution. Birds, fishes, amphibians and even roundworms have a system for delivery of lipids from lipogenic organs to the developing egg that is closely related to the mammalian lipid transport system, and some of the proteins involved are homologous. For instance, the protein involved in lipid transport in these various groups is known as vitellogenin and is related to mammalian apolipoprotein B. The vitellogenin receptor of the chicken oocyte will, in fact, recognize mammalian apolipoprotein B, suggesting a common origin with mammalian lipoprotein receptors. The mammalian lipase gene family (pancreatic, lipoprotein, hepatic and endothelial lipases) is homologous to proteins in egg yolk of *Drosophila* that are presumably involved in related functions.

There are several types of lipoproteins with differing chemical compositions, physical properties and metabolic functions (Table 5.1), but their common role is to transport lipids from one tissue to another, to supply the lipid needs of different cells. The different lipoproteins may be classified in a number of ways depending on their origins, their major functions, their composition, physical properties or method of isolation. They differ according to the ratio of lipid to protein within the particle as well as in the proportions of lipids (triacylglycerols, esterified and non-esterified cholesterol and phospholipids). These compositional differences influence the density of the particles: in general, the higher the lipid to protein ratio, the larger the particle and the lower its density. There is a strong relationship between biological function and the density class into which a particle falls. It is convenient, therefore, to make use of density to separate and isolate lipoproteins by ultracentrifugation, and it is now usual to classify plasma lipoproteins into different density classes. It is almost by good fortune for the lipid biochemist that this classification is also, at least approximately, one of function. From lowest to highest density, these are: chylomicrons, very low density lipoproteins (VLDL),low density lipoproteins (LDL) and high density lipoproteins (HDL). (An additional class of intermediate density lipoproteins, IDL, is sometimes included between VLDL and LDL.) As density increases, so the ratio of triacylglycerols to phospholipids and cholesterol decreases (Table 5.1): chylomicrons and VLDL particles are often grouped as the triacylglycerol-rich lipoproteins, whereas LDL and HDL are more important as carriers of cholesterol.

It is important to emphasize that the classes are not homogeneous: there is a wide variety of particle sizes and chemical compositions within each class, and there is overlap between them. It is also important to realize that lipoprotein particles are not molecules—they are aggregates of lipid and protein molecules with a degree of structural organization. The term particle mass, rather than molecular mass, is used to describe their mass.

Table 3.1 Composition and characteristics of the human plasma lipoproteins

	Chylomicrons	VLDL	LDL	HDL
Protein (% particle mass)	2	7	20	50
Triacylglycerols (% particle mass)	83	50	10	8
Cholesterol (% particle mass; free+ esterified)	8	22	48	20
Phospholipids (% particle mass)	7	20	22	22
Particle mass (daltons)	0.4-30×10^6	10-100×10^6	2-3.5×10^6	1.75-3.6×10^5
Density range (g ml^{-1})	<0.95	0.95-1.006	1.019-1.063	1.063-1.210
Diameter (nm)	80-1000	30-90	18-22	5-12
Apolipoproteins	AI, AII, AIV, B-48, CI, CII, CIII, E	B-100, CI, CII, CIII, E	B-100	AI, AII, AIV, CI, CII, CIII, D, E

The apolipoproteins

The proteins associated with the plasma lipoproteins are known as apolipoproteins. At first a series of letters, A-E, was used to identify apolipoproteins as they were separated by electrophoresis, but it soon became apparent that most of these could be divided further into several individual proteins (Table 3.2). These are usually referred to in abbreviated form: apoAI, apoCIII, etc. The complete amino acid sequences of the nine major human apolipoproteins are now known [AI, AII, AIV, B (-48 and -100), CI, CII, CIII, D and E]. The apolipoproteins other than apoB are often referred to as soluble apolipoproteins. They may exist in lipid free form in the plasma and they may exchange between lipoprotein particles. Apolipoproteins of the groups A, C and E have similar gene structures and some homologous stretches of sequence, and are believed to have evolved from a common ancestral gene, whereas the genes for apoB and apoD have distinct structures. The genes for apoAI, CIII and AIV are arranged in one cluster spanning about 15 kb on human chromosome 11, and the genes for apoCI, CII and E in a cluster on chromosome 19.

The apolipoproteins fulfil two main functions. First, they stabilize the lipid particles in the aqueous environment of the blood, and maintain their structural integrity. Secondly, they are important in 'identifying' the lipoprotein and directing its metabolism in specific ways.

Much interest has centred on the properties of apoB since it is common to the triacylglycerol-rich lipoprotein particles (chylomicrons, secreted from the small intestine, and VLDL secreted from the liver) and cholesterol-rich LDL. It is now realized that apoB synthesis is intimately linked to the assembly of the triacylglycerol-rich lipoprotein particles. In fact the particle forms around one molecule of the protein (Fig. 3.4), which never leaves the particle until it is ultimately catabolized. There are two isoforms of apoB in mammals. They are coded for by the same gene. A full-length transcript produces a very large protein, of 4536 amino acids, known as apoBl00 (since it is 100% of the possible length). A specific enzyme system present in some cells edits the mRNA coding for apoB to introduce a stop codon about half-way along. This transcript codes for a shorter protein,

Table 3.2 Characteristics of the human apolipoproteins

Shorthand name	Molecular mass (daltons)	Ainino acid residues	Function	Major sites of synthesis	Chromosomal location
AI	28 000	243	Cholesterol efflux from cells; activates LCAT	Liver; intestirie	11q13-qter
AII	17 000	154	May inhibit HL activity; inhibits AI/LCAT	Liver; intestine	1p21-qter
AIV	44 500	376	Activares LCAT; some evidence for AN as a satiety factor	Intestine	11q13-qter
B48	241000	2152	Structural component of chylomicrons	Intestine	2p23-p24
B100	513000	4536	Structural component of VLDL; binds to LDL-R	Liver	2p23-p24
CI	6600	57	Activates LCAT	Liver	19q13.2
CII	8800	79	Activates LPL (essential cofactor)	Liver	19q13.2
CIII	8800	79	Inhibits LPL?; may be involved in hepatic uptake of apoE-containing particles	Liver; intestine	11q13-qter
D	22 000	169	Involved in	Brain, adipose	3p14.2-qter

			cholesteryl ester transfer?	tissue (role in brain may not be connected with lipid metabolism)	
E	34000	279	Ligand for LDL-R	Liver (60-80%); other tissues including adipose tissue (remainder)	19q13.2

The functions of many of the apolipoproteins have not been fully clarified. HL: hepatic lipase; LCAT: lecithin-cholesterol acyltransferase; LDL-R: LDL-receptor (B/E receptor); LPL: lipoprotein lipase.

apoB48 (2152 amino acids), so-called because it is 48% of the length of apoBl00. In humans the intestine secretes apoB48 and the liver apoBl00, and these proteins are therefore specific markers of chylomicrons and VLDL, respectively. In the rat, the liver secretes both apoBl00- and apoB48-containing VLDL particles, and the intestine, as in humans, only apoB48.

The different classes of lipoprotein particles transport

Chylomicrons are the largest and least dense of the lipoproteins and their function is to transport lipids of exogenous (or dietary) origin. Because the principal dietary fat is triacylglycerol, this is true also of the chylomicrons. The triacylglycerolcore is stabilized with a surface monolayer of amphipathic molecules: phospholipids, unesterified cholesterol and proteins (Table 3.1). The core lipid also contains some cholesteryl esters and minor fat-soluble substances absorbed with the dietary fats: fat-soluble vitamins, carotenoids and traces of environmental contaminants such as pesticides. It seems that the enterocyte continually secretes particles. Between meals these are small and poor in lipid, which is derived from biliary phospholipids or the lipids of cells shed from the gut mucosa, and from plasma non-esterified fatty acids. When fat is being absorbed the particles increase in size. Thus, increased lipid transport into the circulation after a fat-rich meal is achieved mainly by an increase in the size of the chylomicrons secreted, a mechanism that allows rapid changes in flux through this pathway.

VLDL particles, like chylomicrons, contain predominantly triacylglycerols (Table 3.1) and their function is to transport triacylglycerols of endogenous origin, synthesized in the liver. VLDL are spherical particles with a core consisting mainly of triacylglycerols and cholesteryl esters, with unesterified cholesterol, phospholipids and protein mainly on the surface (Fig. 3.5). As with the chylomicrons, each VLDL particle is assembled around one molecule of apoB (in humans, apoB100), which remains with the particle throughout its lifetime. The particles also contain apoCI, CII, CIII and apoE in variable amounts, decreasing relative to apoB as particle density increases (i.e. as the particles become smaller). The triacylglycerol-fatty acids secreted in VLDL are mobilized from stores within the hepatocyte by an unidentified lipase. They originate from plasma non-esterified fatty acids and triacylglycerol-fatty acids taken up by the liver, and to a variable extent from *de novo* synthesis in the liver. The glycerol 3-phosphate used for their esterification may come either from glucose metabolism or from the phosphorylation of glycerol, taken up by the liver (released in turn mainly from adipose tissue), by the enzyme glycerol kinase.

As in chylomicron biosynthesis in the intestinal mucosa, nascent VLDL particles originate in the lumen of the endoplasmic reticulum where they form around an emerging molecule of newly synthesized apoBl00. They acquire phospholipids and further triacylglycerol, the latter by a process involving the microsomal triacylglycerol transfer protein (MTP). A few individuals have been identified who have mutations rendering MTP inactive; such persons cannot secrete VLDL particles. The resulting syndrome is known as abetalipoproteinaemia (lack of apoBl00 in plasma). The mature particles move to the Golgi apparatus where some of the apolipoproteins are glycosylated. The Golgi vesicles then migrate to the cell surface where VLDL are exported by exocytosis into the subendothelial space, the space of Disse. The regulation of VLDL assembly and secretion is complex. A proportionof all the apoB100 synthesized fails to associate with triacylglycerol and enters a degradative pathway. Some nascent particles fail to acquire sufficient triacylglycerol to become 'competent' for further loading with triacylglycerol, and are also degraded. Thus, increased availability of hepatic triacylglycerol may lead to a greater proportion of nascent VLDL particles reaching sufficient size for secretion.

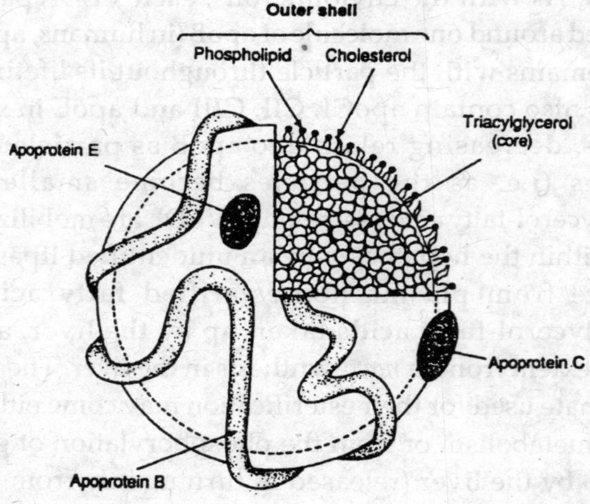

Fig. 3.5 Structure of a VLDL particle. (Other lipoprotein particles are similar in structure.) A monolayer of amphipathic molecules (phospholipids, unesterified cholesterol) stabilizes a droplet of hydrophobic lipids (triacylglycerols and cholesteryl esters) in the core. Apolipoprotein B wraps around the surface of the molecule, and other (soluble/apolipoproteins are present on the surface.

LDL particles are the major carriers of plasma cholesterol in humans although this is not the case for all mammalian species. They are derived from VLDL in the plasma by a series of degradative steps that remove triacylglycerols, resulting in a series of particles that contain a progressively lower proportion of triacylglycerols and are correspondingly richer in cholesterol, phospholipids and protein. The intermediate particles (IDL) are usually present at relatively low concentrations. During the transformations, the apoB component remains with the LDL particles and the apoC and E components are progressively lost. Once only apoB remains, the particle is a mature LDL particle. ApoB has an important role in the recognition of LDL by cells since it must interact with specific cell-surface receptors before the LDL particle can be taken up and metabolized by the cell. Other receptors (for example on macrophages) recognize modified LDL and are responsible for the degradation of LDL particles that cannot be recognized by normal cell-surface LDL-receptors.

There is a variety of types of particle classified as HDL. They are very numerous in plasma: it has been estimated that there are 10-20 times as many HDL particles in the circulation as all other lipoprotein particles combined HDL precursor particles have not been found within cells. Liver cells in culture secrete apoAI, and in vivo it seems that apoAI becomes associated with some phospholipid molecules as it leaves the cell. ApoAI has a very high affinity for phospholipid, a property that is essential to understanding HDL metabolism. The precursor particles in the circulation are described as pre-β HDL (from their migration pattern on electrophoresis). The smallest particles (pre-β1 HDL) are essentially apoAI with some phospholipid molecules. Further addition of phospholipid molecules leads to discoidal particles also called pre-βl HDL. They acquire cholesterol from cells and from other lipoprotein particles and 'mature' into spherical, cholesterol-rich particles. These particles can give up their cholesterol (again by routes described later) and hence can recycle via lipid-poor forms (Fig. 3.6). Most HDL cholesterol is carried by the larger, spherical particles, which are divided by centrifugation into two subclasses, HDL_2 (larger particles) and HDL_3 (smaller).

A further class of lipoprotein particles is found to a variable extent, and is known as lipoprotein(a) (always read as

'Lipoprotein-little a'). Lipoprotein(a), abbreviated Lp(a), is found at a somewhat higher density than LDL in the ultracentrifuge. Its structure is of an LDL particle, to the apoB100 of which is covalently attached a large additional protein, apolipoprotein(a). Apolipoprotein(a) is related to the protein plasminogen, precursor of the enzyme plasmin in the fibrinolytic system. Its structure is characterized by multiple repeats of a 114-amino acid cysteine-rich domain known as a kringle (because its tertiary structure resembles the twisted baked product known as a kringle). The number of kringles varies from person to person and is genetically determined. Variations in the number of kringles give the Lp(a) particles a range of densities. The plasma concentration of Lp(a) particles varies considerably, but is of interest because of a relationship between Lp(a) concentration and risk of atherosclerosis.

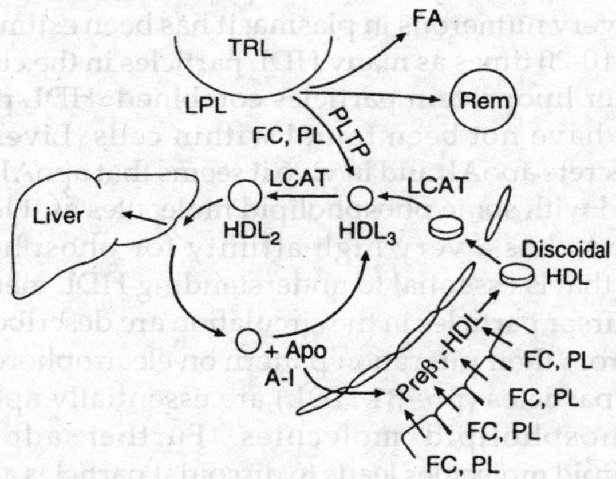

Fig. 3.6 The HDL pathway. Lipid-poor apoAI or pre-βI HDL acquires phospholipids (PL) and free cholesterol (FC) by interaction with cell membranes, forming discoidal HDL in the circulation. Through the action of lecithin-cholesterol acyltransferase (LCAT), and the acquisition of further lipids arising during the action of lipoprotein lipase (LPL) on triacylglycerol-rich particles (TRL) [phospholipid transfer protein, PL TP, is involved in this transfer], these particles swell and become spherical 'mature HOL'; smaller spherical particles are known as HDL3; more lipid-rich particles as HDL2. (Fatty acids, FA, are also produced by the action of LPL, and the smaller particles that remain are known as remnants (rem).) These mature HDL particles can deliver their lipid content to the liver, possibly via the receptor SR-BI, to recycle as lipid-poor apoAI. An

alternative route for delivery of lipid to the liver is through transfer of TRL remnants via the cholesteryl ester transfer protein.

Specific lipoprotein receptors

One function of the apolipoproteins has already been described as targeting of lipoproteins to specific destinations. This is achieved by the interaction of the apolipoproteins with specific cell-surface receptors. The triacylglycerol-rich lipoprotein particles, chylomicrons and VLDL, are largely confined to the vascular compartment by their size until they have undergone extensive lipolysis, allowing them to pass through the fenestrations in the endothelial lining of the capillaries. Some of the receptors involved are therefore expressed on endothelial cells, but most are expressed on parenchymal cells, and part of their selectivity may depend upon the exclusion, by virtue of size, of the larger particles. Before describing the major receptors involved in lipoprotein metabolism in further detail, it is useful to make some general points about membrane-bound receptors.

Membrane receptors

Membrane receptors are proteins designed to recognize specific molecules and bind them in preparation for the initiation of a biological process that takes place in the membrane. The process may be the transport of a molecule (e.g. glucose, LDL) across the membrane or the triggering of a chemical message by a hormone or a growth factor. The recognition step demands that the receptor has specificity, like an enzyme for its substrate. Binding is followed frequently, but not always, by internalization in which the membrane in the vicinity of the receptor forms a vesicle that encapsulates the receptor-ligand complex and enters the cell. Once inside, the receptor complex can be degraded by lysosomal enzymes and the receptor proteins may be recycled to be reinserted into the plasma membrane again. The number of membrane receptors is usually responsive to the availability of the specific ligand. This requires a mechanism for the control of the biosynthesis of the receptor. Receptors are usually glycoproteins, as described for the LDL-receptor below. The functional properties of receptors in membranes appear to be intimately related to the microenvironment provided by membrane lipids. Membrane proteins are on average associated with 30-40 molecules of phospholipid per molecule of peptide. These annular phospholipids are required for functional activity or for the stabilization of protein confermation. The structural and compositional changes of the lipid bilayer, which provides a fluid matrix for proteins, can induce alterations in functional properties of proteins in the intact membrane, for example by allowing changes in protein conformation or the diffusion or position of proteins in the membrane.

This may be an important means by which diet-affecting the nature of the membrane phospholipid fatty acids—can affect hormone action, for instance.

The LDL-receptor

The best-characterized lipoprotein receptor is the LDL-receptor, also known as the B/E receptor because it recognizes (i.e. binds to) homologous regions on apolipoprotein E and on apolipoprotein B100 (but not apoB48). Mutations in this receptor causing loss of function result in marked elevation in the plasma LDL-cholesterol concentration, the condition known as familial hypercholesterolaemia. It is the receptor responsible for the removal of LDL particles from the circulation. The structure of this receptor is shown in Fig. 3.7.

The receptor is synthesized on the rough endoplasmic reticulum as a precursor of molecular mass 120 kDa. About 30 min after its synthesis, the protein is modified to a mature receptor with an apparent mass of 160 kDa. The increase in molecular mass coincides with extensive modifications of the carbohydrate chains. The precursor molecule contains up to 18 acetylglucosamine molecules attached in O-linkage to serine and threonine residues (Fig. 3.7) as well as two high mannose-containing N-linked chains. The latter are modified extensively and the O-linked chains are elongated by addition of one galactose and two sialic acid residues to each N-acetylglucosamine during post-translational processing in the Golgi network.

The receptor is exported to specialized regions of the cell membrane known as coated pits, where the cytoplasmic leaflet is coated with the protein clathrin. After binding of an LDL particle through interaction of its apolipoprotein B100 molecule with the ligand-binding domain of the LDL-receptor, the receptor-LDL complex is internalized by endocytosis of the coated pit (Fig. 3.8). The coated vesicle so formed loses its clathrin coating, and its interior becomes acidified (an endosome). The LDL particle and receptor dissociate in the low pH environment of the endosome, and the receptor is recycled back to the cell surface. The remainder of the particle is transferred to lysosomes, where the cholesteryl esters are hydrolysed by lysosomal acid hydrolases, and the cholesterol adds to the cellular cholesterol pool. Cholesteryl esters may also be resynthesized by the action of acyl-CoA: cholesterol acyltransferase. The size of this cellular cholesterol pool in turn regulates expression of the LDL-receptor so that cellular cholesterol content is strongly regulated.

LDL-receptors are expressed by most cell types. There is some debate, however, about their biological function. They represent a means by which cells can acquire cholesterol exported from the liver. Most cells expressing LDL-receptors, however, also have the enzymic capability of synthesizing cholesterol *de novo*, from acetyl-CoA. The alternative view is that LDL-receptors are there to remove cholesterol from the plasma. The fact that they are highly expressed in steroidogenic tissues such as the adrenal cortex argues for the former. However, in whole-body terms, the main site of LDL-receptor-mediated LDL removal is the liver.

Cholesterol delivered to the liver can be removed from the body either by excretion directly in the bile, or after conversion to bile salts. Increasing LDL-receptor activity with the 'statin' drugs can lead to marked reduction in plasma cholesterol concentration.

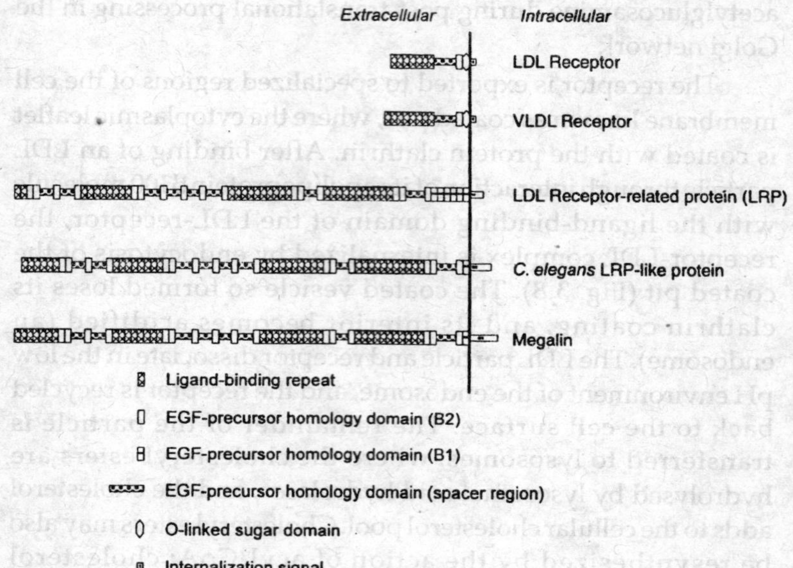

Fig. 3.7 The structure of the LDL-receptor and related receptors. (The apoE receptor 2 is not shown but it is similar to the LDL and VLDL-receptors.) Each of the receptors in this family is built up from repeats of simpler units, some named according to their homology with the epidermal growth factor (EGF) receptor. The ligand binding domains are responsible for binding to the apolipoproteins of the target lipoprotein particles. The internalization signal domain is responsible for internalization of the receptor after the ligand is bound.

The LDL-receptor-related protein

Because of its ability also to bind apoE, the LDL-receptor may play a role in removal of chylomicron remnants from the circulation. But it cannot be the major receptor involved in this process, because humans or animals lacking functional LOL-receptors (as in familial hypercholesterolaemia) do not show accumulation of chylomicron remnants. Recently a related receptor has been identified, known as the LDL-receptor-related protein or LRP. This is the best candidate at present for the chylomicron remnant receptor. The LDL-receptor and LRP, together with the other receptors (described below), form a family of receptors with structural homologies (Fig. 3.7).

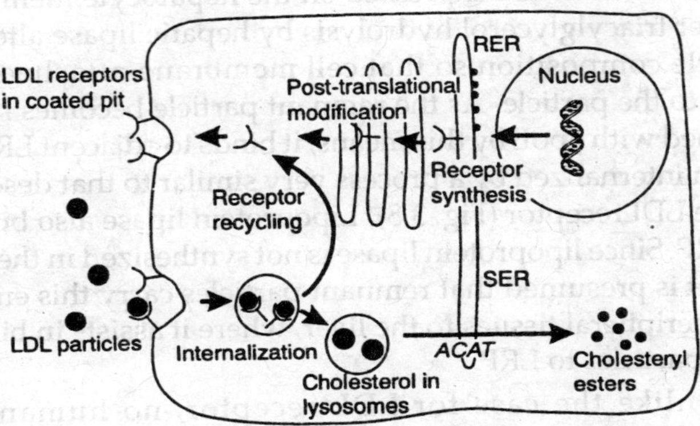

Fig.3.8 Cell biology of the LDL-receptor. Acyl-CoA cholesterol acyltransferase (ACAT) synthesizes cholesteryl esters that may be stored in the cell. The synthesis of the LDL-receptor is closely regulated at the level of gene expression by the cellular cholesterol content.

LRP binds to apoE, in which chylomicron remnants are relatively enriched. It is expressed in most tissues, with the liver, brain and placenta as major sites of expression on a whole-body basis. Its expression is not regulated by cellular cholesterol content as is that of the LDL-receptor. However, in adipocytes, where it is also expressed, its concentration on the cell surface is rapidly increased by insulin. This increase in receptor appearance on the cell membrane is too rapid to be accounted for by an effect of insulin on protein synthesis, leading to the suggestion that the LRP, like the glucose transporter GLUT4, is present within the cells and translocates to the cell membrane when stimulated by insulin. This would represent a mechanism for directing some chylomicron remnants to adipose tissue in the period following a meal, although the major site of chylomicron remnant uptake is still considered to be the liver.

Both hepatocytes and adipocytes also produce apolipoprotein E, which may be expressed on the cell surface. In the liver it has been suggested that chylomicron remnants, once they become small enough via the activity of lipoprotein lipase in the capillary beds of other tissues, may enter the sub-endothelial space (the space of Disse). Here they bind to hepatic lipase, which is also expressed on the hepatocyte membrane. Further triacylglycerol hydrolysis by hepatic lipase alters the particle composition so that cell membrane-attached apoE binds to the particle. As the remnant particle becomes further enriched with apoE by this means, it binds to adjacent LRP and is then internalized by a process very similar to that described for the LDLreceptor (Fig. 3.8). Lipoprotein lipase also binds to the LRP. Since lipoprotein lipase is not synthesized in the adult liver, it is presumed that remnant particles carry this enzyme from peripheral tissues to the liver, where it assists in binding of the particles to LRP.

Unlike the case for LDL-receptor, no human LRP deficiency state has yet been found. Attempts to create mice lacking the LRP have not been successful the phenotype is lethal at an early embryonic stage. It may therefore be an essential protein for life.

A related receptor protein was discovered in 1992 by searching for genes with homology to the LDL-receptor gene. This receptor binds to, and internalizes, VLDL particles particularly avidly. It is expressed mainly in tissues outside the liver that might be expected to have a requirement for fatty acids: heart, skeletal muscle and adipose tissue. Within those tissues the so-called VLDL-receptor is expressed mainly by the endothelial cells, which suits its supposed role as a receptor for large, triacylglycerol-rich particles that cannot cross the endothelial barrier. The precise physiological role of the VLDL-receptor is unknown, however. Its structure is shown in Fig. 3.7.

A further member of the family in mammals is known as megalin or lipoprotein receptor-related protein-2 (LRP-2; Fig. 3.7). It is expressed in absorptive epithelial cells of the proximal tubules of the kidney. It has been suggested that its function is the re-uptake of fat-soluble vitamins that would otherwise be lost by urinary excretion. In 1996 yet another member of the family was described, the apoE receptor 2. This is, again, closely homologous to the LDL and VLDL-receptors. It is highly expressed in the brain and may playa role in lipid transportinto the central nervous system.

This family of receptors is related also to the vitellogenin receptor responsible for delivery of lipid to the developing avian egg yolk, and there are closely related proteins in the fly *Drosophila melanogaster* and the nematode *Caenorhabditis elegans* (Fig. 3.7).

Scavenger receptors

Unrelated structurally to the LDL-receptor family are several receptors with the generic name of 'scavenger receptors'. These were first identified as receptors in macrophages that mediate the uptake of LDL particles that have been chemically modified so that their affinity for the LDL-receptor is reduced. One such modification is oxidation. There are at least three families of scavenger receptors, classes A, B and C. They are characterized by a broad substrate specificity, and they may well have roles in macrophage function in host defence, removal of foreign substances, etc., beyond their role in lipid metabolism. There are two related receptors in class A, the type I and type II scavenger receptors. They are expressed on the macrophage cell surface in clathrin-coated pits, similar to the LDL-receptor (Fig. 3.8). They bind and internalize modified LDL particles, which are then degraded by similar processes to those shown in Fig. 3.8.

However, a crucial difference from the LDL-receptor system is that the cellular cholesterol content does not regulate expression of the scavenger receptors. Thus, their activity can lead to unregulated accumulation of cholesterol in macrophages, an event that may initiate the series of changes that lead to atherosclerosis.

Class B scavenger receptors include the receptor known as Scavenger Receptor Class B type 1 or SRBI, and a protein that had been previously studied as a widely expressed cell-surface antigen given the name CD36. SR-BI was found to bind HDL particles with high affinity, and is now recognized as a long-sought HDL receptor. Its function in this role is covered in more detail in the following section. CD36 appears to have multiple roles. Amongst other physiological roles, it has been suggested as a long-chain fatty acid transporter and also given the name fatty acid translocase (FAT).

The lipoprotein particles transport lipids

It is convenient to describe lipoprotein metabolism as consisting of three pathways, but it will become clear that these are interrelated and interact with one another.

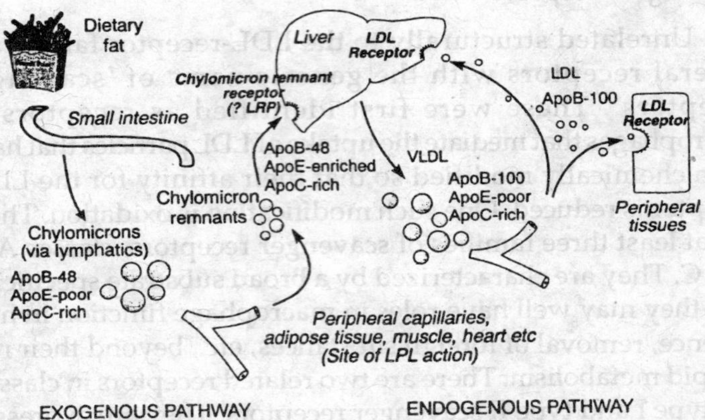

Fig. 3.9 Overview of the exogenous and endogenous pathways of lipoprotein metabolism. Lipolysis of particles by lipoprotein lipase (LPL) in capillaries of extra-hepatic tissues is simplified: VLDL particles, in particular, may go through several cycles of lipolysis, and there is an intermediate class of particle known as Intermediate Density Lipoprotein (IDL) formed before final conversion to LDL particles. VLDL particles themselves may also be removed by VLDL receptors expressed in peripheral tissues. Interactions with HDL are shown in part in Fig. 3.11 but HDL is also an important donor of apolipoproteins (e.g. apoCII to nascent chylomicrons).

The pathway for transport of dietary fat is known as the exogenous pathway (Fig. 3.9). As the tniacylglycerol-rich chylomicron particles, secreted from enterocytes, enter the plasma, they interact with other particles (presumably by simple physical contact) and acquire other apolipoproteins, especially CII, CIII and apoE. ApoCII is essential for their further metabolism. The chylomicron particles come into contact with the enzyme lipoprotein lipase, which is expressed in a number of extrahepatic tissues that can use fatty acids, including adipose tissue, skeletal and cardiac muscle and mammary gland. It is anchored to the luminal aspect of the endothelial cells lining the capillaries in these tissues by binding to heparan sulphate proteoglycans. As chylomicrons pass through the capillaries they bind to lipoprotein lipase and to the proteoglycan chains. LPL, with apoCII as an essential activator (cofactor), hydrolyses the core triacylglycerol in the particle. The rate of lipolysis is rapid, and it has been estimated that somewhere between 10 and 40 lipoprotein lipase molecules must act simultaneously on a chylomicron. The non-esterified fatty acids that are generated can diffuse into the adjacent tissues, either by simple diffusion across the cell membranes or by facilitated diffusion, using one of the recently described fatty acid transport proteins. The triacylglycerol-depleted chylomicron particle shrinks, and as a consequence some of the amphipathic surface monolayer becomes redundant. Surface components dissociate and are acquired by other lipo-proteins, particularly HDL. A phospholipid-transfer protein in the plasma seems to mediate this transfer. The particle dissociates from lipoprotein lipase (the process of lipolysis having taken perhaps a matter of minutes). It is known as a chylomicron remnant. It may undergo a further round of lipolysis in another tissue bed, but after loss of about 80% of its original content of triacylglycerol it will be removed rapidly from plasma. Shrinkage of the particle allows it to cross the endothelial barrier and interact with cell-surface receptors. In addition, conformational changes to the apolipoproteins caused by the shrinkage of the particle expose different regions, which bind to specific cell-surface receptors, probably mainly LRP. The half-life of chylomicron-triacylglycerol in the circulation is about 5 min. The half-life of a chylomicron particle has been estimated at 13-14 min.

The exogenous pathway is regulated mainly by the rate of entry of dietary fat, but the tissue disposition of the dietary fatty acids is regulated by tissue-specific regulation of lipoprotein lipase activity. Adipose tissue lipoprotein lipase activity is up-regulated by insulin and therefore plays an important role in removal of dietary fatty acids in the period following a mixed meal (i.e. a meal that contains both fat and carbohydrate, since the latter will stimulate insulin secretion). The mechanism for activation of adipose tissue lipoprotein lipase by insulin is complex. Whilst insulin increases transcription of the lipoprotein lipase gene in adipose tissue, the major point of regulation in the short term (the hours following a meal) appears to be a diversion within the adipocyte of lipoprotein lipase between a degradative pathway, and secretion in an active form for export to the endothelium. Since the enzymes of triacylglycerol synthesis in adipocytes are also activated in the fed state, the net effect is that dietary fatty acids tend to be readily deposited as adipocyte triacylglycerol. In contrast, lipoprotein lipase in skeletal, and particularly cardiac, muscle is down-regulated by insulin, and increases in activity during fasting, and in skeletal muscle with exercise training. In these conditions, therefore, fatty acids tend to be directed to the tissues that need them for oxidation.

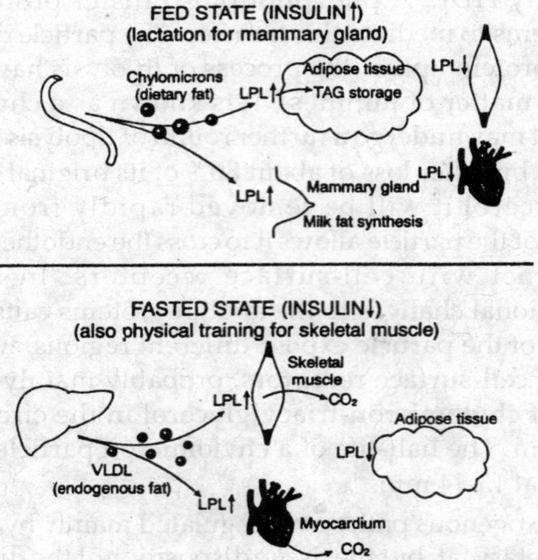

Fig. 3.10 Tissue-specific regulation of lipoprotein lipase (LPL) in relation to nutritional state directs the flow of triacylglycerol-fatty acids to different tissues according to needs.

The secretion of VLDL from the liver, and its further metabolism, is described as the endogenous pathway (Fig.). It is very similar initially to the exogenous pathway. VLDL particles also interact with lipoprotein lipase in capillaries and their core triacylglycerol is hydrolysed. The half-life of VLDL particles in the circulation is typically 12 h, but their fates are heterogeneous. As the particles lose their triacylglycerol through successive interactions with lipoprotein lipase, they also lose their surface apolipoproteins as described above, until they become LDL particles. However, along the way some of the particles are removed intact by endothelial VLDL receptors. LDL particles have an even longer haeflife in the circulation, typically 2-2.5 days, and they are removed by LDL-receptors on cell surfaces in the sub-endothelial space. The major site for removal of LDL particles is normally the liver although the LDL-receptor is a mechanism for delivering cholesterol to most cells. The expression of LDL-receptors is closely regulated to maintain cellular cholesterol homeostasis.

Thus, the endogenous pathway is a means of delivering fatty acids to tissues via lipoprotein lipase and perhaps the VLDL-receptor, and cholesterol via the LDL-receptor.

The third pathway is that of HDL metabolism (Fig. 3.11). As described above, HDL particles may not be secreted as such: rather, the basic components of apo AI and phospholipid are secreted and the particles cycle through various stages in the plasma. If the LDL pathway of cholesterol delivery to tissues is regarded as 'forward cholesterol transport', the HDL pathway represents 'reverse cholesterol transport' or the movement of cholesterol out of tissues and its transport to the liver for ultimate excretion in the bile. (The liver is the only mammalian organ that can excrete cholesterol from the body.)

As described earlier (Fig. 3.6), immature lipid-poor HDL particles (pre-βI HDL) interact with cell membranes to acquire unesterified cholesterol. There has been much debate about whether this involves a specific receptor. Recent advances in the molecular biology of cellular cholesterol efflux have come about through the identification of the gene defect in Tangier disease, a rare inherited disease in which cholesterol accumulates in tissues.

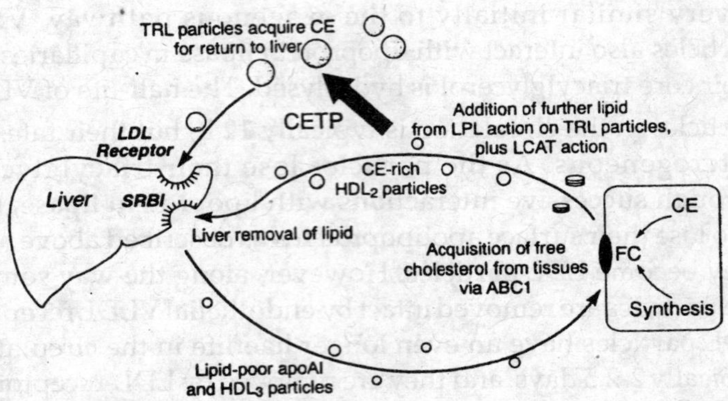

Fig. 3.11 The pathway of reverse cholesterol transport. For more details of HDL metabolism see Fig. 3.5. Lipid-poor apoAI acquires free cholesterol (FC and phospholipid from peripheral tissues via the ATP-binding cassette protein-l (ABCl). The lipid content of the HDL particles is increased by transfer of additional cholesterol from the lipolysis of triacylglycerol-rich lipoprotein (TRL) particles. Esterification of this cholesterol by lecithin-cholesterol acyltransferase (LCAT) produces mature, lipid-rich HDL_2 particles. Thesę can return their cholesterol to the liver in two ways. They may deliver it directly via a hepatic receptor (SR-BI) or they may exchange it for triacylglycerol from the TRL via the action of cholesteryl ester transfer protein (CETP). The TRL particles then carry this cholesterol to the liver for receptor-mediated uptake. (CE denotes cholesteryl esters.)

The gene responsible for Tangier disease encodes a protein that is a member of a family of cell-membrane-associated transporter proteins that have the property of binding ATP on their cytoplasmic side (through the so-called A TP-binding cassette, ABC: this particular protein has now become known as ABC1). Hydrolysis of ATP can then power the transfer of substances across membranes. In the case of cholesterol efflux, ABCl seems to mediate the transfer of cholesterol from 'cholesterol rafts' in the cell membrane to the lipidpoor HDL particles (where it binds to apoAI). The unesterified cholesterol acquired by the particles becomes esterified with a long-chain fatty acid through the action of the enzyme lecithin-cholesterol acyltransferase (LCAT), which is associated with HDL particles. This enzyme catalyses the transfer of a fatty acid from phosphatidylcholine to cholesterol to form a cholesteryl ester

(Fig. 3.12). The phospholipid substrate for the reaction is present in the HDL particle. The hydrophobic cholesteryl ester moves to the core of the particle, which swells and becomes spherical as more cholesterol is acquired and esterified. The other product, lysophosphatidylcholine, is transferred to plasma albumin from which it is rapidly removed from blood and reacylated. This is probably the origin of the bulk of HDL-cholesterol in plasma, although some is also acquired from the surface components of chylomicrons and VLDL particles when their triacylglycerol is hydrolysed by lipoprotein lipase, as described above.

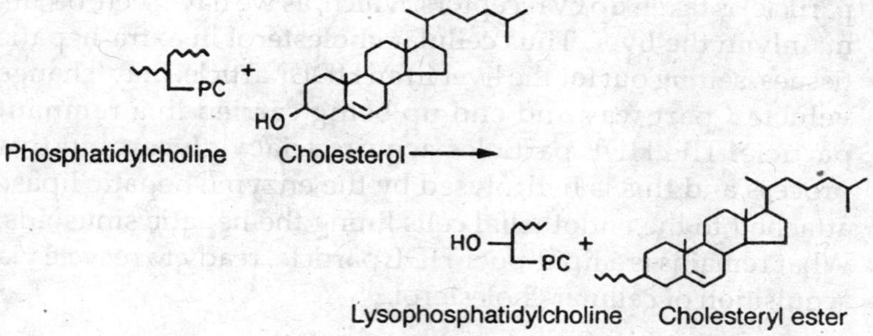

Phosphatidylcholine + Cholesterol → Lysophosphatidylcholine + Cholesteryl ester

Fig. 3.12 The lecithin-cholesterol acyltransferase (LCAT) reaction.

To complete the pathway of reverse cholesterol transport, HDL-cholesterol must reach the liver. There are at least two routes by which this can occur. First, the scavenger receptor SR-BI is expressed in the liver and also in steroidogenic tissues (e.g. adrenal gland and ovary). 'Docking' of HDL particles with SR-BI is followed by off-loading of their cholesteryl ester content. The cholesteryl esters enter the cellular pool and may be hydrolysed by lysosomal acid hydrolases as shown for LDL-receptor-mediated uptake (Fig. 3.8). This process is fundamentally different from the uptake of LDL particles by the LDL-receptor, however, and has been called 'selective lipid uptake'. The difference is that the particle itself is not internalized and the cholesterol-depleted particle leaves the receptor to re-enter the cycle of the HDL pathway.

The second mechanism for delivery of HDLcholesterol to the liver brings us to an important way in which the lipoprotein pathways interact. Human plasma contains a protein known as cholesteryl ester transfer protein (CETP), which is secreted from the liver and from adipose tissue. CETP mediates the exchange of 'neutral lipids' (triacylglycerol and cholesteryl esters) between particles, according to concentration gradients. When HDL particles become enriched in cholesteryl esters, they may exchange these esters for triacylglycerol carried in the triacylglycerol-rich lipoproteins, i.e. chylomicrons and VLDL. Thus, chylomicrons and VLDL acquire cholesterol that stays with the particle since it is not removed by lipoprotein lipase, and it is eventually removed when the remnant particle (or LDL particle) is taken up by receptors, which, as we have seen, occurs mainly in the liver. Thus cellular cholesterol in extra-hepatic tissues, setting out for the liver in an HDL particle, may 'change vehicles' part way and end up being carried in a remnant particle. The HDL particles acquire triacylglycerol in this process and this is hydrolysed by the enzyme hepatic lipase attached to the endothelial cells lining the hepatic sinusoids. What remains is a lipid-poor HDL particle, ready to recycle via acquisition of cellular cholesterol.

Species differ quantitatively

In some species, like man, guinea-pig and pig, lipoproteins of the LDL type (in which apolipoprotein B predominates) account for more than 50% of the total substances of density f $<1.21\ gml^{-1}$. They are the LDL mammals. In the majority of mammals, however, HDL are the predominant class and may account for up to 80% of plasma substances of density $< 1.21\ gml^{-1}$. Herbivorous species, with the exception of guinea-pigs, camels and rhinos, and carnivores are HDL mammals. It is worth noting that although rats are often used for the study of lipid biochemistry, their lipoprotein pattern is of the HDL type and very different from that of man. Another distinct difference is that the rat liver secretes VLDL particles that contain either apoB48 or apoB100 (whereas in humans the liver secretes only apoB100-containing particles). This almost certainly relates to the fact that their LDL-cholesterol levels are so relatively low, since apoB48-containing particles tend to be cleared from the circulation rapidly (as with chylomicrons in man). Caution needs to be exercised in extrapolating results on experimental animals to the human situation.

Whichever species one is studying, it is often necessary to quantify various aspects of the lipoprotein system.

THE CO-ORDINATION OF LIPID METABOLISM IN THE BODY

Several genes involved in lipid metabolism are themselves regulated by lipid or lipid-related molecules, and adipose tissue itself secretes proteins that regulate food intake and fat storage.

The flux of fatty acids through the pathways described in this chapter can vary enormously. The fat content of the diet varies in different parts of the world, between different people within one country, and even from day to day in one individual. The rate at which we transport fatty acids through the plasma varies considerably during a normal day. For non-esterified fatty acids it is high at night during fasting, suppressed after meals, and stimulated to very high levels during aerobic exercise; for triacylglycerol-fatty acids it is roughly the converse. This requires co-ordination between metabolic pathways in different organs, in relation both to the influx of dietary lipids and the body's requirement for lipids. We have seen how this co-ordination can be brought about by hormones, and in particular by insulin whose secretion from the pancreas is stimulated in the 'fed' state. Insulin acts both by affecting the activity of enzymes on a short-term basis (e.g. by reversible phosphorylation/dephosphorylation) and on a longer term basis by regulation of gene expression. However, the main stimulus for insulin secretion is a rise in the plasma glucose concentration. It would seem sensible that the body should also have means for homeostasis of lipid metabolism that do not depend upon the simultaneous ingestion of carbohydrate in appropriate amounts.

Over the past few years there have been great advances in our understanding of how this is achieved. There are several, possibly interrelated, systems that regulate gene expression to control the flux through various pathways of lipid metabolism, and whose regulators are themselves lipids or products derived from lipids.

The sterol regulatory element binding protein (SREBP) system controls pathways of cholesterol accumulation in cells and may also control fatty acid synthesis

The American biochemists and 1985 Nobel Prize winners, Mike Brown and Joseph Goldstein, showed that cellular cholesterol content is regulated by two parallel mechanisms. When the content of unesterified cholesterol in cells increases, the expression of the LDL-receptor protein decreases. In addition, the key enzymes of cholesterol biosynthesis (hydroxymethylglutaryl(HMG)-CoA synthase, HMG-CoA reductase, squalene synthase, farnesyl diphosphate synthase) are repressed. Thus, any further increase in cellular cholesterol is minimized. Conversely, if the cellular unesterified cholesterol content falls, these pathways are activated (Fig. 3.13). The molecular basis for this regulation is now clear. The genes for all these proteins contain an upstream sequence known as the sterol regulatory element-l (SRE-l). The protein that binds to SRE-l, and activates expression of the gene, is part of a larger protein called sterol regulatory element binding protein (SREBP). SREBP is normally localized in the endoplasmic reticulum. In the absence of cholesterol a specific protease releases a peptide from SREBP, which migrates to the nucleus and binds to SRE-l, activating transcription. When cellular cholesterol levels are high, this protease is inhibited and gene expression is repressed. The natural regulator of this system appears not to be cholesterol itself but a hydroxylated derivative of cholesterol, an oxysterol. The working of the system is illustrated in Fig. 3.14.

There are three known isoforms of SREBP: SREBP-1a, SREBP-1c and SREBP-2. The first two arise from differential splicing of the transcript from one gene, whereas SREBP-2 is the product of a separate gene. It seems that SREBP-1a and 1c are concerned more with regulation of genes involved in fatty acid metabolism including fatty acid synthase, acetyl-CoA carboxylase and stearoyl-CoA (Δ9) desaturase, whereas SREBP-2 regulates transcription of genes involved in cholesterol metabolism. SREBP-1c is also expressed in adipose tissue (where it is also known as adipocyte determination and differentiation factor-1, ADD-1), and has been shown to regulatestu fatty acid synthesis by increasing the expression of fatty acid synthase in adipose tissue as well as liver. Leptin production and secretion are also increased. Expression of SREBP-1c is itself under regulation by insulin. Thus, when insulin levels are high, more SREBP-lc is produced and fatty acid synthesis is up-regulated. Since this also requires the involvement of cholesterol (or a derivative), it might be that the regulation really works more in the negative way: in times of fasting (low insulin), SREBP-lc expression is severely reduced, and fatty acid synthesis cannot be activated by the SREBP pathway. Some of the genes whose expression is regulated by the SREBP system are shown in Table. 3.3.

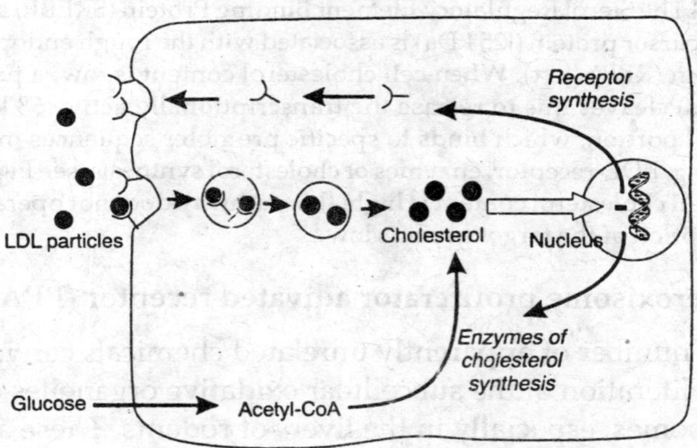

Fig. 3.13 Regulation of the cellular cholesterol content. When the content of unesterified cholesterol in cells increases, the expression of the LDL-receptor protein and the enzymes of cholesterol biosynthesis is repressed.

The SREBP system has been manipulated to alter cholesterol metabolism. Certain fungal metabolites, compactin from *Penicillium* spp. and mevinolin from *Aspergillus terreus*, were found in the 1970s to inhibit HMG-CoA reductase.

Synthetic derivatives of these molecules are now widely used as drugs (the 'statins') to lower plasma cholesterol concentrations. The main mode of action is that inhibition of hepatic cholesterol synthesis reduces hepatocyte unesterified cholesterol concentrations, and this results in up-regulation of hepatic LDL-receptor expression: hence more LDL particles are removed from the circulation.

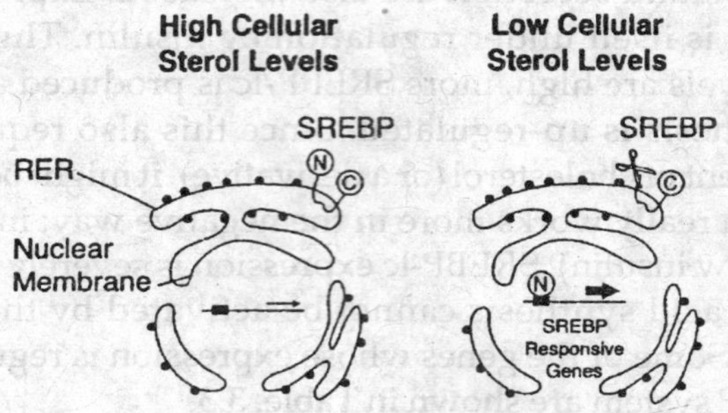

Fig. 3.14 The Sterol Regulatory Element Binding Protein (SREBP) system. The precursor protein (125 kDa) is associated with the rough endoplasmic reticulum (RER) (left). When cell cholesterol content is low, a protease (scissors) cleaves this to release the transcriptionally active 68 kDa N-terminal portion, which binds to specific promoter sequences in target genes (e.g. LDL-receptor, enzymes of cholesterol synthesis; see Fig. 3.13). When cell cholesterol content is high, this pathway does not operate and transcription of the target genes is low.

The peroxisome proliferator adivated receptor (PPAR)

A number of apparently unrelated chemicals can induce the proliferation of the subcellular oxidative organelles called peroxisomes, especially in the livers of rodents. These agents were suggested to work through a common receptor called the peroxisome proliferator activated receptor (PPAR). It is now realized that there is a family of PPARs and that a major role is to regulate fatty acid metabolism. Like the SREBP system, the PP ARs are activated by lipids - in this case derivatives of fatty – acids and act upon specific response elements, peroxisome proliferator response elements (PPRE), in the regulatory region of. genes concerned with lipid metabolism (Table 3.4).

PPAR-α is expressed mainly in the liver. It is activated by long-chain fatty acids, amongst other factors. The actual ligand is thought to be a mmm-

Table 3.3 Genes whose expression is increased by the sterol regulatory element binding protein (SREBP) system

Gene	Function in cell
LDL-receptor	Import of LDL particles
HMG-CoA synthase	*de novo* cholesterol synthesis
Squalene synthase	*de novo* cholesterol synthesis
Farnesyl diphosphate synthase	*de novo* cholesterol synthesis
Lanosterol synthase	*de novo* cholesterol synthesis
Acetyl-CoA carboxylase	*de novo* fatty acid synthesis
Fatty acid synthase	*de novo* fatty acid synthesis
Stearoyl-CoA desaturase	Synthesis of oleate
Glycerol 3-phosphate acyltransferase	Triacylglycerol and phospholipid synthesis
Lipoprotein lipase	Import of lipoprotein triacylglycerol fatty acids
PPAR-γ	Induction of adipocyte differentiation
Leptin	Secretion from adipocyte to signal fat store size

Stimulation occurs in response to low cellular cholesterol content. These functions have been shown in hepatocytes and adipocytes. Nutritional and hormonal regulation of enzymes in fat synthesis: studies of fatty acid synthase and glycerol 3-phosphate acyltransferase gene transcription. Fatty acids: links between genes involved in fatty acid and cholesterol metabolism.

bolic derivative of a fatty acid such as a member of the prostaglandin family. One candidate is 15-deoxy-$(\Delta)^{12,14}$-prostaglandin J_2, which has been shown to be particularly active in cellular systems. The activated PPAR-α then dimerizes with another receptor, the retinoid-X receptor, which must itself be activated by binding 9-*cis*-retinoic acid. The heterodimer migrates to the nucleus where it binds to the PPRE and activates gene expression. Some genes whose regulation is altered by PPAR-α are listed in Table 3.4. In general PPAR-α activation can be seen to up-regulate the secretion of the apolipoproteins forming HDL, and to increase hepatic fatty acid oxidation.

Another isoform, PPAR-γ, is expressed mainly in adipose tissue and its major role in that tissue is the regulation of fat storage. Activation of PPAR-γ occurs in a similar way through the binding of a fatty acid or a fatty acid derivative, and after dimerization with the activated retinoid-X receptor, PPAR-γ binds to PPREs in the adipocyte nucleus to activate expression of specific genes. In this case the genes include those of adipocyte differentiation from precursor cells (immature adipocytes), so activation of PPAR-γ causes the recruitment of new adipocytes, as well as increased fatty acid deposition by up-regulation of the expression of genes regulating fat storage. These include components of the pathway of glucose uptake and synthesis, glucose transporter-4 (GLUT4) and phosphoenolpyruvate carboxykinase (an enzyme of gluconeogenesis). Presumably their function is to increase availability of glycerol 3-phosphate for triacylglycerol synthesis. Therefore, in times of excess fatty acid availability, this system ensures the co-ordinated storage of the excess fatty acids in adipose tissue.

There is cross-talk between the SREBP and PPAR pathways. One of the SREBP isoforms, SREBP-1c (ADD-1), is expressed in adipocytes. As described above, SREBP-1c activation in adipose tissue increases expression of fatty acid synthase, and thus potentially the generation of ligands for PPAR-γ. But there is a more direct link in that SREBP-1 activation directly increases PPAR-γ expression.

Just as the SREBP system has been manipulated pharmacologically to alter cholesterol metabolism, so the PPAR system is the target for drug intervention. A group of agonists of the PPAR-α receptor, known as fibric acid derivatives or fibrates, are useful drugs in the treatment of patients with elevated plasma triacylglycerol concentrations and low HDL-cholesterol concentrations. Their effects are apparently brought about by up-regulation of lipoprotein lipase and hepatic fatty acid oxidation, thus allowing the body to dispose of excess fatty acids, and also secretion of apolipoproteins AI and AII, leading to increased HDL-cholesterol concentrations. More recently a new group of drugs, the thiazolidinediones or 'glitazones', have been developed, which appear to improve the sensitivity of the body to insulin. These are finding a place in the treatment of type 2 (maturity-onset) diabetes, which is usually associated with obesity and in which the ability of tissues to respond to insulin is reduced. It has been discovered that the thiazolidinediones are potent agonists of PPAR-γ in adipose tissue. How activation of PPAR-γ in adipose tissue should lead to a widespread improvement in response to insulin is not clear, although it is interesting that an early effect of thiazolidinedione treatment is a reduction in the plasma concentration of non-esterified fatty acids.

Other nuclear receptors

It seems likely that new mechanisms for the regulation of gene expression by lipid-related molecules will continue to be uncovered. One example is the nuclear receptor/transcription factor known as Hepatic Nuclear Factor-4α (HNF-4α). Fatty acyl-CoA esters appear to be an important natural ligand for HNF-4α, which in turn activates expression of a number of genes including apolipoprotein CIII, and pyruvate kinase. It has been suggested that the down-regulation of apoCIII expression by PPAR-α is mediated by competition with HNF-4α.

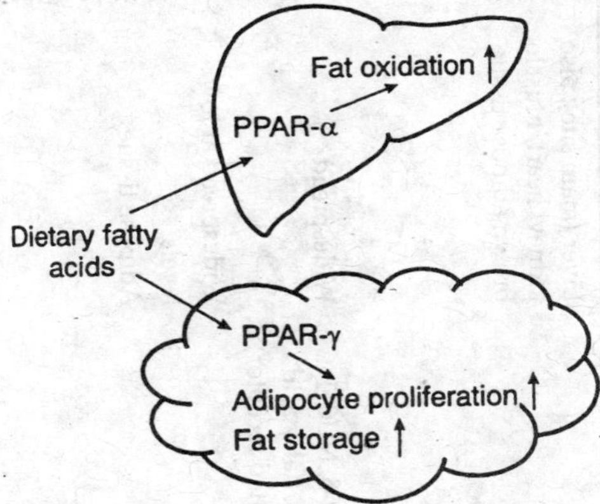

Fig. 3.15 The peroxisome proliferator activated receptor (PPAR) system and regulation of fatty acid disposition in liver and adipose tissue.

Secretion of harmones by Adipose tissue

In recent years it has become clear that adipocytes are also active in secretion of a number of peptides and other

Table 3.4 Peroxisome proliferator activated receptors (PPARs): tissue distribution and effects of activation

Receptor	Other names	Main tissue distribution	Genes whose expression is increased by PPAR activation	Genes whose expression is suppressed by PPAR activation
PPAR-α		Liver (main site); also kidney, heart, muscle, brown adipose tissue	Apolipoprotein AI; apolipoprotein AII; enzymes .of peroxisomal fatty acid oxidation; liver FABP; CPT-1; enzymes of mitochondrial fatty acid oxidation	Apolipoprotein CIII
PPAR-δ	PPAR-δ, NUC-1, FAAR (fatty acid activated receptor)	Widespread	Not known although HDL concentrations increase with activation	Not known
PPAR-γ1		Widespread at low levels		
PPAR-γ2		Adipose tissue	Factors involved in adipocyte differentiation; adipose tissue FABP (also known as aP2); lipoprotein lipase; fatty acid transport protein; acyl-CoA synthase; GLUT4; phosphoenolpyruvate carboxykinase	Leptin

compounds that may in turn help to regulate both lipid metabolism and other functions. Some of these are listed in Table 3.5. The hormone, leptin is a protein of 167 amino acids. It was discovered in 1994 by positional cloning of the *ob* gene responsible for gross, spontaneous obesity in a mutant strain of laboratory mice. Only mice homozygous for the mutation *(ob/ ob* mice) show the obese phenotype. Painstaking work from the laboratory of Jeffrey Friedman at Rockefeller University in New York led to the identification of the gene and the realization that it coded for a novel hormone, expressed almost exclusively in white adipose tissue. The hormone was called leptin (Greek leptos, thin). When purified recombinant leptin was injected into *ob/ob* mice, they lost weight and became normal. The weight loss was the result of both decreased food intake, and increased energy expenditure. In normal animals, and humans, leptin secretion increases with increasing fat stores. Leptin receptors are widespread, and there are several isoforms produced by alternative splicing of transcripts from a single gene. A short, membrane-spanning form, believed to be responsible for leptin transport, is found in the choroid plexus, the brain region that governs transport of peptides between the bloodstream and the hypothalamus. In the hypothalamus itself, a brain region that controls many bodily functions including appetite and the activity of the autonomic nervous system, the long-form, signalling isoform of the leptin receptor is expressed. Leptin carries a signal from adipose tissue to the brain that the fat stores are growing and that it is appropriate to reduce food intake and increase energy expenditure. It also signals to other organs and tissues, and the widespread role of leptin in regulation of energy metabolism is now being discovered. One important role is that of a signal to the reproductive system that the fat stores are sufficient to support pregnancy and birth. Female and male *ob/ob* mice are sterile, and their fertility returns on treatment with recombinant leptin. This is probably the explanation also for the common observation that extreme thinness (arising, for instance, either from eating disorders or from regular physical training) is associated with amenorrhoea (absence of menstruation).

Although the main factor regulating leptin secretion from the adipocyte is cell size (i.e. triacylglycerol content), there is additional modulation by the short-term feeding state. This appears to be mediated largely by insulin, which acts through the adipocyte determination and differentiation factor-1, ADD-1 (or SREBP-1c), to increase leptin expression. Thus, leptin secretion increases on feeding (acting to reduce further food intake) and is

Table 3.5 Proteins and other factors secreted by adipocytes

Secreted product	Potential role
Proteins	
Lipoprotein lipase	Transport to endothelium; uptake of circulating triacylglycerol fatty acids.
Complement components D (adipsin), B, C3	Formation of C3a-desarg or acylation stimulating protein; stimulation of fatty acid storage.
Cholesteryl ester transfer protein	Systemic participation in reverse cholesterol transport; adverse role in generation of atherogenic lipoprotein phenotype in obesity / insulin resistance.
Apolipoprotein-E	Lipoprotein metabolism.
Angiotensinogen	Regulation of blood pressure (after conversion to angiotensin-II).
Leptin	Hormone signalling size of fat stores, playing widespread role in metabolic regulation.
Plasminogen activator inhibitor-1 (PAI-1)	Increases coagulating ability of blood.
Tissue factor	Coagulation pathway.
Tumour necrosis factor-α	Cytokine, perhaps reducing fat storage; systemically, may reduce sensitivity to insulin (although there is some doubt about adipose tissue's contribution to systemic concentration).
Interleukin-6	Cytokine with role in inflammatory processes.

Prostaglandins PGE1, PGI_2 (prostacyclin)	Probably local role in regulation of blood flow and lipolysis
Non-esterified fatty acids, glycerol	Products of lipolysis; distri-bution to other tissues.

This tabulation is not exhaustive and is expanding all the time.

suppressed during fasting (so acting to increase food intake).

The leptin system also operates in humans, although it became clear soon after the discovery of leptin that the vast majority of obese humans have high, not low leptin concentrations in their circulation, just as appropriate for their enlarged fat stores. The message is that the majority of human obesity does not result from a failure of adipose tissue to secrete leptin, but from a failure of the brain to respond to it. The first clinical trials of recombinant leptin as a treatment for human obesity have recently been reported. Obese patients treated with high doses of leptin, by subcutaneous injection once a day, showed some weight loss, although it was not marked—perhaps not surprisingly in view of their 'leptin resistance'. However, we can be certain that the leptin system plays an important role in humans. A small number of families has been identified in which there are one or two individuals who have shown remarkable obesity from a very early age. Sequencing of the leptin gene has shown that these families carry a mutation rendering it ineffective, and that some of the rare cases of extreme precocious obesity are homozygous for this mutation or for a mutation in the leptin receptor. One pair of grossly obese cousins studied by Stephen O'Rahilly and his group at the Department of Medicine in Cambridge, UK, has displayed an almost incredible craving for food and failed ever to lose weight despite periods' in hospital to try to restrain their eating. They are both homozygous for a particular frameshift mutation in the leptin gene. They have now been treated with recombinant human leptin, which has normalized their eating behaviour and led for the first time to weight loss. There appears to be little effect on energy expenditure, so in humans the leptin system probably regulates food intake more than energy expenditure. These may be rare accidents of nature, but they show that the leptin system is crucial for normal regulation of feeding in humans.

Other proteins secreted from adipocytes that have a role in lipid metabolism include lipoprotein lipase (secreted for transport to the capillary endothelium), apolipoprotein E and cholesteryl ester transfer protein. Adipocytes also secrete a number of peptide components of the complement system, which is important in host defence against infection. It has recently been realized that these include the factors known as D, B and C3, all components of the alternative (as opposed to the classical) complement pathway. Factors D (also known as adipsin, to reflect its origin), B and C3 interact to produce a peptide of 76 amino acids that is known to immunologists as C3a-desarg (formed by removal of a C-terminal arginine from the fragment C3a). C3a turns out to be identical to a protein that had been identified by Allan Sniderman and Katherine Cianflone, of McGill University in Montreal, as a potent stimulator of fatty acid esterification in adipocytes and termed by them acylation stimulating protein. Production of ASP by adipocytes is stimulated by the presence of chylomicrons. This suggests a co-ordinated system of regulation, whereby the arrival in adipose tissue capillaries of dietary fat in the form of chylomicron-triacylglycerol triggers ASP secretion, which in turn stimulates the storage of dietary fatty acids as adipocytetriacylglycerol after their release from chylomicrons by lipoprotein lipase (Fig. 3.16).

DISEASES INVOLVING CHANGES OR DEFECTS IN LIPID METABOLISM

Defects in a variety of lipid metabolic pathways can give rise to inappropriately high (or sometimes low) concentrations of lipoproteins in the blood. This in turn may be associated with a range of chronic conditions including atherosclerosis, obesity and diabetes.

In earlier chapter we described deficiency diseases that arise because the diet contains insufficient of several types of lipids that the body is unable to synthesize for itself. Here, the discussion is devoted to diseases that involve a defect, or defects, in a metabolic pathway, usually because an enzyme or some other protein is deficient or not functioning correctly. This may be because of a mutation in the protein, or a mutation in another protein that in some way regulates the protein we are interested in, or (more commonly) a collection of mutations (or polymorphisms), each of which may be insignificant on its own, but which act together with lifestyle factors (such as excessive energy or fat intake, or lack of exercise) to produce the observed phenotype.

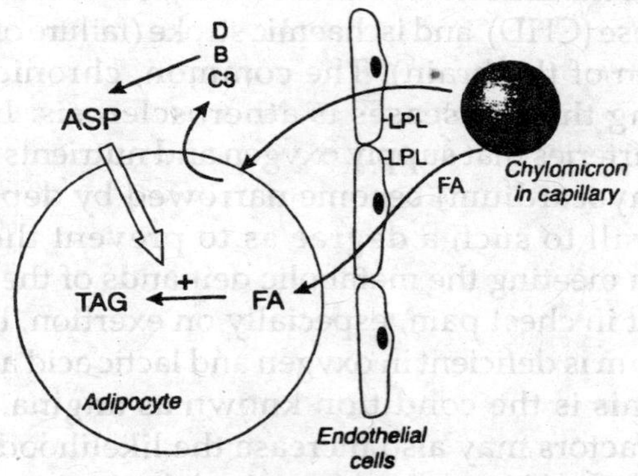

Fig. 3.16 The acylation stimulating protein (ASP) system in adipose tissue. Arrival of a chylomicron particle in the capillary sends a signal to the adipocyte (probably by transfer of a protein carried by the chylomicron to a specific receptor on the adipocyte surface), which increases secretion of the three proteins D, B and C3; components of the alternative complement pathway. These interact to form the protein C3a-desarg or ASP. ASP acts via a receptor (not yet characterized) to stimulate triacylglycerol (TAG) synthesis from fatty acids (FA) released by the action of lipoprotein lipase attached to the capillary endothelial cells on the chylomicron-TAG.

The common outcome of many of these conditions is an increased risk of atherosclerosis, a process that involves accumulation of fatty deposits in arteries. This condition will therefore be described first, followed by a consideration of the disorders of lipid metabolism and how they lead to generation of atherosclerosis. Finally, we will consider briefly some diseases of lipid storage that may have different clinical outcomes.

Atherosclerosis

Atherosclerosis involves the build-up of deposits in arterial walls, characterized by high concentrations of lipids that derive from plasma lipoproteins; lipids are also involved in the formation of thrombi that may lead to the blockage of blood vessels narrowed by atherosclerosis.

Cardiovascular disease is a broad term for a number of conditions involving pathological changes in blood vessels associated mainly with the heart and brain. The major causes of death are ischaemic heart disease (IHD), often called coronary heart disease (CHD), and ischaemic stroke (failure of blood flow to a region of the brain). The common, chronic condition underlying these diseases is atherosclerosis. In IHD, the coronary arteries that supply oxygen and nutrients to the heart muscle (myocardium) become narrowed by deposits in the arterial wall to such a degree as to prevent the coronary circulation meeting the metabolic demands of the heart. This may result in chest pain, especially on exertion, because the myocardium is deficient in oxygen and lactic acid accumulates locally. This is the condition known as angina. Local and systemic factors may also increase the likelihood of platelet aggregation and blood clot (thrombus) formation at the site of narrowing. This can result in complete blockage of an artery, a heart attack (myocardial infarction) or stroke if it is in the brain. The immediate effect is failure to supply a region of the myocardium, or brain, with oxygen. If the blockage is of a major vessel, this may be fatal. Peripheral vascular disease involves blood vessels in, for example, the limbs and although it is not a major cause of mortality, it is associated with considerable morbidity and distress for the patient.

Atherosclerosis is an irregular thickening of the inner wall of the artery that reduces the size of the arterial lumen particularly near junctions in the arterial tree. The thickening is caused by the accumulation of atherosclerotic plaques, consisting of a proliferation of smooth muscle cells, connective tissue, mucopolysaccharides, fat-filled foam cells, in which the predominant lipid is cholesteryl ester, and deposits of calcium. The artery wall is locally thickened and loses elasticity. The term atherosclerosis was coined in 1904, from the Greek athere, meaning porridge or gruel and referring to the soft consistency of the core of the plaque, and sclerosis, hardening.

The first stage in atherosclerosis is the fatty streak, a yellowish, minimally raised spot (the spots later merging into streaks) in the arterrial wall. Forty-five per cent of infants coming to post mortem examination during the first year of life have fatty streaks in their aorta, so this is an early development in life, and the development of the mature atheromatous plaque must be a long-term process. The origin of the fatty streak has mostly been studied in animal models of hyperlipidaemia, usually rabbits with a genetic predisposition to high blood cholesterol levels, fed on a high-fat diet containing predominantly saturated fatty acids and cholesterol. It is preceded by the adhesion of monocytes and T-lymphocytes to an area of endothelium. The monocytes migrate into the subendothelial space where they differentiate into macrophages, which then begin to engulf large amounts of lipid via the scavenger receptor pathway. Current understanding is that the main source of this lipid is LDL particles that have become chemically modified, probably through the process of lipid peroxidation, which leads in turn to peroxidation of the apoB100. This reduces the affinity of the LDL particles for the LDL-receptor, and instead makes them ligands for the family of scavenger receptors. Lipid uptake by the scavenger receptor pathway is not subject to feedback regulation by cellular cholesterol content; therefore, the macrophages may engulf large amounts of lipid, giving them a foamy appearance under the microscope. These are so-called foam cells, and are characteristic of the atherosclerotic plaque.

Further development of the atherosclerotic plaque involves proliferation of smooth muscle cells of the arterial wall and the elaboration of a connective tissue matrix, forming a fibromuscular cap to the lesion. Within the lesion there may be breakdown of dead macrophages and release of their contents, with formation of a semi-liquid pool of extracellular lipid. The lid of the lesion may remain firm, in which case the lesion may protrude into the arterial lumen, obstructing flow, but not causing acute damage. However, for reasons that are not understood, some plaque caps become unstable and are damaged, and the contents of the plaque are exposed, resulting in the normal response to damage to a vessel wall–thrombus formation. The processes involved in the development of ischaemie heart disease and myocardial infarction, and the possible involvement of lipids, are illustrated in Fig. 3.7.

Examination of the lipid contents of atherosclerotic plaques shows cholesterol and fatty acids that appear to have been derived from the circulation. This has been shown in animal experiments using radioactive labelling, and more recently in humans by analysis of the fatty acid content of atherosclerotic plaques and showing a relationship to dietary fatty acids. In addition, plaques can be shown to contain apoBl00. This all points to circulating lipoprotein particles as the major source of plaque lipid.

Several aspects of lipid metabolism relate to the risk of atherosclerosis and CHD. The first is the LDL-cholesterol concentration. Over many years, epidemiological observations of the differences in CHD rates between different countries have shown a strong relationship with average (across the population) serum LDL-cholesterol levels. In addition, people with a defect in LDL removal, usually through a mutation in the LDL-receptor (the condition of familial hypercholesterolaemia) have a very high risk of developing CHD at an early age. Only 25 years ago the 'lipid hypothesis' of CHD–that is, the hypothesis that elevated serum LDL-cholesterol concentrations playa direct role in CHD–was the subject of considerable controversy. Now it is more widely accepted. A major reason for this acceptance has been that several major prospective clinical trials have shown that lowering LDL-cholesterol concentrations with the 'statin' drugs, inhibitors of HMG-CoA reductase, leads to marked reductions in mortality from CHD, in proportion to the reduction in LDL-cholesterol achieved.

Despite widespread acceptance of the role of LDL-cholesterol, it remains true that many patients suffering an early myocardial infarction have relatively normal levels of LDL-cholesterol (Fig. 3.18). Increasingly, other abnormalities of lipid metabolism are being seen as associated with development of atherosclerosis. The clearest relationship is with the serum HDL-cholesterol concentration, which, in epidemiological studies, is strongly negatively related to CHD risk. This is understandable if we consider HDL-cholesterol concentrations to reflect activity of the reverse cholesterol transport pathway, removing excess cholesterol from peripheral tissues, such as macrophages, to the liver for ultimate disposal. Again, epidemiological studies have now been reinforced by clinical trials. Fibrate drugs that raise HDL-cholesterol concentrations by acting as PPAR-α agonists reduce mortality from CHD.

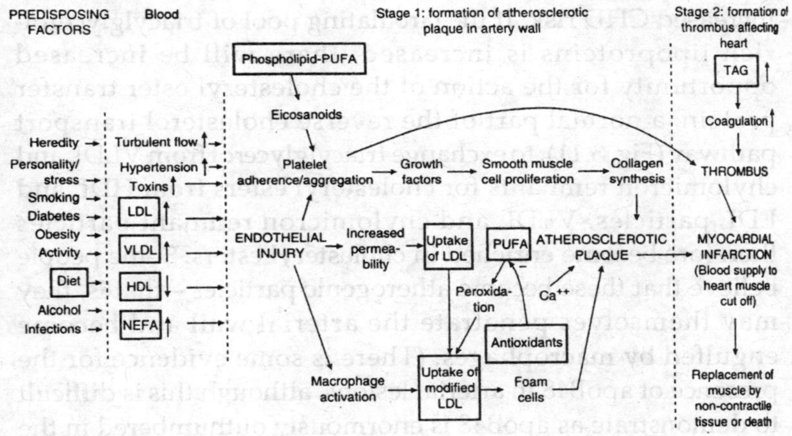

Fig. 3.17 Possible processes involved in the development of ischaemic heart disease and myocardial infarction. Boxes indicate points at which lipids may be involved.

Beyond the positive relationship of CHD with elevated LDL-cholesterol, and its negative relationship with HDL-cholesterol, there are other more subtle effects. Increasingly there is recognition of a phenotype that predisposes to CHD, known by a variety of names, including the atherogenic lipoprotein phenotype, the metabolic syndrome, syndrome X, or hyperapobetalipoproteinaemia (a high concentration of apoB in the circulation). The characteristics usually associated with the atherogenic lipoprotein phenotype are listed in Table 3.6. A brief description of its origin in terms of lipid metabolism may be given as follows. It appears to stem from a disturbance of the metabolism of the triacylglycerol-rich lipoproteins, with overproduction of VLDL from the liver. That, in turn, may result from increased delivery of non-esterified fatty acids to the liver. An increased circulating concentration of VLDL particles will .compete for clearance by lipoprotein lipase when chylomicrons become enriched with triacylglycerol in the postprandial period. This will lead to a prolonged residence of partially lipolysed chylomicron and VLDL remnants in the circulation, something that has often been noted to be associated with increased CHD risk. If the circulating pool of triacylglycerol-rich lipoproteins is increased, there will be increased opportunity for the action of the cholesteryl ester transfer protein, a normal part of the reverse cholesterol transport pathway (Fig. 3.11), to exchange triacylglycerol from VLDL and chylomicron remnants for cholesteryl esters from HDL and LDL particles. VLDL and chylomicron remnant particles therefore become enriched in cholesteryl esters. Some people believe that these become atherogenic particles—that is, they may themselves penetrate the arterial wall and become engulfed by macrophages. (There is some evidence for the presence of apoB48 in arterial lesions, although this is difficult to demonstrate as apoB48 is enormously outnumbered in the circulation by apoBlOQ.) But in addition, LDL and HDL particles thereby lose cholesteryl esters.

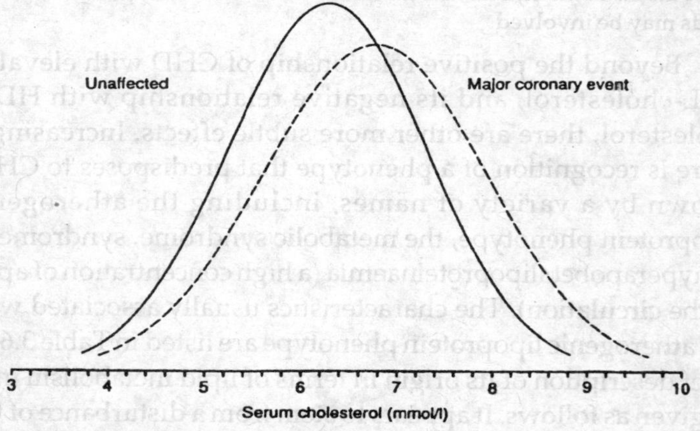

Fig. 3.18 Distribution of serum cholesterol concentrations in 438 men who

had a major coronary event and 7252 unaffected men. Note that there is considerable overlap between the distributions. An elevated serum cholesterol concentration is ertainly a major risk factor for development of coronary heart disease, but nevertheless many people experiencing the disease have relatively normal cholesterol concentrations.

Table 3.6 Features of the atherogenic lipoprotein phenotype

Feature	Comment
Hypertriglyceridaemia	May be relatively mild, but usually is more marked after meals.
Low HDL -cholesterol concentration	Reflects impairment of triacylglycerol-rich lipoprotein metabolism.
Small, dense LDL particles	Total LDL-cholesterol concentration may be normal, hence number of particles must be increased.
Hyperapobetalipoproteinaemia (high concentration of apoB100 in plasma)	Reflects increased number of LDL particles, as above.
Insulin resistance	These features are often associated with impaired action of insulin on metabolic processes.

The triacylglycerol that they have gained in exchange can be hydrolysed by hepatic lipase, which acts preferentially on these smaller particles. Therefore the particles have lost both triacylglycerol and cholesterol. The result is a low HDL-cholesterol concentration (associated with increased risk of CHD as discussed above) and a predominance of small, relatively lipid-poor LDL particles. In the past few years these small, dense LDL particles have attracted much attention as the probable atherogenic particles of the atherogenic lipoprotein phenotype. There is always a population of LDL particles. of different sizes in the circulation, but some people tend to have a predominance of large, buoyant cholesterol-rich LDL particles (known as pattern A) and others a predominance of smaller, denser relatively lipid-poor LDL particles (pattern B). There is some evidence for a genetic component to this LDL phenotype, but it is clearly modifiable by diet and by other influences on the lipoprotein phenotype. Pattern B is associated with much higher risk of CHD than is pattern A, even though the total LDL-cholesterol concentration may be the same. The increased number of LDL particles that must be present in pattern B (to give the same total LDL-cholesterol concentration) accounts for the term hyperapobetalipoproteinaemia, which covers the same, or a closely related, phenotype. It is postulated that smaller LDL particles may more readily cross the endothelial wall and hence become substrates for macrophage uptake. In addition, there is considerable experimental evidence that small, dense LDL particles are more susceptible to peroxidation than are larger, more buoyant particles, and hence again more likely to be substrates for macrophage uptake.

Once the atherosclerotic plaques have developed, further events must follow to lead to myocardial infarction or ischaemic stroke. These are aggregation of platelets, probably at the site of rupture of the atherosclerotic lesion, and then the formation of a thrombus. These processes may be related to aspects of lipid metabolism. Platelet aggregation, an early step, is propagated by the formation of the eicosanoid, thromboxane A_2 from arachidonic acid in platelets. Release of thromboxane A_2 by activated platelets leads to a cascade effect on aggregation. Formation of throrhboxane A_2 is inhibited by aspirin (an inhibitor of the cyclo-oxygenase path way), and this is the reason that low doses of aspirin can markedly reduce the risk of re-infarction in people who have already suffered one myocardial infarction. Platelet aggregation is also inhibited by *n*-3 polyunsaturated fatty acids. This is probably mediated via the generation of eicosanoids of the 3-series rather than the 2-series. The blood coagulation pathways are also related to lipid metabolism. High circulating concentrations of the activated form of coagulation factor VII, often known as VIIa (or, if measured by assay of its coagulant activity, as VIIc), are associated with increased risk of CHD. Generation of factor VIIa is now known to relate to triacylglycerol metabolism. There is a relationship between circulating concentrations of VIIa and of triacylglycerol, and the elevation of plasma triacylglycerol concentrations that occurs after a meal is associated with activation of factor VII.

Myocardial infarction may lead to death. If it does, this is not usually because of complete obstruction of blood flow to the myocardium, but because the electrical rhythm of the heart becomes grossly disturbed, leading to the critical conditions of ventricular fibrillation or cardiac arrest. Then blood supply to the rest of the body, including the brain, ceases. The stress reaction that sets in at the onset of myocardial infarction can lead to very high plasma concentrations of non-esterified fatty acids, and there is considerable evidence that these may themselves be a potent cause of ventricular dysrhythmias. It has been postulated that the stresses associated with modern living lead to consistent elevations of the plasma non-esterified fatty acid concentration that are now inappropriate, and may lead to — or potentiate — ventricular dysrhythmias. The UK biochemist Eric Newsholme has put forward this view, and suggests that in earlier times we would have oxidized the excess fatty acids in muscular activity; hence the hazards of stressful, sedentary living. There is an interesting recent development in this story. The American nutritionist Alexander Leaf has shown that, amongst fatty acids, the *n*-3 polyunsaturated fatty acids appear to have a unique ability to stabilize the heart rhythm. In several studies in experimental animals he has' shown that acute administration of *n*-3 polyunsaturated fatty acids can prevent death from coronary artery obstruction. There is convincing support now from human dietary trials. In the Diet and Reinfarction (DART) study, conducted in the UK, survivors of myocardial infarction were randomized into two groups, one of which was given advice to eat oily fish (and if this advice was not followed, they received supplementation with fish-oil capsules). Over the next two years, the fisheating group suffered the same rate of new myocardial infarctibns as did the control group; but their death rate was significantly reduced, by 29%.

It has been proposed that this reflects protection against the fatal dysrhythmias that often accompany myocardial infarction. More recently, the Lyon Dief-Heart Study also investigated the effects of dietary modification in people who had suffered from a myocardial infarction. In this case the 'test' group was asked to follow a Mediterranean-style diet, which was particularly rich in α-linolenic acid (18:3*n*-3). The control group was given conventional dietary advice. The trial was stopped prematurely, after an average of 2.5 years on the diet, because the results were so clearly significant, with lower death rates in the Mediterranean-diet group. A four-year follow-up of the participants showed that even after the formal trial had finished, the subjects had largely kept to their diets, and the difference in all-cause mortality between the groups was marked (about half in the test group compared with the control group). For 'cardiac deaths' the ratio was 3:1.

Risk factors for CHD and the effects of diet

From the preceding description, it will be clear that the major lipid risk factors for CHD are: an elevated circulating LDL-cholesterol concentration; a low HDL-cholesterol concentration; a predominance of small, dense LDL particles; and elevated circulating triacylglycerol concentrations (particularly in the postprandial period). The total cholesterol concentration in serum is often listed as a risk factor, but this reflects the fact that LDL-cholesterol is the major component of total serum cholesterol. These risk factors are modified to a great extent by genetic and environmental factors but they are also modulated to a considerable extent by dietary factors, which will be reviewed briefly here.

The realization of a connection between serum cholesterol concentra tions and CHD risk began in the 1950s with the work of the celebrated American nutritionist, Ancel Keys. Keys had been travelling in the Mediterranean countries, investigating the apparently very low incidence of chronic diseases including CHD. He questioned whether this might be related to aspects of the diet in those areas. This led to the foundation of the Seven Countries Study, an international study of CHD and associated factors in a number of countries (it soon expanded beyond seven) with widely differing incidences of CHD. There were two major, early findings. Comparing average values in one country with another, there was a strong positive relationship between serum cholesterol concentration and incidence of CHD, and a strong positive relationship between consumption of saturated fat and serum cholesterol. This led to the recognition that the nature of the fatty acids in the diet is an important influence on the serum cholesterol concentration: saturated fatty acids tend to raise it, polyunsaturated fatty acids to lower it. For some years it was felt that monounsaturated fatty acids were neutral in this respect. However, the answer to any experimental trial is dependent upon how the experiment is conducted, and it is now realized that if monounsaturated fatty acids replace saturated fatty acids in the diet, they reduce serum cholesterol concentration. Amongst these fatty acid classes, not all fatty acids have the same effect. Of the saturated fatty acids, for instance, those with chain length below 12C seem to have no effect; 12C, 14C and 16C lengths raise cholesterol, whereas stearic acid, 18:0, may slightly reduce cholesterol (perhaps because of rapid desaturation to oleic acid). In contrast to the effects of dietary fatty acids, the effect of dietary cholesterol on serum cholesterol concentration is relatively weak over any reasonable range of intake.

This can be understood if we remember the powerful systems that regulate cholesterol accumulation in cells. The relative effects of fatty acids and cholesterol have been combined by many investigators to produce equations that predict the change in serum cholesterol in response to any dietary manipulation. The mechanism for the effects of different dietary fatty acids on serum cholesterol concentration is not clearly understood. It appears to reflect a shift in the hepatic cholesterol pool between esterified and unesterified forms, with up-regulation of hepatic LDL-receptors (and a consequent reduction in serum LDL-cholesterol concentration) when polyunsaturated fatty acids predominate.

The *n*-3 polyunsaturated fatty acids have a special role. They are relatively neutral in terms of serum cholesterol, but they have a potent effect in reducing serum triacylglycerol concentrations. They also reduce platelet aggregation, and may stabilize the myocardial rhythm, as described above.

The double bonds in dietary unsaturated fatty acids are mostly of the *cis* geometrical configuration. However, some foods contain significant quantities of isomeric fatty acids in which the double bonds are in the *trans* configuration . These are often produced as a result of hydrogenation of unsaturated fatty acids. This occurs in the reducing environment of the bovine rumen, so that dairy products and beef fat contain small proportions of the *trans* fatty acids. It also occurs during the hydrogenation of vegetable oils to produce 'harder' fats, e.g. in the production of margarine. A number of epidemiological studies have shown a relationship between *trans* fatty acid intake and cardiovascular disease, and controlled feeding studies suggest that dietary *trans* fatty acids raise serum cholesterol concentrations to a very similar extent to saturated fatty acids. Given the similarity of their molecular configurations, this is perhaps not surprising. The content of *trans* fatty acids is being reduced by many food manufacturers as a result of this finding.

In most naturally occurring *cis*-polyunsaturated fatty acids, the double bonds are separated by a methylene bridge; e.g. the most common form of linoleic acid in nature is (*cis*-9, *cis*-12) 18:2. However, a large number of isomers of linoleic acid is found. Some of these have the double bonds between consecutive pairs of carbon atoms [the most common is (*cis*-9, *trans*-11) 18:2, also known as rumenic acid]. This arrangement of double bonds is similar to that seen in benzene, and is known as conjugated. The electrons become delocalized over conjugated double bonds, and this can confer unusual chemical properties. The group of isomers of linoleic acid with this configuration has become known (collectively) as conjugated linoleic acid (CLA). CLA is formed in the rumen of ruminant animals, and is found in milk fat, cheese and beef. It has come to prominence because of claims from animal studies that CLA can protect against some forms of cancer. Dietary CLA has also been shown to alter body composition in mice, with a loss of body fat, and in cultured adipocytes to reduce the activity of lipoprotein lipase. More recently, it has been claimed that dietary CLA may protect against atherosclerosis. However, the evidence in this respect is not clear-cut: there have also been demonstrations that high levels of CLA fed to rodents can predispose to the formation of fatty streaks in the aorta (presumed to be the precursors of atherosclerotic lesions). Various mechanisms for the potential beneficial effects have been proposed, and may differ for the different isomers. For instance, some isomers are potent agonists of PPAR-α.

The total quantity of dietary fat has an important influence on serum lipoprotein concentrations, beyond the nature of the fatty adds it contains. It is a common observation that a change to a diet containing a low proportion of energy from fat, and a correspondingly higher proportion from carbohydrate, is associated with a reduction in serum HDL-cholesterol concentration and an elevation of serum triacylglycerol concentration. The reduction in serum HDL-cholesterol concentration may reflect reduced flux of fat through the exogenous lipoprotein pathway, which involves the transfer of surface components from the chylomicrons, as they are lipolysed by lipoprotein lipase, to HDL particles. The elevation of serum triacylglycerol concentration is not clearly understood. It is thought to represent increased secretion of VLDL-triacylglycerol from the liver, perhaps because of a change in the metabolic partitioning of fatty acids in the liver between oxidation (favoured when fat levels in the diet are high) and esterification (favoured when carbohydrate is plentiful, and insulin levels are high). A recent study using isotopic methods to assess hepatic fatty acid metabolism has suggested, however, that the elevation of triacylglycerol concentration on a high-carbohydrate diet may reflect impaired removal of triacylglycerol from the circulation in peripheral tissues. These observations have led to questions about the safety of low-fat diets, since both these changes (depression of HDL-cholesterol, elevation of triacylglycerol concentrations) appear to be deleterious in terms of CHD risk. Such questioning is important, but many nutritionists feel that a shift towards lower fat diets may play an important role in reducing the incidence of obesity. The answers are not yet clear. We need to know: (1) Are the changes in lipoprotein particles that lead to elevation of triacylglycerol concentrations the same as those seen in subjects at increased risk of CHD? (2) Whether the changes in lipid concentrations are maintained long term (there is some evidence that they are not, but there are few long-term studies) (3) Whether the changes in lipid concentrations depend upon the nature of the dietary carbohydrate: in many short-term experimental studies, much of the dietary carbohydrate is in the form of simple sugars, which may have a particularly marked triacylglycerol-raising effect.

Hyperlipoproteinaemias

Diseases that involve elevation of circulating lipid concentrations (hyperlipoproteinaemias or hyperlipidaemias) are often associated with increased incidence of atherosclerosis and myocardial infarction. They are often sub-divided into primary and secondary hyperlipoproteinaemias. Primary hyperlipoproteinaemias are considered to arise directly from a genetic cause, while the secondary condition arises from some other medical or environmental cause such as diabetes or obesity. If the diabetes or obesity is controlled or cured, the hyperlipoproteinaemia will disappear or lessen. However, the simplicity of this classification is now under question as we recognize increasingly that most diseases reflect interactions between genome and environment. For instance, even in familial hypercholesterolaemia, caused by a defect in the LDL-receptor, the degree of elevation of plasma cholesterol concentration is dependent upon other lifestyle and medical factors such as diet and thyroid hormone status.

A different classification is based upon the observed lipid phenotype. The classification normally used is that proposed in 1967 by the American clinician and biochemist, Donald Fredrickson (Table 3.7). Fredrickson believed this was a classification of familial (i.e. primary) hyperlipidaemias, but it is now recognized to include many conditions that we would regard as secondary.

There are a few well-described single-gene mutations that lead to 'classic' primary hyperlipoproteinaemias, but for many conditions the background is polygenic, and the phenotype may be correspondingly more variable according to environmental factors. A given phenotype may result from different causes. For instance, amongst single-gene mutations, those that abolish activity of the LDL-receptor lead to familial hypercholesterolaemia, but an identical phenotype (Fredrickson Type IIa) is produced by rarer mutations in apoBlOO that affect its ability to bind to the LDl- receptor (the condition is known as familial defec. tive apolipoprotein B.100 or FDB, but in most clinical situations would not be distinguished from familial hypercholesterolaemia). Complete lack of activity of lipoprotein lipase leads to familial chylomicronaemia syndrome or Fredrickson Type I hyperlipoproteinaemia, but again an identical phenotype is caused by mutations in apoCII that abolish its ability to activate lipoproteinlipase. For clinical purposes it may not be important to distinguish these causes, although genetic analysis in increasingly being applied to these conditions.

Secondary hyperlipoproteinaemias may be associated with many conditions, but amongst the most common are diabetes and obesity (see the following section), thyroid disease (low thyroid hormone concentrations lead to hypercholester olaemia and hypertriglyceridaemia), over-consumption of alcohol (hypertriglyceridaemia),and liver and kidney diseases. Some drug treatments for other medical conditions may also lead to hyperli poproteinaemias. A prominent example that is causing concemat present is a marked hyper triglyceridaemia that is commonly observed in patients with human immunodeficiency virus. (HIV) infection treated with viral protease inhibi tors.

Obesity and diabetes

Mortality from cardiovascular disease (mainly coronary heart disease and ischaemic stroke) is considerably increased in obese compared to lean individuals. Almost all this increase can be accounted for by alterations in plasma lipid and lipoprotein concentrations. Statistically, it is some times claimed that obesity *per se* is not a significant risk factor for cardiovascular disease, but this merely implies that it has no residual effect beyond what would be expected from the alterations in lipid concentrations. The typical alterations in lipid concentrations in obesity are those described earlier as the atherogenic lipoprotein phenotype (Table 3.6). Because concentrations of some lipoprotein species (particularly HDL) may be decreased, this is often described as a dyslipidaemia rather than a hyperlipidaemia.

Table 3.7 The classification of hyperlipidaemias according to phenotype

Type	Plasma cholesterol	Plasma triacylglycerol	Particles accumulating	Usual underlying defect	Treatment
I	+	+++	Chylomicrons	Lipoprotein lipase deficiency; apolipoprotein CII deficiency.	Low-fat diet; medium-chain triacylglycerols to replace long-chain fatty acids.
IIa	++	N	LDL	LDL-receptor defect or LDL overproduction.	Reduce dietary saturated fat and cholesterol; resins; 'statin' drugs (HMGCoA reductase inhibitors).
IIb	++	++	VLDL, LDL	VLDL or LDL overproduction or impaired clearance.	Reduce dietary saturated fat and cholesterol; reduce weight; fibrate drugs (PPAR-α activators); statins.
III	+	++	Chylomicron remnants and VLDL remnants	Impaired remnant removal; may be due to particular isoform of apolipoprotein E, or apo-E deficiency.	Reduce dietary saturated fat and cholesterol; fibrates; statins.
IV	Nor+	++	VLDL	VLDL overproduction or clearance defect.	Reduce weight if obese, reduce alcohol intake; *n*-3 PUFA supplementation; fibrates.

V	+	+++	Chylomicrons, VLDL and remnants	Lipoprotein lipase defect (not complete absence) or apolipoprotein CII deficiency.	Reduce weight if obese, reduce alcohol intake; *n*-3 PUFA supplementation; fibrates.

This is known as the Fredrickso classification. N: normal; +: mildly raised; ++: moderately raised; +++: severely raised. Resins are non-absorbed compounds that bind bile salts and cholesterol in the gastrointestinal tract so that they are excreted.

It is a long-standing observation that the health risks of obesity vary according to the distribution of fat over the body. The observation of different patterns of fat distribution is an old one; a 25 000-year-old carved figurine shows a female, perhaps thought to be especially fertile, with large fat deposits around her hips and breasts; Shakespeare in *Comedy of Errors* has Dromio referring to Nell, who clearly carried her fat around her midriff, with the words: 'No longer from head to foot than from hip to hip: she is spherical, like a globe'. This subject was first investigated systematically by the French physician Jean Vague in Marseilles in the 1940s.

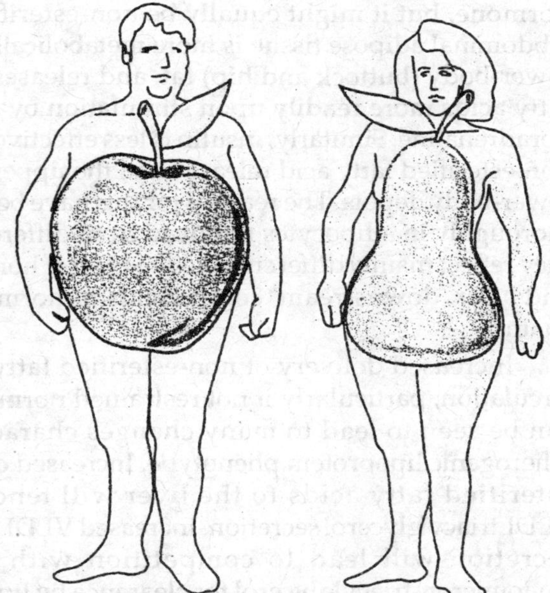

Fig. 3.19 Characterization of the two extremes of body fat distribution as apple- and pear-shaped.

(Jean Vague is still active and publishing today.) In a classic paper published in 1947

Vague drew attention to two broad patterns of fat distribution, one typically female with fat mainly on the lower body (often thought of as pear-shaped) and one typically male with abdominal and upper-body fat accumulation (often envisaged as apple-shaped; Fig. 3.19). He pointed out that the male pattern can also occur in women, and vice versa, and that irrespective of sex, the upper-body (apple) type of fat distribution was associated with a much higher incidence of several diseases including coronary heart disease and risk of developing diabetes.

One widely accepted hypothesis for the link between obesity, especially of the upper-body variety, and dyslipidaemia predisposing to coronary heart disease, suggests that it is mediated by the abdominal fat releasing some factor or factors that affect both lipid metabolism, and sensitivity of tissues to insulin. This factor might be a cytokine or other hormone, but it might equally be non-esterified fatty acids. Abdominal adipose tissue is more metabolically active than is lower-body (buttock and hip) fat, and releases non-esterified fatty acids more readily upon stimulation by adrenaline and noradrenaline. Similarly, insulin is less effective at suppressing non-esterified fatty acid release from the upper-body than the lower-body depots. The reasons for this have been investigated thoroughly in adipocytes removed from different depots, and they reflect mainly differential expression of hormone receptors and *other* 'downstream' components of hormone-signalling chains.

Increased delivery of non-esterified fatty acids into the circulation, particularly if not restrained normally by insulin, can be seen to lead to many changes characteristic of the atherogenic lipoprotein phenotype. Increased delivery of non-esterified fatty acids to the liver will tend to stimulate VLDLtriacylglycerol secretion. Increased VLDL-triacylglycerol secretion will lead to competition with the influx of chylomicron-triacylglycerol for clearance by lipoprotein lipase in the period following meals. Thus, hypertriglyceridaemia and particularly increased postprandial hypertriglyceridaemia may be explained. It was described earlier how this can lead to depression of HDL cholesterol concentrations and the production of small, dense LDL particles through the action of cholesteryl ester transfer protein (CETP). It is interesting also that the activity of CETP is itself increased by high concentrations of non-esterified fatty acids.

With the advent, in the last few decades, of imaging techniques such as computer-assisted tomography and magnetic resonance imaging, it has been appreciated that upper-body fat accumulation is associated with increased deposition of adipose tissue within the abdominal cavity, surrounding the visceral organs - so-called visceral fat. Two of the larger intra-abdominal depots, those associated with the greater omentum and the mesentery, have the property that the blood passing through them is conducted to the hepatic portal vein. It has also been found in many studies that adipocytes isolated from these depots (taken at abdominal surgery) is even more metabolically active than is abdominal fat from the subcutaneous depot. Therefore some people now consider that liberation of non-esterified fatty acids from these depots, reaching the liver directly (where they may influence VLDL-triacylglycerol secretion), plays a particular role in the dyslipidaemia of upper-body obesity. This remains to be confirmed because it is very difficult to study the metabolic pattern of these adipose tissue depots within the body.

Increased circulating concentrations of non-esterified fatty acids have effects in addition to (those on VLDL secretion from the liver. The supply of fatty acids to the liver is a major determinant of the rate of release of glucose from the Jiver. This may arise because fatty acids are readily oxidized in the liver, and this provides a source of energy for the process of gluconeogenesis (glucose synthesis from other precursors). Therefore increased non-esterified fatty acid concentrations may lead to hepatic over-production of glucose. Fatty acids are a preferred metabolic luel for many tissues including skeletal muscle, where they displace glucose as a fuel for oxidation. This can lead to inefficiency of glucose disposal, particularly in the postprandial period when skeletal muscle glucose uptake is normally stimulated by insulin. These changes in glucose metabolism together tend to raise concentrations of glucose in the plasma. The rise may be counteracted by an increased secretion of insulin from the pancreas. This is considered by many to be a plausible scenario for the development of insulin resistance, a situation in which insulin concentrations are high and do not regulate glucose metabolism normally. It was described earlier (Table 3.6) as a feature of the atherogenic lipoproteinemphenotype, and is a very typical finding in upper-body obesity.

In some people, increasing insulin resistance can be matched by increasing insulin secretion from the pancreas. In this case it is accompanied by the dyslipidaemia described above, but glucose metabolism remains relatively normally controlled. In others, however, presumably those with a genetic predisposition, the pancreatic β cells fail to match the demand and become 'exhausted', so that insulin concentrations in plasma no longer match the glucose concentrations, which may then rise to high levels. This is the condition of diabetes mellitus, diagnosed by measurement of increased plasma glucose concentrations, fasting or after an oral glucose dose.

There are two major types of diabetes mellitus. The type described above, very often associated with insulin resistance, usually coming on in later life (after the age of 40), and associated with marked insulin resistance, is called Type II diabetes mellitus. In older literature it is called non-insulin-dependent diabetes mellitus (because insulin is not usually necessary for treatment) or maturity-onset diabetes. The other major type is Type I, or insulin-dependent or juvenile-onset diabetes. This almost always develops in childhood and requires insulin for treatment. People with Type I diabetes are often thin, although insulin resistance is also present (less marked than in Type II). Type I diabetes mellitus arises because of an autoimmune process that destroys the insulin-secreting β-cells of the pancreas and thus its aetiology is not directly concerned with lipid metabolism.

Both types of diabetes are associated with considerably increased risk of cardiovascular disease, including coronary artery disease, peripheral vascular disease and ischaemic stroke. About 75% of people with diabetes will eventually die from coronary heart disease, an incidence that is about twofold to sixfold greater than that for non-diabetic people matched for age, sex and for major risk factors including LDL-cholesterol concentration. The reasons for this are not entirely clear. In Type II diabetes there is clear association with the atherogenic lipoprotein phenotype (Table 3.6). In Type I diabetes the alterations in lipid metabolism may be more subtle and involve changes in the composition of lipoprotein particles that have not been fully delineated. Increased oxidative stress is one unifying hypothesis that has attracted some attention, since peroxidation of LDL particles may precede their uptake by macrophages in the arterial wall.

The treatment of diabetes is not strictly within the scope of this book except in so far as it relates to lipid metabolism. Type I diabetes requires injections of insulin. Type II diabetes is usually managed with drugs that boost insulin secretion (sulphonylureas) or apparently increase sensitivity to insulin by an unknown mechanism (metformin, a biguanide drug). Recently, however, a new class of agents has been discovered and is being introduced for clinical use in Type II diabetes. These are the thiazolidinediones or 'glitazones', agonists of PP AR-γ.

Hypolipoproteinaemias

These conditions in which lipoprotein concentrations are below normal or there is a complete absence of one lipoprotein class are much rarer than those characterized by raised lipoprotein concentrations (Table 3.7). There are some 'classic' single-gene mutations in this group.

Chylomicron retention disease (Anderson's disease) is usually diagnosed in childhood as a result of poor growth and intestinal problems. The enterocytes are unable to secrete chylomicrons, and although other lipoprotein species are present in plasma their concentrations are low and their composition abnormal. It is a familial disease which is recessively expressed, and occurs mainly in males. The precise defect is not known and may differ between different families. It is not a defect in microsomal triacylglycerol transfer protein (MTP) although people with that defect may exhibit some impairment of fat absorption.

Abetalipoproteinaemia is a complete absence of apoB100 in plasma, i.e. there are no VLDL or LDL particles. It is an autosomal recessive disease. There may be signs of fat malabsorption, and neurological and visual impairment. A characteristic finding is abnormally shaped red blood cells (often star-shaped) which reflects abnormal membrane phospholipid composition, secondary to the changes in lipoprotein concentration and composition. The molecular defect is in MTP, which plays a critical role in VLDL assembly (Fig. 3.4). There are also conditions in which levels of apoB100 are low but not absent (hypobetalipoproteinaemia). This may represent the heterozygous form of abetalipoproteinaemia or it may be a separate condition, associated with a mutation causing a truncation of apoB100.

There are several diseases associated with low concentrations of HDL, but the classic single-gene condition associated with absence of HDL is Tangier disease, named after a small island off the coast of Virginia, USA, where it was first described by Donald Fredrickson in 1961. Lack of HDL is associated with cholesterol accumulation in tissues, especially macrophages that accumulate in lymphoid tissue such as the tonsils. Since these macrophages have accumulated their lipid initially by taking up LDL particles, and since LDL particles also carry fat-soluble pigments such as β-carotene, the tonsils in Tangier disease are typically orange in colour, a hallmark of the disease. It encodes the cholesterol efflux regulatory protein, responsible for efflux of cholesterol from tissues. Paradoxically, sufferers are not at greatly increased risk of cardiovascular disease.

A further defect leading to low, but not absent concentrations of HDL-cholesterol is a defect in lecithin-cholesterol acyltransferase (LCAT). LCAT is responsible for the esterification of cholesterol acquired by HDL particles either by interaction with cell membranes or during the shedding of excess surface components as triacylglycerol-rich lipoproteins are hydrolysed by lipoprotein lipase. A striking feature of this disease is corneal opacity, caused by lipid infiltration of the cornea. This led Swedish investigators who described one variant, to give it the name Fish Eye Disease (because the opacified corneas were thought to resemble the eyes of boiled fish). Plasma lipoproteins are abnormal in composition in LCAT deficiency, with an excess of unesterified cholesterol as expected. As in Tangier disease, the patients are unexpectedly not at greatly increased risk of atherosclerosis.

SUMMARY

In this chapter we looked at how dietary lipids are taken up into the body, how lipids are transported around the body, and how these systems are regulated according to supply and demand. In addition, diseases relating to lipid transport have been considered. Emphasis has been given to processes in the human body, but differences between humans and other mammals have been touched upon.

Humans and some other mammals take in a large part of their dietary energy in the form of fat. This dietary fat is a source of energy, of fuel for storage, and a precursor for many important biological molecules. In order for the body to use it, it has first to be taken up from the intestinal tract. This involves emulsification, to produce a large surface area upon which various lipolytic enzymes can act to cause hydrolysis of ester bonds. Simpler molecules (e.g. monoacylglycerols, fatty acids) are released and taken up into absorptive cells of the small intestine. Here, larger molecules (e.g. triacylglycerols, cholesteryl esters) are resynthesized and packaged for export to the circulation. The distribution of water-insoluble lipid molecules around the body in the aqueous environment of the blood plasma has been achieved early on in evolution by the production of emulsion-like particles in which hydrophobic molecules are surrounded by an amphipathic shell, mainly of phospholipids. These are the lipoprotein particles. There are various classes of lipoproteins, distinguished by size, composition and metabolic function. The largest and most lipid-rich are the chylomicrons, synthesized in the small intestine and carrying dietary fat (mainly triacylglycerol) into the circulation. The triacylglycerol in these particles is delivered to tissues in the form of fatty acids after hydrolysis by the enzyme lipoprotein lipase, attached to the capillary endothelium. The activity of lipoprotein lipase is regulated in different tissues according to the need for fatty acids. Thus, dietary fatty acids are distributed according to the metabolic needs of the various tissues.

During fasting, no dietary triacylglycerol enters the system but the liver secretes triacylglycerol-rich particles, the very low density lipoproteins. They also deliver fatty acids to tissues via lipoprotein lipase. The smaller particles that remain after loss of most of the triacylglycerol are known as low-density lipoproteins. They are the main transporters of cholesterol in the circulation. They may be taken up by tissues via the cell-surface low-density lipoprotein receptor, thus delivering cholesterol from liver to other tissues. Tissues can also export excess cholesterol through the pathway of reverse cholesterol transport. This involves the high-density lipoprotein particles, which can pick up cholesterol from cell surfaces and transport it to the liver where it can be excreted in the bile, either intact or after conversion to bile acids.

Lipoprotein and lipid metabolism are regulated in a number of ways. Lipoproteins are targeted by the proteins (apolipoproteins) that they carryon their surface, which interact with specific receptors on cell surfaces. The expression of these receptors, especially the low-density lipoprotein receptor, is regulated according to the cell's lipid status. Fatty acids and cholesterol, and derivatives of them, can directly regulate gene expression to control the synthesis of lipids and of receptors.

Excessive concentrations of lipids in the circulation are associated with a number of diseases, especially cardiovascular disease in which the blood supply to various organs is impaired. This can lead to partial or complete loss of function, e.g. of the brain (stroke), heart (coronary heart disease and myocardial infarction) or limbs (peripheral vascular disease). A common underlying process is that of atherosclerosis, the build-up of fatty deposits in the arterial wall. Circulating lipid concentrations are responsive to the diet. In particular, the nature of the predominant dietary fatty acids regulates the serum cholesterol concentration (saturated fatty acids tend to raise it, polyunsaturated to lower it).

Chapter 4

Lipids in Cell Organelles

One of the most important general functions for lipids is their role as constituents of cellular membranes. These membranes not only separate cells from the external environment, but also compartmentalize cells and provide a special milieu for many important biochemical processes. Here we shall describe the different types of lipids found and their distribution in various cellular membranes, as well as aspects of their function there.

CELL ORGANELLES

Lipids contribute to the structures of a variety of membranes in different parts of cells and in different organisms. The composition of these lipids varies considerably depending on the type and function of the membrane.

Eukaryotic cells characteristically contain a range of different cell organelles. All of these are separated from the cytosol (and usually from each other) by membranes. Some organelles such as mitochondria or chloroplasts are surrounded by two distinct membranes whereas others, such as microbodies, have a single membrane only. The relative numbers and detailed structure of most of these organelles vary from cell to cell. Moreover, in some cases, for example chloroplasts, the morphology will change depending on the developmental stage of the cell and will even be affected by the environment and metabolic state of the organelle.

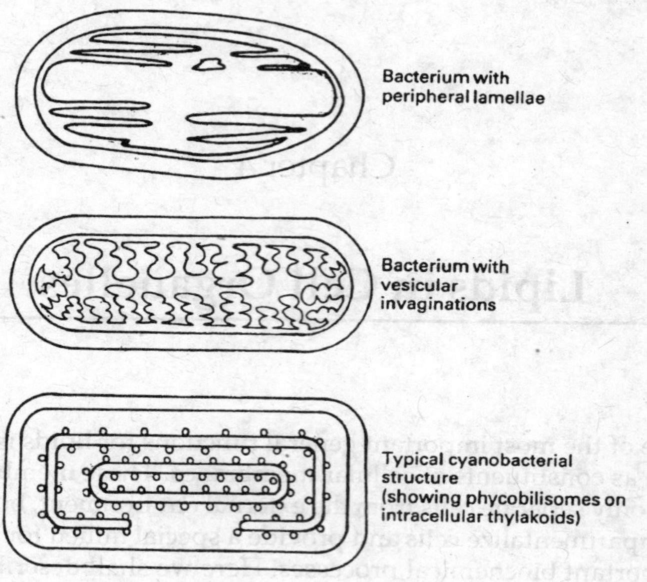

Fig. 4.1 Diagrammatic representation of internal membrane systems of bacteria and cyanobacteria.

Even in prokaryotes, many species contain a complex internal membrane structure. This is especially true in photosynthetic organisms such as purple or green bacteria and cyanobacteria (Fig. 4.1). Moreover, in some organisms membrane (or organelle) formation can be induced (Table 4.1) – thus providing a convenient experimental system from which to isolate membranes and in which to study membrane biogenesis.

Since organelles and their membranes carry out such a wide range of different functions, it is scarcely surprising that their lipid compositions vary. Moreover, the percentage of lipid and protein in different membranes can range from 80% lipid in myelin to only about 25% in mitochondria (Table 4.2). Almost all membrane lipids are amphipathic molecules such as phospholipids or glycolipids; neutral lipids such as triacylglycerols have little part in membrane structure. In organisms that make or utilize sterols, these tend to be concentrated in the external surrounding membrane (compare chloroplast or mitochondria with erythrocyte or myelin membranes in Table 4.2).

Table 4.1 Experimental systems for studying membrane induction

Organism	Stimulus	Morphological change
Photosynthetic bacteria	Starvation, light	Photosynthetic internal membranesformed
Euglena	Starvation, light	Chloroplasts produced
Yeast	Glucose in medium reduced	Mitochondria produced
Etiolated plants	Light	Etioplasts converted to chloroplasts
Animals	Fed barbiturates	Liver and endoplasmic reticulum proliferate
Animals	Fed clofibrate	Liver peroxisomes proliferate

Before describing the composition, structure and function of membranes, it is necessary to detail the molecules that make up the lipid components.

GLYCEROLIPIDS

Glycerolipids are the most widespread of membrane lipids. Phosphoglycerides dominate in higher animals whereas glycosylglycerides are important in plants.

Three important classes of membrane lipid are widely distributed – glycerolipids, sphingolipids and steroids. Of these, glycerolipids are quantitatively by far the most important group. They can be conveniently divided into two main groups – those containing phosphorus (phosphoglycerides) and those without phosphorus but containing a sugar constituent (glycosylglycerides). Confusingly, some compounds (e.g. sphingomyelin) can be classified in more than one group.

Table 4.2 Lipid composition of different membranes

	Membrane (% total lipid)				
	Chloroplast (spinach)	Protoplast (B. *megaterium*)	Mitochondrion (rat)	Erythrocyte (rat)	Myelin (rat)
Lipid:protein (wt/wt)	1:1	1:3	1:3	1:3	3:1
Phospholipid	12	48	90	61	41
PC	tr	0	40	34	12
PE+PI+PS	tr	19	41	11	26
PG	12	26	–	–	–
DPG	–	3	7	–	–
SPH	–	–	2	16	3
Phospholipid	80	52	–	11	42
MGDG	41	–			
DGDG	23	–			
SQDG	16	–			
Sterol, sterol ester	tr	0	tr	28	17
Acylglycerols	–	–	10	–	–
Pigments	8	–	–	–	–

PC = phosphatidylcholine; PE = phosphatidylethanolamine; PS = phosphatidylserine;
PG = phosphatidylglycerol; SPH = sphingomyelin;
MGDG = monogalactosyldiacylglycerol; DGDG = digalactosyldiacylglycerol;
SQDG = sulpholipid; DPG = diphosphatidylglycerol;
tr = trace. The glycolipids in rat membranes are sphingolipids.

Phosphoglycerides

The phosphoglycerides comprise a very widespread and diverse group of structures. In most membranes they are the main lipid components and, indeed, the only general exceptions to this statement are the photosynthetic membranes of plants, algae and cyanobacteria and the archaebacterial membranes.

Usually, phosphoglycerides contain fatty acids esterified at positions 1 and 2 of glycerol. They are, thus, diacylphosphoglycerides. These lipids are named after the moiety which is attached to the phosphate esterified at position 3 of glycerol. Thus, the compounds can be thought of as derivatives of diacylglycerols in which the hydroxyl on carbon atom 3 is esterified with phosphoric acid, which in turn is esterified with a range of molecules – organic bases, amino acids, alcohols.

The simplest phosphoglyceride contains only phosphoric

Table 4.3 Structural variety of different diacylglycerophospholipids

X	Name of phospholipid	Source	Remarks
H	Phosphatidic acid	Animals, higher plants, microorganisms.	Only small amounts normally found. Main importance as a biosynthetic intermediate.
$OH{\cdot}CH_2CH_2N^+(CH_3)_3$ Choline	Phosphatidyl choline (lecithin)	Animals; first isolated from egg yolks; higher plants; rare in micro-organisms.	Most abundant animal phospholipid
$OH{\cdot}CH_2CH_2NH_3^+$ Ethanolamine	Phosphatidyl ethanolamine	Animals; higher plants; micro-organisms.	Widely distributed but in small amounts. Minor components of old 'cephaline' fraction. *N*-acetyl derivatives in brain; fatty amides in wheat flour, peas.
$OH{\cdot}CH_2 \cdot CH(COO^-) \cdot NH_3^+$ Serine	Phosphatidylserine	Animals; higher plants; micro organisms.	Widely distributed but in small amounts. Minor component of old 'cephaline' fraction. Serine as L isomer. Lipid usually in salt form with K+, Na+, Ca^{2+}

Contd.....

Table 4.3 Contd.

X	Name of Phospholipid	Source	Remarks
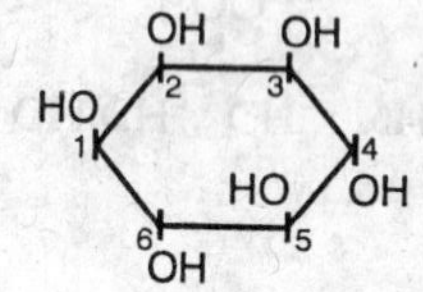 Myo-inositol	Phosphatidylinositol	Animals, higher Plants, micro organisms.	The natural lipidis found as a derivative of *myo*-inositol-1-phosphate only. The major inositol-containing phosphoglyceride (aout 80% of total in animals). 3-Phosphate derivatives of inositol phosphoglycerides present in very small amounts.
Inositol-4-phosphate	Phosphatidylinositol phosphate	Animal; small amounts in yeast; plants.	Mainly nervous tissue, but also plasma membranes of other cells.
Inositol-4, 5-*bis*phosphate	Phosphatidylinositol *bis*phosphate	Animals; small amounts in yeast; plants.	Distribution as above. Both compounds have very high rates of turnover.
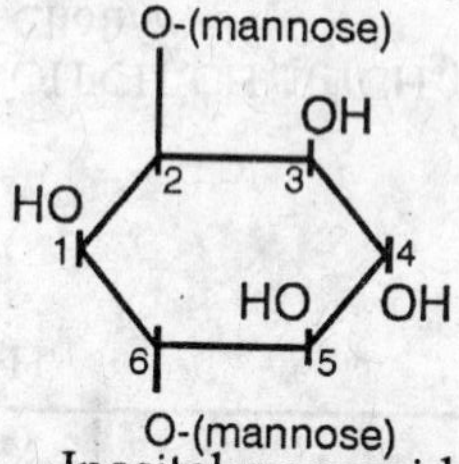 Inositol mannoside	Phosphatidylinositol mannoside X=0, monomannoside X=1, dimannoside etc.	Micro-organisms (*M.phlei*,*M. tuberculosis*).	

Contd.....

Table 4.3 Contd.

X	Name of Phospholipid	Source	Remarks
CH_2OH. CHOH; CH_2OH Glycerol	Phosphatidylglycerol	Mainly higher plants and micro-organisms.	'Free' glycerol has opposite stereochemical configuration to the acylated glycerol, i.e. 1,2-diacyl-1-*sn*-glycerl.Probably most abundant phospholipid because of its prevalence in photosynthetic membranes.
$CH_2OC(=O)R^1$ $CHOC(=O)R^2$ $CH_2OP(=O)(O^-)—OCH_2CH(OH)—CH_2OH$ Phosphatidylglycerol of	Diphosphatidylglycerol (*bis*phosphatidylglycerol, cardiolipin)	Animal; higher plants; micro-organisms.	Major component of many bacteria, localized in inner mitochondrial membrane eukaryotes.

acid attached to diacylglycerol and is called phosphatidic acid (Table 4.3). Where additional 'X' groups are esterified to the phosphate moiety the lipids are called phophatidyl-X. Major types of such diacylphosphoglycerides are shown in Table 4.3 where relevant comments about their distribution and properties are made also.

Phosphonolipids

In 1953, Rouser and his associates first identified a phosphonolipid in biological extracts—from a sea anemone. Subsequent work has shown that such lipids, which contain a C-P bond, are significant constituents of lower animals such as molluscs, coelenterates and protozoa. Phosphonolipids can be detected in bacteria and mammals, but only in very low quantities. Two types of glycerophosphonolipids have been found (Fig. 4.2) and there are also other structures that contain a sphingosine (and, sometimes, a galactosyl) residue instead of glycerol. Usually, organisms tend to accumulate phosphonolipids based on either glycerol or on sphingosine rather than both types. In protozoa such as *Tetrahymena,* the glycerophosphonolipids are concentrated in the ciliary membranes. This is perhaps because the phosphonolipids are particularly resistant to chemical as well as enzymic attack.

Glycosylglycerides

Glycolipids that are based on glycerol have been found in a wide variety of organisms. However, whereas in animals they are only found in very small quantities, they are major constituents of some micro-organisms and are the main lipid components of the photosynthetic membranes of algae (including the blue-greens or cyanobacteria) and plants (Table 4.4). Their structure is analogous to that of glycerophospholipids with the sugar(s) attached glycosidically to position 3 of glycerol and fatty acids esterified at the other two positions.

Since the membranes of higher plant chloroplasts are the most prevalent on land, and the photosynthetic membranes of marine algae are the most common in the seas and ocean, it follows that these glycosylglycerides are the most abundant membrane lipids in Nature, in spite of the sparse attention paid to them in most standard biochemistry textbooks! The two galactose-containing lipids – monogalactosyldiacylglycerol and digalactosyldiacylglycerol – represent about 40% of the dry weight of photosynthetic membranes of higher plants. Position 1 of the galactose ring has a β-link to glycerol whereas, in digalactosyldiacylglycerol there is an α,1-6 bond between the sugars. Whereas galactose is almost the only sugar found in the glycosylglycerides of higher plants, other sugars such as glucose may be found in algae, particularly marine species. In bacteria several combinations of residues may be found in diglycosyldiacylglycerols (Table 4.5). The most common combinations are two glucose, two galactose or two mannose residues linked α, 1-2 or β,1-6. Such glycosylglycerides do not form a large proportion of the total lipids in bacteria, but are found more frequently in the Gram-positives or photosynthetic Gram-negatives. In addition, bacteria may contain higher homologues with up to seven sugar residues.

Apart from the galactose-containing lipids, a third glycosylglyceride is found in chloroplasts (Table 4.4). This is the plant sulpholipid. It is more correctly called sulphoquinovosyldiacylglycerol and contains a sulphonate constituent on carbon 6 of a deoxyglucose residue. This sui phonic acid group is very stable and also highly acidic so that the plant sulpholipid is a negatively charged molecule in Nature. Although this sulpholipid occurs in small amounts in photosynthetic bacteria and some fungi, it is really characteristic of the photosynthetic membranes of chloroplasts and cyanobacteria.

All three chloroplast glycolipids usually contain large amounts of α-linolenic acid (Table 4.6). In fact, monogalactosyldiacylglycerol may have up to 97% of its total

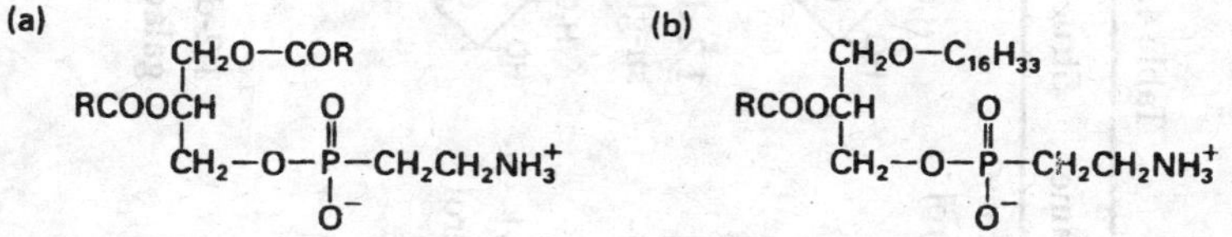

(a)

$CH_2O{-}COR$
$RCOOCH$
$CH_2{-}O{-}P(=O)(O^-){-}CH_2CH_2NH_3^+$

(b)

$CH_2O{-}C_{16}H_{33}$
$RCOOCH$
$CH_2{-}O{-}P(=O)(O^-){-}CH_2CH_2NH_3^+$

Fig. 4.2 Glycerophosphonolipids: (a) phosphatidylethylamine; (b) an analogue, which is a derivative of chimyl alcohol rather than a diacylphosphonolipid. It is named l-hexadecyl-2-acyl-*sn*-glycero(3)-2-phosphonoethylamine.

Table 4.4 Major glycosylacylglycerols of plant and algal photosynthetic membranes

Common name	Structure and chemical name	Source and fatty acid composition
Monogalactosy diacylglycerol (MGDG)	1,2-diacyl-[β-D-galactopyranosyl-(1→3)–*sn*-glycerol	Especially abundant in plant leaves and algae; mainly in chloroplast. Contains a high proportion of polyunsaturated fatty acids. *Chiarella vulgaris* MGDG has mainly 18:1, 18:2 when dark grown but 20% 18:3 when grown in the light. *Euglena gracilis* MGDG has 16:4. Spinach chloroplast MGDG has 25% 16:3, 72% 18:3. Also found in the central nervous systems of several animals in small quantity.
Digalactosyl-diacylglycerol (DGDG)	1,2-diacyl-[α-D-galactopyranosyl-(1′→6)–β-D galactopyranosyl-(1′→3)]-*sn*-glycerol	Usually found together with MDGD in chloroplasts of higher plants and algae. Not quite so abundant as MGDG. Also has high proportion of polyunsaturated fatty acids, especially 18:3. In both lipids the glycerol has the same configuration as in the phospholipids.

Continued

Plant sulpholipid (sulphoquinovosyl-diacylglycerol; SQDG)	$H_2C—SO_3H$, O, OH, HO, OH, O—CH_2, $CHOCOR^1$, CH_2OCOR^2 1,2-diacyl-[6-sulpho-α-D-quinovopyranosyl-(1′→3)]-*sn*-glycerol. [D-quinovose is 6-deoxy-D-glucose. Note the carbon-sulphur bond]	Usually referred to as a 'sulpholipid' as distinct from a 'sulphatide', which is reserved for cerebroside sulphates. Typical lipid of chloroplast membranes but present elsewhere in some marine algae. Also found in cyanobacterial photosynthetic membranes and, to a lesser extent, in purple photosynthetic bacteria. Contains more saturated fatty acids (mainly palmitic) than the galactolipids, e.g. spinach leaf sulpholipid has 27% 16:0, 39% 18:2, 28% 18:3.

acyl groups as this one component in some plants! The reason for this exceptional emichment is not known although speculations have been made. Moreover, a unifying theory connected with photosynthesis is not possible since many cyano-bacteria do not make α–linolenate and marine algae contain little of the acid. In the plant sulpholipid the most usual combination of acyl groups is palmitate/α-linolenate. Interestingly, unlike animal lipids most palmitate is esterified at position 2 and most of the α-linolenate at position 1. This distribution of fatty acids stems from the special features of plant fatty acid metabolism and because the sulpholipid is assembled via the 'prokaryotic pathway' of metabolism. Since the position at which saturated and unsaturated fatty acids are attached to the glycerol has a marked effect on their melting properties, fatty acid distribution may have functional significance. It has also been speculated that the different distribution of acyl moieties between plants and animals may be related to specific interactions with membrane proteins.

Bacteria contain a number of phosphatidylglycolipids. These compounds are confined to certain types of Gram-positive organisms such as the streptococci or mycoplasmas. In addition, different species of algae and bacteria (including archaebacteria) contain various sulphoglycolipids

Table 4.5 Some glycosylglycerides found in bacteria

Glyceride	Structure of glycoside moiety	Occurrence
Monoglucosyldiacyl-glycerol	α-D-Glucopyranoside	*Pneumonoccus, Mycoplasma*
Diglucosyldia-cylglycerol	β-D-Glucopyranosyl-(1→6)-O-β-D-glucopyranoside	*Staphylococcus*
Diglucosyldia-cylglycerol	α-D-Glucopyranosyl-(1→2)-O-α-D-glucopyranoside	*Mycoplasma, Streptococcus*
Dimannosyl-diacylglycerol	α-D-Mannopyranosyl-(1→3)-O-α-D-mannopyranoside	*Micrococcus lysodeikticus*
Galactofuranosyldia-cylglycerol	β-D-Galactofuranosid	*Mycoplasma, Bacteroides*
Galactosylglucosy-ldiacylglycerol	α-D-Galactopyranosyl-(1-2)-O-α-D-glucopyranoside	*Lactobacillus*
Glucosylgalactosyl-glucosyldiacylglycerol	α-D-Glucopyranosyl--(1-6)-O-α-D-galactopyranosyl-(1→2)-O-α-D-glucopyMranoside	*Lactobacillus*

usually with the sulphur present in a sulphate ester attached to the carbohydrate moiety. Lipids containing taurine seem to be significant components of some protozoa.

Betaine lipids

Betaine ether-linked glycerolipids are naturally occurring lipids which are important in many lower plants (e.g. bryophytes), algae, fungi and protozoa. Because of their structure (Fig. 4.3) and properties, they are often envisaged as substituting for phosphatidylcholine in membranes. A number of structures has been identified but diacylglyceryltrimethylhomoserine (DGTS) is the most common. Typically, it represents 10-20% of all membrane lipids but, in some species of mosses, it may be around 40% of the total.

Ether-linked lipids and their bioactive species

Apart from the betaine ether-linked glycerolipids, there are many lipids that have ether links to carbon position 1 (and sometimes position 2) of the glycerol backbone. In mammals,

Table 4.6 Fatty acid compositions of glycosylglycerides in two plants

Plant leaf		(% total fatty acids)				
		16:0	16:3	18:1	18:2	18:3
Spinach	MGDG	trace	25	1	2	72
('16:3-plant')	DGDG	3	5	2	2	87
	SQDG	39	-	1	7	53
Pea	MGDG	4	-	1	3	90
('18:3-plant')	DGDG	9	-	3	7	78
	SQDG	32		2	5	58

'16:3-plants' contain hexadecatrienoate in their monogalactosyldiacylglycerol while '18:3-plants' contain α-linolenate instead. The reason for this is provided by the differences in fatty acid metabolism between these two types of plants.

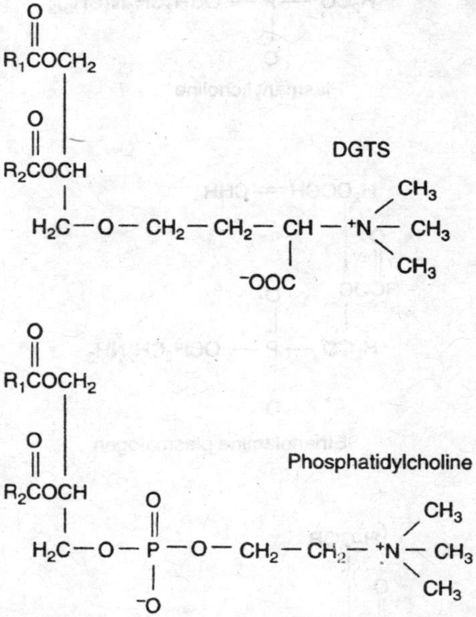

DGTS

R_1COOCH$_2$ – R_2COOCH – H$_2$C–O–CH$_2$–CH$_2$–CH(COO$^-$)–$^+$N(CH$_3$)$_3$

Phosphatidylcholine

R_1COOCH$_2$ – R_2COOCH – H$_2$C–O–P(=O)(O$^-$)–O–CH$_2$–CH$_2$–$^+$N(CH$_3$)$_3$

Fig. 4.3 Comparison of the structure of the betaine lipid, DGTS with phosphatidylcholine. DGTS = 1,2-diacylglyceryl-3-0-4' (*N,N,N*-trimethyl)-homoserine.

there are almost exclusively ether linkages in the choline or ethanolamine glycerolipid class. If the ether link is unsaturated (Fig. 4.4) then the lipid is termed a plasmalogen. Indeed, of the ether-linked phosphoglycerides, ethanolamine plasmalogen is the most widespread. It is espedally abundant in nervous tissues and in white blood cells and platelets. The ether lipids are not confined to mammals but are also found in lower animals, some plants and many bacteria. Particular interest has recently been paid to the biologically active platelet activating factor (PAF) (Fig. 4.4) and its derivatives. PAF has a similar structure to the ether derivative of phosphatidylcholine (plasmanylcholine) except that the fatty acid at the *sn*-2 position is replaced by an acetyl group.

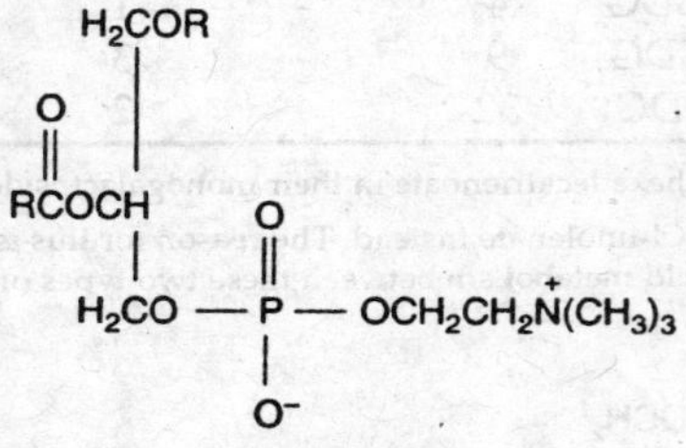

H_2COR
$RC(=O)OCH$
$H_2CO-P(=O)(O^-)-OCH_2CH_2\overset{+}{N}(CH_3)_3$

Plasmanyl choline

$H_2COCH{=}CHR$
$RC(=O)OC$
$H_2CO-P(=O)(O^-)-OCH_2CH_2\overset{+}{N}H_3$

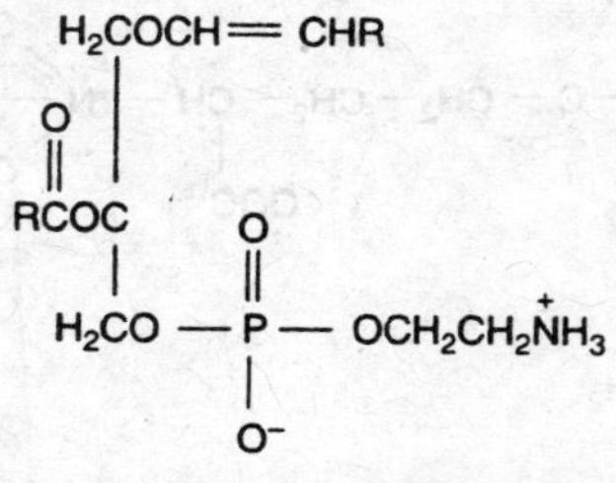

Ethanolamine plasmalogen

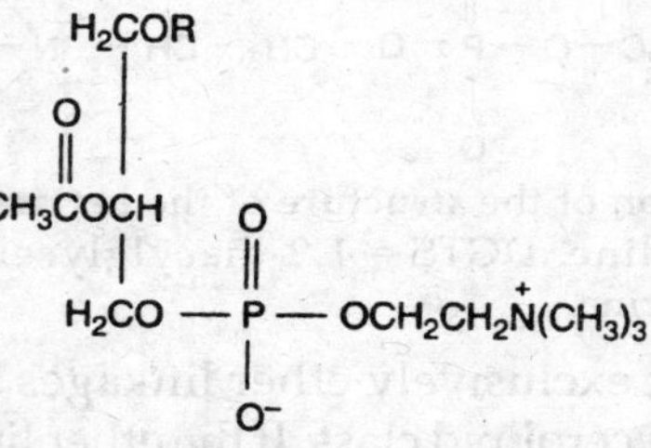

H_2COR
$CH_3C(=O)OCH$
$H_2CO-P(=O)(O^-)-OCH_2CH_2\overset{+}{N}(CH_3)_3$

Platelet activating factor (PAF)

Fig. 4.4 Some important ether lipids.

SPHINGOLIPIDS

Lipids based on a sphingosine rather than a glycerol backbone are also widespread in membranes but are particularly abunaant in nervous tissue. Generally, they tend to be concentrated in the outer leaflet of the plasma membrane.

The lipids were based on a glycerol backbone. However, another important group of acyl lipids have sphingosine-based structures. Both glycolipids and phospholipids are found, with some compounds capable of dual classification, i.e. phospholipids containing sugar residues. More than 300 different types of complex sphingolipids have been reported and this does not take account of variations in the ceramide backbone!

Sphingosine (D-erythro- 2-amino-trans-4-octadecane-1,3-diol or 4D-sphingenine) is a long-chain amino alcohol (Fig. 4.5 and 4.6). Several other sphingosylalcohols are also found in naturally occurring sphingolipids and these differ mainly in the number and position of double bonds and whether they are hydroxylated or not. (Some examples are given in Fig. 4.6) Both the amino and the alcohol moieties of sphingosine can be substituted to produce the various sphingolipids (Fig. 4.5).

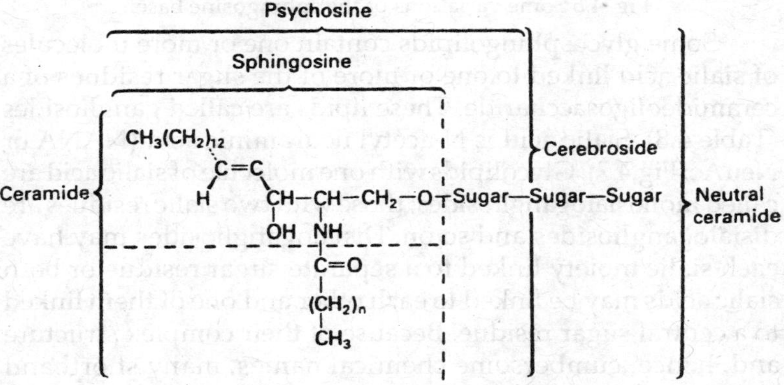

Fig. 4.5 Basic structure of sphingolipids.

Attachment of an acyl moiety to the amino group yields a ceramide. This acyl link is resistant to alkaline hydrolysis and, therefore, can be easily distinguished from the O-esters found in glycerol-based acyl lipids. The simplest glycosphingolipids are the monoglycosyl ceramides (or cerebrosides). In animals galactosylcerebroside is the most common. Further attachment of hexosides to glucosylcerebroside yields the neutral glycosylsphingolipids (or neutral ceramides) (Table 4.7). These are usually written with a shorthand nomenclature, e.g. Glc-Gal-Glc-Cer would be glucosyl-(1-4)galactosyl-(1-4)glucosyl-ceramide. The deacylated product of galactosylcerebroside, osphingosyl-galactoside is called psychosine.

Structure	Name	Abbrev.
$CH_3(CH_2)_{12}$(H)C=C(H)–CH(OH)–CH(NH_2)–CH_2OH	D-*erythro*-sphingosine (D-*erythro*-4-sphingenine)	(d18:1)
$CH_3(CH_2)_{12}$–CH_2–CH_2–CH(OH)–CH(NH_2)–CH_2OH	D-*erythro*-sphinganine	(d18:0)
CH_3·$(CH_2)_{12}$·CH_2–CH(OH)–CH(OH)–CH(NH_2)–CH_2OH	4-Hydroxy-D-*erythro*-sphinganine	(t18:0)

Fig. 4.6 Some variations of the sphingosine base.

Some glycosphingolipids contain one or more molecules of sialic acid linked to one or more of the sugar residues of a ceramideoligosaccharide. These lipids are called gangliosides (Table 4.8). Sialic acid is N-acetyl neuraminic acid (NANA or NeuAc; Fig.4.7). Glycolipids with one molecule of sialic acid are called monosialogangliosides; those with two sialic residues are disialogangliosides and so on. Disialogangliosides may have each sialic moiety linked to a separate sugar residue or both sialic acids may be linked to each other and one of them linked to a central sugar residue. Because of their complex structure and, hence, cumbersome chemical names, many shorthand notations have been employed. One of the most commonly used is that introduced by the Swedish scientist, Svennerholm. The parent molecule is denoted as G_{MI} and other derived structures are as shown in Table 4.8. The subscripts M, D, T and Q refer to monosialo-, disialo-, trisialo and quatra (tetra)sialogangliosides, respectively. A more recent nomenclature system, due to the German, Wiegandt, is also included in Table 4.8. The latter method has the advantage that, once the symbols have been learnt, the structure can be worked out from the shorthand notation and it can be applied to non-sialic acid-containing glycolipids as well. However, it has yet to supersede the Svennerholm notation, which will be used in this book.

Instead of glycosylation of the alcohol moiety of the sphingosine base, esterification with phosphocholine can take place. The phospholipid produced thus is called sphingomyelin.

Three structures constitute the main bases in animals with sphinganine and sphingosine being present in the sphingomyelin of most tissues, while a trihydroxylated derivative of sphinganine, called phytosphingosine (t18:0 in Fig. 4.6) is found in kidney. This base is also prominent in phytoglycolipid, which is a sphingolipid exclusive to plants that contains inositol as well as several other monosaccharide residues.

Physical techniques have revolutionized the analysis and identification of sphingolipids. In many cases a combination of gas-liquid chromatography, mass spectrometry and nuclear magnetic resonance spectroscopy provides sufficient information to elucidate a complete structure. The position of attachment of the sugar residues can be worked out from the use of glycosidase enzymes or by permethylation analysis. On the other hand the molecular weight and sequence of some quite complex glycosphingolipids can often be established in one step by the use of the gentler techniques of mass spectrometry such as fast-atom bombardment.

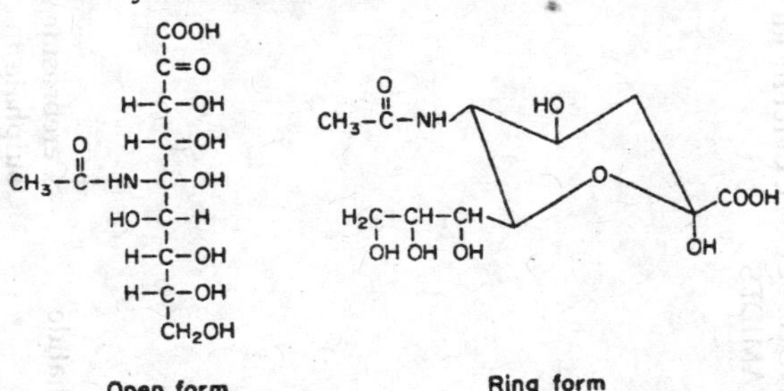

Fig. 4.7 N-acetylneuraminic acid (NeuAc) or sialic acid.

Table 4.7 The structure of some glycosyl ceramides [Cer = ceramide; glc = glucose; gal = galactose; GalNAc = N-acetyl galactosamine]

Accepted nomenclature	Old trivial name	Structure	Description	General remarks
MONOGLYCOSYL CERAMIDES	Cerebroside	Cer-gal Cer glc In all glycosyl ceramides and gangliosides there is an O-glycosidic linkage between the primary hydroxyl of sphingosine and C-1 of the sugar	'Cerebroside' originally used for galactosyl ceramide of brain but now widely used for monoglycosyl ceramides. Sugar composition depends largely on tissue. Brain cerebroside mainly galactoside while serum has mainly glucose. In animals the highest concentration is in brain. Monogalactosyl ceramide is largest single component of myelin sheath of nerve. Intermediate concentrations in lung and kidney. Also found in liver, spleen, serum with trace amounts in almost all tissues examined.	*Fatty acids:* cerebrosides containing galactose are characterized by having large concentrations of (a) hydroxy acids, (b) long chain odd and even fatty acids, in comparison with other lipids. The hydroxy acids include α-hydroxy acids formed by α-oxidation. This oxidation mechanism is probably also responsible for odd-chain acids. Typical acids are: 22:0 (behenic); 24:0 (lignoceric); 24:1 (nervonic); α-OH-24:0 (cerebronic)
Sulphatide	Cerebroside sulphate	Cer-gal-3-sulphate	Very generally distributed like cerebrosides. Fatty acid and base composition similar.	

Continued

Table 4.7 Contd.

Accepted nomenclature	Old trivial name	Structure	Description	General remarks
DIGLYCOSYL CERAMIDES	Cytosides		Major diglycosyl ceramide. Widely distributed. Key substance in glycosyl ceramide metabolism. It may accumulate, but as it is a precursor for both ceramide oligosaccharides or gangliosides it may be present in trace quantities only.	Base:d18:1 (sphingosine) is dominant. Smaller amounts of d18:0 (sphinganine) (especially leaves, wheat flour). Hydroxylated bases (e.g. t18:0) also occur (see Fig. 4.6).
(a) Lactosyl ceramide	Cytolipin H	Cer-glc-(4←1)-gal		
(b) Digalactosyl ceramide		Cer-gal-(4←1)-gal	Minor diglycosyl ceramide. Found especially in kidney (human and mouse).	
TRIGLYCOSYL CERAMIDES				
Digalactosyl glucosyl ceramide		Cer-glc-(4←1)-gal-(4←1)gal	Source: kidney, lung, spleen, liver. Most analyses have been on hunnan tissue.	In general, each *organ* has a dominant type of glycolipid but its nature may depend also on the *species*, e.g. monoglycosyl ceramdes are dominant in brain; trihexosides and aminoglycolipids in red cells.
TETRAGLYCOSYL CERAMIDES				
(a)	Aminoglycolipid globoside	Cer-glc-(4←1)-gal-(4←1)-gal-(3←1)-β-*N*-acetyl galactosamine	Most abundant lipid in hunnan erythrocyte stroma.	
(b)		Cer-glc-(4←1)-gal-(4←1)-gal-(3←1)-α-*N*-acetyl galactosamine	So-called 'Forssman Antigen'.	A ceramde with a 30C wi-hydroxy fatty acid is a major skin component, where it is largely responsible for water
(c)	Asialoganglioside	Cer-glc-(4←1)-gal-(4←1)-galNAc-(3←1)-gal		

Continued

Table 4.7 Contd.

Accepted nomenclature	Old trivial name	Structure	Description	General remarks
			Basic ganglioside structure without N-acetyl neurawc acid (sialic acid) residues. Internnediate in ganglioside biosynthesis (rat, frog, brain).	impe-reability. Modern analytical techniques are revealing that most types are widespread among tissues. Cerebrosides are widespread, but nninor connponents of higher plants, the best characterized being those of bean leaves and wheat flour where glucose is probably the only sugar. Sphingolipids are 'rare in microorganisnns.

As a generalization, sphingolipids have a rather specific distribution in cells where they are concentrated in the outer (external) leaflet of the plasma membrane.

STEROLS

Sterols are common in the membranes of eukaryotic organisms, where they are important in providing stability, but are rare in prokaryotes. In the membranes of higher animals cholesterol is the almost exclusive sterol constituent whereas in plants, other sterols, such as β-sitosterol, predominate.

Major sterols

Cholesterol and the functionally related sterols of fungi and plants are significant components of many organisms, especially of their external cellular membranes. However, they are not needed by all types of organisms nor, indeed, are they components (or, therefore, necessary) for all classes of membranes even in organisms with an appreciable sterol content.

Table 4.8 Structure of some gangliosides

Class	Structure	Wiegandt shorthand	Svennerholm shorthand	Composition and occurrence
MONOSIALOGANGLIOSIDE	β β β (a) Cer-glc-(4←1)-gal-(4←1)-galNAc-(3←1)-gal 3 (↑) 2 NANA	$G_{GNT}1$	G_{M1}	*Major bases:* C_{18} and C_{20} sphingosines. Minor amounts of dihydroanalogues. *Fatty acids:* large amounts of 18:0 (86-95% in brain). *Occurrence:* mainly in grey matter of brain but also in spleen, erythrocytes, liver kidney. Modern analytical techniques have shown them to be present in a much wider range of tissues than previously realized. Main gangliosides of human brain are G_{GNT} 1, 2a, 2b, 3a. Gangliosides appear to be confined to the animal kingdom. In man, cattle, horse, main ganglioside outside brain is $G_{LACT}1$. N-glycolyl-neuraminic acid is chief sialic acid in erythrocyte and spleen
Tay-Sachs Ganglioside	β β (b) Cer-glc-(4←1)-gal-(4←1)-galNAc 3 (↑) 2 NANA	$G_{GNTrII}1$	G_{M2}	
Haematoside	(c) Cer-glc-(4^β←1)-gal-(3^β←2)-NANA	$G_{LACT}1$	G_{M3}	
DISIALOGANGLIOSIDE	β β β (a) Cer-glc-(4←1)-gal-(4←1)-galNAc-(3←1)-gal 3 3 (↑) (↑) 2 2 NANA NANA	$G_{GNT}2a$	G_{D1a}	

Continued

Table 4.8 Contd.

	(b) Cer-glc-(4←1)-gal-(4←1)-galNAc-(3←1)-gal (β linkages at each ←1); NANA-(8←2)-NANA linked 2→3 to the inner gal	$G_{GNT}2b$	G_{D1b}	
TRISIALOGA-NGLIOSIDE	Cer-glc-(4←1)-gal-(4←1)-galNAc-(3←1)-gal (β linkages at each ←1) inner gal 3 (↑) 2 NANA-(8←2)-NANA (above) inner gal 3 (↑) 2 NANA-(8←2)-NANA (below) terminal gal 3 (↑) 2 NANA	$G_{GNT}3a$	G_{r1}	gangliosides of horse and cattle. *Physical properties:* insoluble in non-polar solvents; form micelles in aqueous solution.

Abbreviations in column 2 are the same as in Table 4.7. NANA = N-acetylneuraminic acid (sialic acid). Wiegandt abbreviations: G = ganglioside. Subscript denotes sialic-free oligosaccharide.

Arabic numeral gives number of sialic residues. T = basic tetraose; Tr = triose; Lact = lactose. Trisaccharides and disaccharides that can be derived from the basic tetraose are distinguished by I II, etc., I describing the sugar that originated from the non-reducing end of the tetrasaccharide; a, b refer to positional isomers with respect to sialic residues.

Sterols are common in eukaryotes, but rare in prokaryotes. Whereas vertebrates synthesize cholesterol, invertebrates rely on an external sterol supply.

Yeast and fungi, in the main, have sidechainalkylated compounds, while plants and algae contain sitosterol and stigmasterol as their most abundant sterols (Fig.). However, it should be emphasized that, as with most situations in biology, too many generalizations should not be made. For example, cholesterol, which in many animal tissues comprises over 95% of the sterol fraction, cannot be regarded as exclusively the animal sterol. Thus, marine invertebrates, whether or not they are sterol auxotrophs, discriminate much less than the mammalian intestine in the sterols that they absorb and, therefore, have compositions dependent on the general availability of sterols. Furthermore, whereas plants typically have C24-ethyl sterols such as stigmasterol, some species of red algae contain only cholesterol.

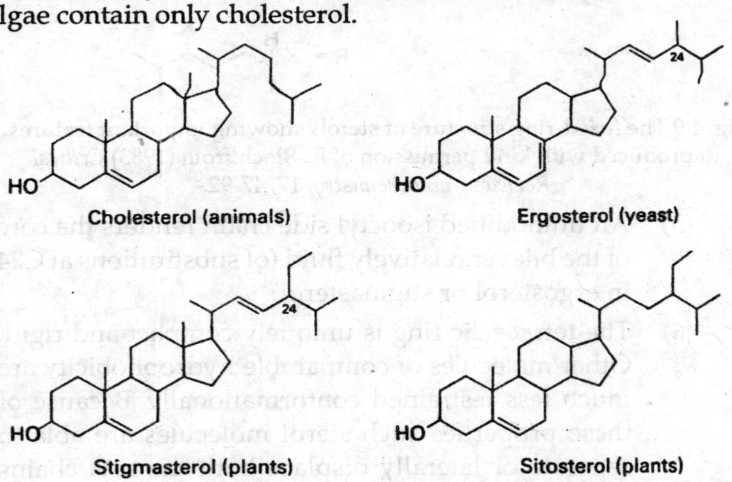

HO
Cholesterol (animals)

24
HO
Ergosterol (yeast)

24
HO
Stigmasterol (plants)

HO
Sitosterol (plants)

Fig. 4.8 Major types of membrane sterols.

It is generally believed that the main function for membrane sterol is in the modulation of fluidity. The latter is mediated by the interaction of sterol with the glycerolipid components. For that purpose, clolesterol seems to have been optimally designed. Any changes in the structure appear to reduce its effect on membranes (Fig. 4.9). Points to note are:

(1) All *trans*-fusion of the rings gives a planar molecule capable of interacting on both faces.

(2) The 3-hydroxy function permits orientation of cholesterol in the bilayer.

(3) In naturally occurring sterols, the side chain is in the (17β, 20*R*) configuration, which is thermodynamically preferred and permits maximal interaction with phosphoglycerolipids.

(4) Methyl groups at C10 and C18 are retained and provide bridgeheads.

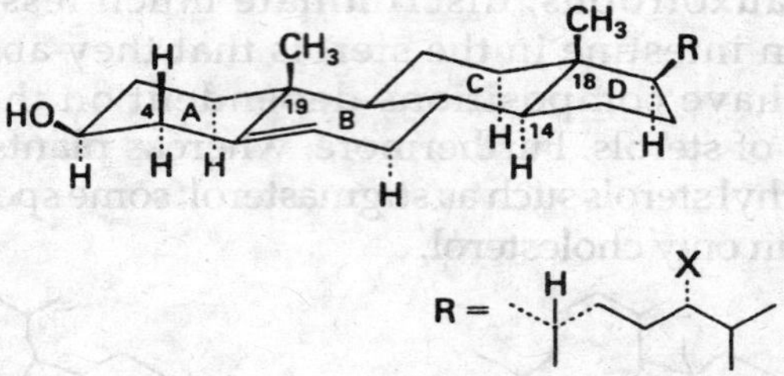

Fig. 4.9 The fused-ring structure of sterols showing important features. Reproduced with kind permission of K. Bloch from (1983) *Critical Reviews in Biochemistry*, 17, 47-92.

(5) An unmodified isooctyl side chain renders the core of the bilayer relatively fluid (of substitutions at C24 in ergosterol or stigmasterol).

(6) The tetracyclic ring is uniquely compact and rigid. Other molecules of comparable hydrophobicity are much less restrained conformationally. Because of these properties such sterol molecules are able to separate or laterally displace both the acyl chains and polar head groups of membrane phospholipids.

As mentioned above, cholesterol is, by far, the most important sterol in mammalian tissues. So far as plants and algae are concerned, the major structures are sitosterol (~70%), stigmasterol (~20%), campesterol (~5%) and cholesterol (~5%). Yeasts can accumulate large amounts of sterols (up to 10% of the dry weight) and phycomycetes contain almost exclusively ergosterol. This compound is the major sterol of other yeasts and mushrooms except the rust fungi in which it is absent and replaced by various C29 sterols. With very few exceptions, sterols are absent from bacteria.

The 3-hydroxyl group on ring A can be esterified with a fatty acid. Sterol esters are found in plants as well as animals. Although sterols are present in most mammalian body tissues, the proportion of sterol ester to free sterol varies markedly. For example, blood plasma, especially that of humans, is rich in sterols and like most plasma lipids they are almost entirely found as components of the lipoproteins. About 60-80% of this sterol is esterified. In the adrenals, too, where cholesterol is an important precursor of the steroid hormones, over 80% of the sterol is esterified. By contrast, in brain and other nervous tissues, where cholesterol is an important component of myelin, virtually no cholesterol esters are present.

The 3-hydroxyl group can also form a glycosidic link with the 1-position of a hexose sugar (usually glucose). Sterol glycosides are widespread in plants and algae and the 6-position of the hexose can be esterified with a fatty acid to produce an acylated sterol glycoside.

Other steroids

A full description of the various types of steroids, their metabolism and function, is beyond the scope of this book. Cholesterol is the essential precursor for bile acids, corticoids, sex hormones and vitamin D-derived hormones, as has been well established for all vertebrates.

MEMBRANE STRUCTURE

The detailed structure of membranes differs considerably according to their location. However a general model for most membranes is consistent with an extended bilayer of amphiphilic lipids with hydrophobic moieties directed to the centre and hydrophilic head groups at the two surfaces. Ionic or hydrophobic interactions occur between lipids and membrane proteins. The latter are located at the surface or embedded within the membrane.

Early models

It has been known for a long time that membranes were composed basically of lipid and protein. The question was how were these constituents arranged in the membrane and how could their special properties be reconciled with their possible physiological functions?

In 1925, the Dutch workers Gorter and Grendel extracted the lipids from erythrocytes and calculated the area occupied by the lipid from a known number of cells when it was spread as a monolayer on a *Langmuir Trough*. There was sufficient, they claimed, to surround a red cell in a layer two molecules thick. (Actually these workers made two mistakes which happened to exactly cancel each other out. First, they didn't know how hard to compress the monolayer and, hence, estimate the area aceurately. Secondly, they made no allowance for the contribution of proteins.) Nevertheless, the estimates made by Gorter and Grendel turned out to be quite accurate and were accepted at the time to indicate that a bimolecular lipid leaflet surrounded cells. This organization, which provides a permeability barrier between various cellular compartments, was further refined by Danielli and Davson. They were unaware of Gorter and Grendel's paper and proposed a similar model in 1935, but realized that the measured surface tension of the membane was too low to be accounted for by lipid and proposed that a layer of protein was present at each surface.

With the advent of electron microscopy came pictures of membranes that seemed to support the Danielli model. For the purposes of electron microscopy, the specimen has to be dehydrated, stained (usually with osmium tetroxide or potassium permanganate – so-called positive staining) and embedded in a plastic material such as an epoxy resin. The sample sections are then cut and examined under the microscope. Much of the interpretation of the structures observed depends on how much shrinkage occurs during dehydration of the sample, and what regions of the membrane take up the stain; it is now generally accepted that osmium tetroxide accumulates at the polar regions of lipid and protein. In just about every membrane examined in this way a triple-layered structure, was seen on the micrographs and on this basis Robertson put forward the unit membrane hypothesis, which held that every membrane had a basic structure consisting of a bimolecular lipid leaflet sandwiched between a layer of protein on one side and glycoprotein on the other. Further support for the hypothesis came from X-ray studies on myelin by Finean. Myelin is an ideal material for this purpose; it can be easily in a pure state, has a simple composition and regular repeating features providing good diffraction patterns.

In Table 4.2 we have tried to indicate that from the point of view of composition at least, there is no such thing as a typical membrane. With such widely differing compositions it is unreasonable to expect a universal structure. Myelin, the structure of which lent most support to the unit membrane theory, seems to be the most atypical.

The lipid-globular protein mosaic model

Implicit in the bimolecular leaflet model is the idea that the protein is spread as an extended sheet (β-conformation) over the ionic heads of the phospholipids and that the binding between phospholipid and protein is essentially electrostatic. Several observations about membrane lipids and proteins were not consistent with this model.

First, the protein component of many membranes represents well over 50% of the bulk of the membrane material. The average amino acid composition of membrane proteins shows no marked preponderance of ionic or hydrophobic residues. Thus, if the protein were extended as a sheet, a significant proportion of the hydrophobic groups would have to be in contact with the water.

Secondly, increasing the ionic strength of the medium surrounding the membrane does not dissociate a large fraction of membrane proteins from lipids as would be expected if electrostatic interactions were predominant.

Thirdly, when the lipid moiety is removed, the proteins are not very soluble in water; in fact they tend to interact hydrophobically with one another.

Lastly, most membrane lipids are zwitterionic rather than

having a net charge and would have no strong electrostatic interactions. Those that are charged tend to be acidic and preferably interact with predominantly basic proteins, which are not common in membranes.

Subsequent observations such as the known rapid lateral diffusion of lipid and protein in the plane of the membrane and knowledge that proteins are often inserted into and through the lipid matrix have been added to the points made above. These considerations have been allowed for in the proposal by Singer and Nicholson in 1972 of their fluid mosaic model for membrane structure (Fig. 4.10). In the diagram an asymmetric lipid bilayer forms the basis of the membrane structure with proteins spanning the membrane or embedded into the hydrophobic core region.

Two techniques in the electron microscopist's armoury have helped to encourage a reappraisal of membrane structure. One is negative staining, in which the sample is not fixed and embedded but dispersed in an aqueous solution of the negative stain (phosphotungstate) and dried down on a support film. Stain accumulates in the hydrophilic regions. The other method, drastically different from the staining methods and therefore useful to give independent corroboration, is the freeze-etching technique.

The freeze-etching technique allows a three dimensional view of membrane surfaces to be made. Furthermore, the analogous freeze-fracture method permits visualization of the interior of the membrane bilayer. With both techniques intrinsic membrane proteins can be seen that penetrate into the bilayer or pass through it, just as proposed in the Singer/Nicholson model (Fig. 4.10).

Membrane structure

The ability of many membrane lipids to form the basic bilayer structure is caused by a number of properties, the most important of which is their amphipathic character. Amphipathicity is caused by the lipids having a polar or hydrophilic head group region and a non-polar or hydrophobic part. Thus, such molecules will naturally orientate themselves to ensure that the polar groups associate with water molecules while the hydrophobic tails interact with each other. Provided the molecule in question is roughly cylindrical in dimension and has no net charge then bimolecular planar leaflets will be the most stable configuration in an aqueous system. However, it is also true that many major lipid components of membranes do not form bilayers when isolated and placed into aqueous systems in vitro. Some of the packing characteristics of different lipids and the structures they form are shown in Fig. 4.11.

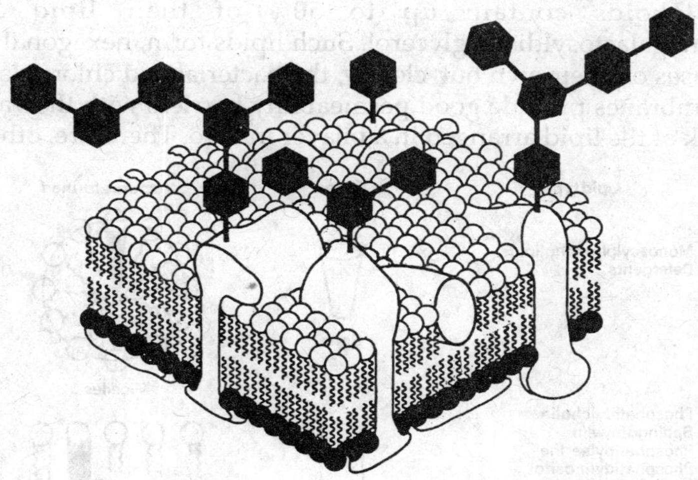

Fig. 4.10 The Singer and Nicholson fluid-mosaic model of membrane structure. The figure shows the topography of membrane protein, lipid and carbohydrate in the fluid-mosaic model of a typical eukaryotic plasma membrane. Phospholipid asymmetry results in the preferential location of phosphatidylethanolamine and phosphatidylserine in the cytosilic monolayer. Carbohydrate moieties on lipids and proteins face the extracellular space.

Israelachvili, in a masterly exposition in 1978, has discussed the theoretical shapes of these structures and the consequences for cells. In some naturally occurring membranes, a very high proportion of the lipid is capable of adopting the hexagonal II phase (inverted micelles) rather than bilayers. In this arrangement cylinders of lipids are formed with an aqueous interior in contact with the head groups of the lipids (Fig. 4.11). The cylinders interact with each other by hydrophobic forces and this can be clearly seen in freeze-fracture micrographs as a regular corrugated pattern.

In general, biological membranes contain an appreciable fraction (up to 40 mol%) of lipid species that, individually, prefer the hexagonal II arrangement. But it has been discovered that, depending on the acyl chain composition, temperature and head group size and charge, complete bilayer stabilization can be achieved by the addition of 10-50 mol% of bilayer-forming lipid species. Moreover, these properties are displayed without any stabilizing effects from membrane proteins. Thus, many bacteria contain unsaturated phosphatidylethanolamine and diphosphatidylglycerol in large amounts while chloroplast thylakoids contain up to 50% of their lipid as monogalactosyldiacylglycerol. Such lipids form, hexagonal II phases on their own but, clearly, the bacterial and chloroplast membranes provide good permeability barriers with the vast bulk of the lipid arranged in a bilayer in vivo. Therefore, other

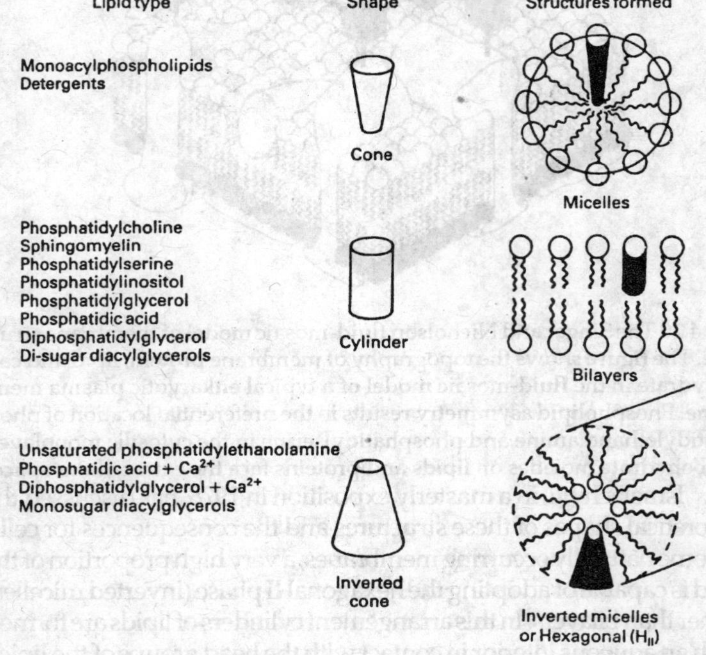

Fig. 4.11 Lipid shapes and their packing characteristics.

factors such as the ionic environment, interaction with other lipids and the contributions of membrane proteins are obviously important.

For certain purposes, it seems important for non-bilayer phases to be present in membranes. A good example is in membrane fusion or in the budding of vesicles.

The fluidity of membranes is dependent on the nature of the hydrophobic moieties of lipids. When lipids are isolated the acyl chains of each type undergo transitions from a viscous gel to a fluid state at a certain temperature. Above the transition temperature, the molecules existin a liquid-crystalline state where the acyl chains, but not the head group, are fluid. This is the normal state of membrane lipids and ensures proper functioning of the membrane proteins. When the temperature of a cell is lowered, three phenomena can contribute to impairment of membrane function. First, ice crystals formed in the aqueous compartments can cause physical damage. Second, transition of the acyl chains into the gel phase leads to changed enzyme activity and altered transport. Third, temperature lowering can lead to the membrane lipids becoming phase-separated. Concentration of particular lipid types by this phase separation can cause the formation of non-bilayer structures.

Further remarks

We emphasized that different membranes contain quite different lipid compositions. Because such compositions are maintained generally within quite strict limits, one must presume that a membrane's complement of lipids is particularly adapted to its needs. However, the overall lipid composition of a membrane is itself a simplification which needs some qualification. First, individual lipid classes (e.g. phosphatidylcholine) consist of a large number of molecular species where the combinations and positional distribution of acyl chains differ. The nature of these individual molecular species is important for the physical properties, membrane function and further metabolism of the lipids. Second, lipids are not distributed evenly in membranes. For example, all naturally occurring membranes, which have been examined, show sidedness, i.e. two halves of the bilayer are dissimilar. In addition, lateral segregation has been shown in a few instances or implied from the specific association of certain lipids with particular membrane proteins.

Transbilayer lipid asymmetry

The first natural membrane to be examined for asymmetry was that from erythrocytes. Such cells are very convenient for membrane studies since erythrocytes can be easily purified in large quantities and contain only a single membrane. In order to study the two leaflets of such membranes, intact cells were probed (or labelled) in order to establish those components that were in the outer leaflet (i.e. that were accessible).

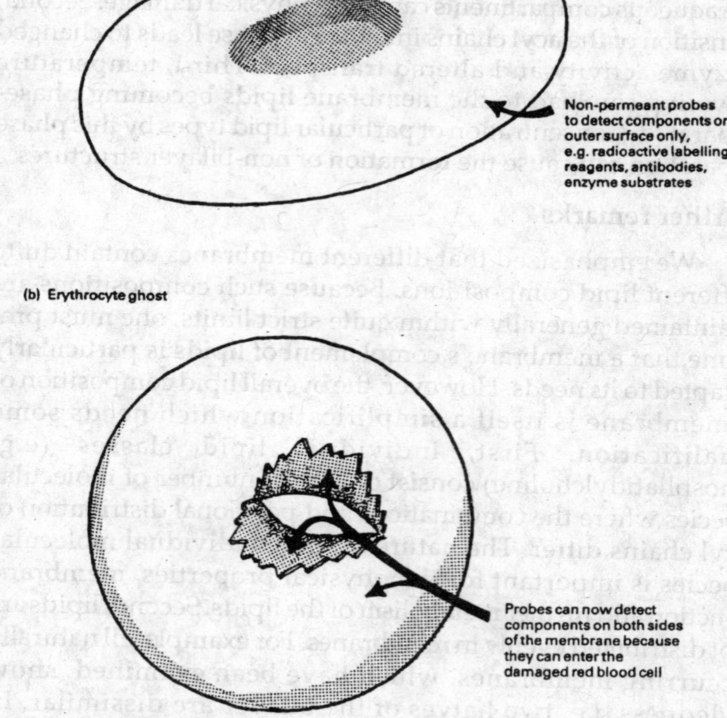

Fig. 4.12 Design of an experiment to establish transbilayer composition in erythrocytes (red blood cells.)

In a second experiment, the erythrocytes were broken by a suitable method (e.g. osmotic shock, sonication) to allow access of the probe to the inner surface as well (Fig. 4.12). Control experiments are also necessary to evaluate the total membrane proteins and lipids, since not all of these may be accessible to the probes used.

Various methods have been used to evaluate the lipid composition of inner and outer membrane leaflets. These are listed in Table 4.9. Enzymic digestion is probably the most common method since it is easily carried out and allows analysis of almost all membrane lipids. A drawback with the method, however, is that removal of lipids from the outer leaflet enhances transfer of replacement lipids from the inner leaflet (transbilayer movement). In addition, the products of lipid digestion may be detergents, which could themselves alter membrane structure (e.g. lysophospholipids). In some instances, similar methods (ie. radiolabelling, enzyme digestion, antibody labelling) may be used to analyse membrane proteins and, in addition, if the latter are enzymes, then they can be measured with their substrates.

An important caveat for all experiments to probe membrane sidedness is that the membrane itself is not permeable to the probe, otherwise the latter will, of course, label both sides of the bilayer.

All membranes that have been examined so far show very different lipid (and protein) compositions for their two leaflets. A common feature of mammalian cell membranes is that the aminophospholipids and phosphatidylinositol are concen-trated in the cytosolic leaflet (Fig. 4.13). The sphingolipids are highly concentrated in the external leaflet of the plasma membrane. By contrast, in the plant membrane example shown in Fig. 4.13 (castor bean glyoxysome) while phosphatidy-lethanolamine was mainly in the cytosolic (i.e. outside) leaflet, phosphatidylinositol and phosphatidylserine were mainly in the opposite side. This latter example emphasizes that generalization should not be made from system to system without firm evidence. It should also be noted that the distribution calculated may be biased if significant amounts of lipids are inaccessible to probes and, therefore, cannot be located.

Table 4.9 Methods used to determine membrane sidedness

Method	Notes
Antibody labelling	Antibodies can be raised to specific proteins or certain lipids. Distribution of antibodies is measured with a radiolabel $C^{125}I$ tag, a linked enzyme assay or by immunogold visualization.
Enzymic digestion	Specific enzymes are the best to use since large-scale phospholipid or protein digestion can destroy the membrane's structure. Measurement of the decrease in given constituents allows an estimation of their accessibility and, hence, distribution.
Chemical (or radio) labelling	Various compounds that react with primary amino groups (phosphatidylethanolamine, phosphatidylserine, proteins) most used.
Lipid exchange proteins	These allow exchange of specific lipids between the outer leaflet and liposomes in the suspension medium. Thus, the natural membrane composition is retained. By initially (radio) labelling the lipid in liposomes, the amount of the same substance in the outer leaflet can be estimated. Unfortunately, such exchange proteins ate few and not widely available.
Assay of enzymic function	Only useful for a limited number of proteins.
NMR	Signals measured in the absence and presence of paramagnetic ions – these act as chemical shift or broadening agents. Results are difficult to interpret and not very specific.

Nevertheless, lipid asymmetry in membranes raises an important question as to how it arises and also has obvious implications for membrane function. The preservation of a symmetry clearly depends on a very low frequency of exchange from one surface to another or a controlled reciprocal exchange.

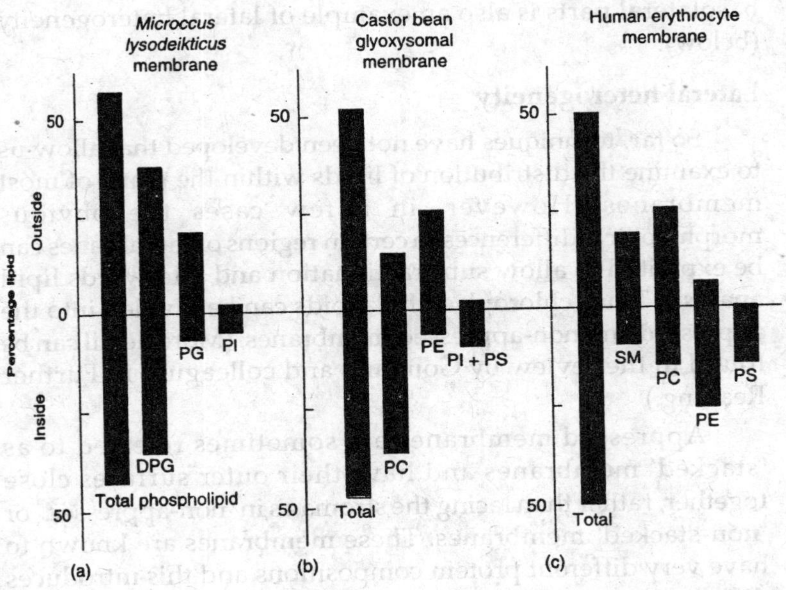

Fig. 4.13 Asymmetry of lipid distribution in a bacterial, plant and an animal membrane. PG = phosphatidylglycerol; PI = sphingomyelin; DPG = diphosphatidylglycerol.

An interesting example of the importance of lipid asymmetry has emerged recently. All body cavities in mammalian organisms are lined with epithelia. In most cases (e.g. the gastrointestinal and urinary tracts) a single layer of cells is involved and these cells have their plasma membrane divided into two parts. An apical portion, which borders the lumen, is connected to adjacent cells through tight junctions. The remainder of the plasma membrane faces the underlying tissue and is called the basolateral portion. Several studies have shown that the apical portion is significantly less fluid (and, hence, less passively permeable angmore mechanically stable) than the basolateral membranes. It turns out that much of the rigidity is due to a very high proportion of glycosphingolipids in the outer leaflet. Such lipids are much more highly ordered than phospholipids owing to their extensive hydrogen-bonding capacity. Division of the plasma membrane into apical and basolateral parts is also an example of lateral heterogeneity (below).

Lateral heterogeneity

So far, techniques have not been developed that allow us to examine the distribution of lipids within the plane of most membranes. However, in a few cases the obvious morphological differences in certain regions of membranes can be exploited to allow sub-fractionation and afterwards lipid analysis. Thus, chloroplast thylakoids can be divided into the appressed and non-appressed membranes. (More detail can be found in the review by Gounaris and colleagues in Further Reading.)

Appressed membranes are sometimes referred to as 'stacked' membranes and have their outer surfaces close together, rather than facing the stroma as in 'non-appressed' or 'non-stacked' membranes. These membranes are known to have very different protein compositions and this introduces different charges on to the membrane surfaces. The membrane vesicles deriving from appressed or non-appressed regions can then be separated using two-phase partition systems. Analysis of such membranes showed that, in contrast to the very different protein contents, the lipids are rather similar. The ratio of lipid to protein is, however, very different with non-appressed regions being enriched 3-4 fold in lipid compared to appressed membranes.

In photosynthetic bacteria, where the cytoplasmic membrane becomes infolded to produce photosynthetic membranes (Fig. 4.1) the two regions can be separated. Here, there is a clear difference in lipid composition with the photosynthetic membrane being enriched in phosphatidylglycerol compared to the cytoplasmic membrane. In addition, the carotenoid and bacteriochlorophyll pigments are exclusive to the photosynthetic membrane though this probably reflects their binding to specific proteins rather than a special property of the lipids themselves.

Micro-lateral heterogeneity is present in membranes where particular lipids are found tightly bound to individual proteins. The clearest examples of this are the pigment-protein complexes of photosynthetic organisms. However, certain purified proteins appear to be highly enriched in specific lipids which are very tightly bound and survive solubilization and purification steps in the presence of detergents. Thus, chloroplast ATPase from spinach and the green alga *Dunaliella salina* have been reported to contain highly saturated molecular species of sulphoquinovosyldiacylglycerol. In addition, cytochrome c from mitochondria is a very basic protein that in vivo has been suggested to interact with diphosphatidylglycerol. However, in many other cases, what were thought to be specific interactions with proteins turn out to be general properties of lipids, which can be replaced by artificial compounds or even detergents while still preserving functional activity. The functions of lipids in providing permeability barriers interacting with proteins, aiding membrane fusion and in drug delivery are discussed later.

Physical examination

Fluidity in membranes is an easily visualized phenomenon and is a term that is widely used. However, it can be misleading. For example, it is commonly assumed (and stated) that making membrane lipids more saturated or adding cholesterol makes a membrane less fluid. The assumption is that such alteration will reduce the speed of movement of lipids. However, introduction of cholesterol into phosphatidylcholine model membranes has no effect on lateral movement and may actually increase rotational diffusion rates. What cholesterol and increased saturation do is to increase the order in the hydrocarbon matrix and this is what can be measured easily by NMR or ESR order parameters.

Spectroscopic studies have demonstrated that lipid molecules in membranes normally show considerable freedom of movement. Lipids can rotate, bob up and down in the plane of the bilayer, show fluid movement down the hydrocarbon chain and, to a small extent, undergo flip flop (i.e. movement from one leaflet to the other). Flexing of the hydrocarbon chains occurs fastest with the methyl ends of the chains undergoing the most movement. Rotation of the molecule is also fast (say about every 10^{-7} s in artificial systems) but flip-flop is much slower (Fig. 4.14) and preserves the bilayer asymmetry of natural membranes.

Although it is accepted generally that membrane lipids show considerable movement, their rate of flip-flop, in particular, can be influenced by proteins. For example, when the cytoskeleton proteins of erythrocytes are modified their transverse asymmetry is changed. Moreover, the sided distribution of lipids in artificial membranes can be changed by posing a transmembrane pH gradient and this redistribution may also occur in vivo. Furthermore, specific proteins, such as an aminophospholipid translocase, have been identified that assist the movement of specific lipids across bilayers.

GENERAL FUNCTIONS OF MEMBRANE LIPIDS

Detailed discussion of membrane properties and functions is beyond the scope of this book but numerous detailed reviews are available. Nevertheless, a few remarks centred on the role of lipids can be made. In the previous sections, we discussed current ideas for membrane structure and the concept that, although lipids are capable of adopting non-bilayer structures when isolated, by and large they participate in bilayers in vivo.

Under normal conditions the membrane bilayer is in a 'fluid' state. Membrane proteins can migrate within the plane of the membrane with diffusion coefficients of about 10^{-10} cm^2 S^{-1} while lipids diffuse with coefficients of about 10^{-8} cm^2 5^{-1}. Overall behaviour might be considered, therefore, in thermodynamic terms. But generalized deductions relating the fluidity of the membrane to enzyme activity are difficult to make for several reasons. For example, motion in a given lipid molecule may include rapid rotations, but slow lateral movement. Also, increased disorder in a bilayer may not correlate with increased translational motion. Moreover, all membranes so far examined have shown transbilayer asymmetry while there is evidence in several cases for at least small areas of concentration of certain lipids, i.e. micro-lateral heterogeneity. These sorts of consideration complicate the interpretation of experiments designed to show how the bilayer lipids affect membrane proteins at a molecular level.

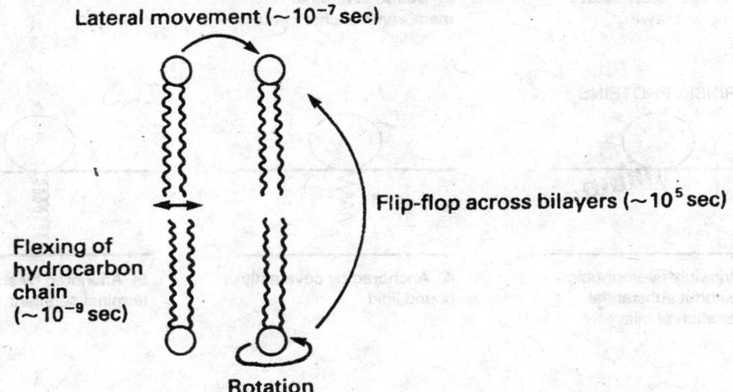

Fig. 4.14 Lipid movement in an artificial membrane.

Enzymologists frequently study how reaction rates, substrate binding, etc. change under a variety of experimental conditions. Parameters that may be changed include pH, ionic concentration and the addition of water-soluble inhibitors. However, the membrane protein is only probed in regions where it is exposed to bulk water or at the interfacial region where such water meets the bilayer. This often only represents a small part of the whole membrane protein. Lipids can interact with proteins in several ways. First, there are interactions with the hydrophobic proteins that penetrate into or through the bilayer.

These are interactions with the membrane's integral or intrinsic proteins. Secondly, the water-soluble extrinsic proteins can interact electrostatically with the charged head groups of the membrane (phospho) lipids. Thirdly, covalent lipid modifications may take place and these will help proteins to associate with and/or become anchored to the membrane. Some basic types of arrangement of membrane proteins are indicated in Fig. 4.15.

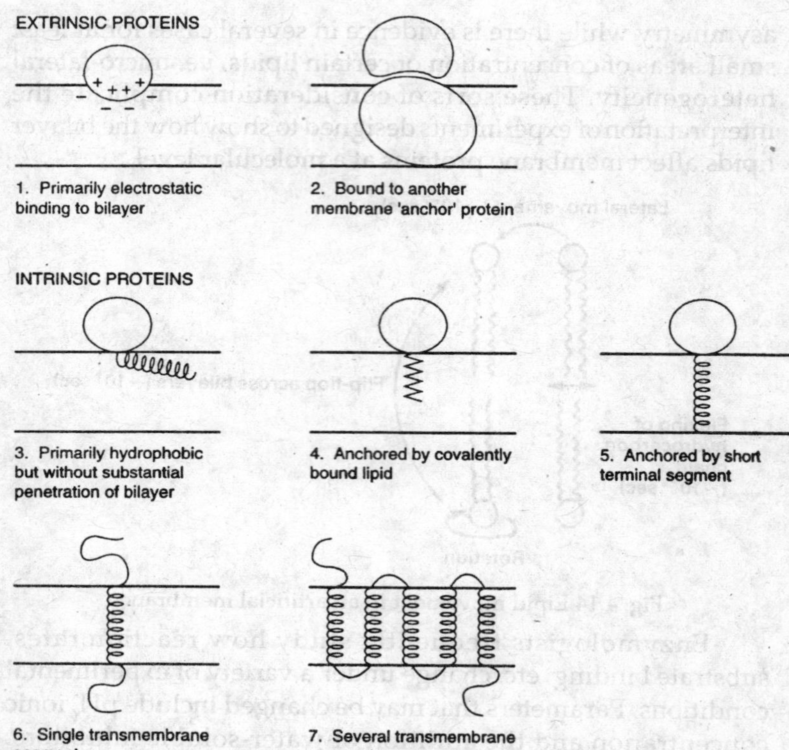

Fig. 4.15 Arrangement of proteins in membranes.

Interactions of lipids with a variety of extrinsic proteins have been studied. The proteins that have been studied (e.g. cytochrome c, spectrin) are usually basic molecules that interact with negatively charged lipids. Even if no negatively charged lipid is tightly bound to these proteins, they induce a time-averaged enrichment of negatively charged lipids in their vicinity. However, this has not been shown to lead to the formation of crystalline domains via local fluidity changes. Indeed, there may be an increase in fluidity due to a disturbance of the hydrophobic interactions of the acyl chains in the bilayer. In addition, divalent cations compete with basic proteins for the negatively charged head groups. In fact, both calcium and basic proteins are able to induce the formation of hexagonal II phases in micelles of diphosphatidylglycerol. This property may explain the ability of cytochrome c to induce non-bilayer structures in diphosphatidylglycerol-containing systems and also to translocate rapidly across bilayers such as the inner mitochondrial membrane (which contains high amounts of diphosphatidylglycerol).

To remove intrinsic proteins from a membrane it is necessary to solubilize the molecules using detergents that can break the hydrophobic interactions with lipids. The lipids adjacent to the proteins experience a different environment from the bulk lipid of the membrane and this has led to the concept of boundary lipids. ESR (electron-spin resonance) studies showed that lipids residing at the lipid-protein interface exhibited increased-order parameters (restricted motion) in the acyl chain region. Recent experiments, however, paint a different picture. First, the lipid-protein interactions in general appear to be relatively non-specific with most isolated integral proteins capable of being reactivated by a whole host of membrane lipids. Secondly, the evidence for a long-lived boundary of lipid is less convincing when further experiments are considered. When NMR (nuclear magnetic resonance) data are used, boundary-bulk lipid exchange appears rapid. This reflects the time-scale of NMR (~10^{-6} s) versus ESR (~10^{-10} s) measure ments.

Although the general concept of long-lived boundary lipids surrounding intrinsic proteins does not appear to be viable at present, there are well-documented cases where particular lipids are associated with individual proteins. For example, the purple membranes of *Halobacterium halobium,* apart from having a rigidly defined protein structure, also contain a unique lipid composition. The lipids appear to be interdigitated between the protein rods and have a composition quite distinct from the rest of the membrane.

Membrane proteins that clearly have an impor tant part spanning the bilayer are those concerned with transport. These molecules must have at least two portions: (1) a domain, such as a channel, that allows the movement of the substrate vectorallyacross the membrane; (2) a binding site for the substrate that confers selectivity and also, when occupied, initiates transport. There may also be regulatory domains allowing the opening of the channel gate. Each of these three parts could be modulated by lipids in bilayer.

An example of the type of information obtained for lipid-protein interactions with a transporter is that derived from the experiments conducted by Melchior and Carruthers on the erythrocyte sugar transporter. The protein is particularly useful because it can be purified and reconstituted into lipid vesicles of different composition and the exact number of reconstituted transporters determined through the binding of radiolabelled cytochalasin B, a specific inhibitor. Furthermore, the kinetic properties, complete amino acid sequence and arrangement of the protein in erythrocyte membranes are known. Comparison of transport rates with the number of reconstituted transporters allowed estimation of the turnover number (T_n), and allowed comparison between different lipids used for the reconstitution (Table 4.10). It was found that several factors influenced transport rates. In order of importance these were: lipid head group> lipid acyl chain length and saturation index> lipid backbone>>bilayer fluidity. Negatively charged phosphoglycerides such as phosphatidylserine were the most active for reconstruction and suggested that the normal tertiary structure of the translocating domain might need a high surface potential. In fact, for the best lipids, such as phosphatidylserine, the fluidity of the bilayer had little effect on transport rate.

In contrast, for a lipid showing slower turnover numbers, such as phosphatidylcholine, transporter T_n increased markedly when the bilayer changed from the crystalline to the gel state. Thus, the nature of the lipid head group(s) influenced the degree to which bilayer fluidity altered transport activity.

The length of the acyl chains (and hence membrane thickness) was also important - not surprisingly because the sugar transporter is known to contain 12 membrane-spanning hydrophobic domains. In effect the thickness of the membrane has to be right for the correct arrangement of these hydrophobic domains.

Table 4.10 Factors influencing the activity of the erythrocyte sugar transporter in reconstituted lipid vesicles

Factor and effect	Experimental result
Nature of the phospholipid head group (alters surface charge/potential)	Vesicles composed of single phospholipids had the following order of activity: Ps>PA>PG>>Pc.
Acyl hainl ength (important to be correct for size of transmembrane protein portions)	For saturated chain acyl chains: C18>C20>C16>C14. For C18 chains: stearoyl>elaidoyl>oleoyl.
Bilayer fluidity (affects immediately environment of hydrophobic domains)	Relatively little change in activity on going from gel to crystalline phases except for micelles containing PC.
Cholesterol inclusion (affects lipid lateral mobility)	Lipid packing alterations correlate with transporter activity changes.

Of course, natural membranes contain not just a single lipid species but a whole host of different molecular types. Thus, experiments to determine the effect of lipids on membrane proteins in vivo are extremely difficult to interpret. In animal membranes, cholesterol is frequently present in addition to phosphoglycerides. Addition of cholesterol to a lipid bilayer induces complex alterations in lipid packing. McConnell and colleagues at Stanford showed that these alterations in lipid packing were associated with altered lateral mobility within the bilayer. In experiments with the sugar transporter referred to above, changes in transporter activity correlated closely with cholesterol-induced alterations in lipid packing. They did not, however, apparently result from cholesterol-induced changes in bilayer lipid order. Experiments with other systems (such as the Ca2+-$Mg2^{+}$-ATPase from endoplasmic reticulum) have also shown no consistent correlation in bilayer fluidity and enzyme activity. Once more, the length of the acyl chains seemed a more important parameter.

Membrane fluidity at low temperatures

We have already seen that membrane lipids can exist in the gel - or liquid - crystalline phases, depending on the prevailing temperature. Whereas isolated lipid samples show a sharp gel-liquid phase transition, naturally occurring lipids show broad non-cooperative transitions due to the heterogeneity of the acyl chains. For a series of molecular species of, say, phosphatidylcholine, the temperature (T_c) for the gel-liquid phase transition depends on the fatty acid composition (with *cis*-unsaturated fatty acids lowering T_c) and even on the position of acyl esterification.

It was thought that the membranes of natural organisms could not function when their lipids were in the gel-crystalline phase and that changes in growth or other physiological parameters seen in prokaryotes on lowering temperature could be explained simply by proposing that phase transitions had taken place. However, it is now clear that while a correlation can sometimes be seen between, for example, enzyme activity changes and the temperature for membrane lipid phase changes, this relationship is by no means simple nor the correlation universal. Nevertheless, as a general rule it can be stated that once significant areas of a natural membrane are in the gel state then function is significantly impaired. It is known, for example, that as temperatures are lowered phase separation of certain lipid species takes place and membrane proteins are squeezed out of such areas leading to the formation of protein-free areas, which can be seen by electron microscopy. Studies with various fatty acid-requiring mutants of *Escherichia coli* have shown that if less than half of the lipids of the cytoplasmic membrane are fluid then it becomes non-functional and the bacterium dies.

Thus, poikilotherms (organisms unable to regulate their own temperatures), which grow at low temperatures, have a pattern of membrane lipids possessing lower phase transition temperatures than those of the same organisms existing under hotter conditions. Furthermore, following changes in environmental temperature, an alteration in the composition of membrane lipids is seen. Although an increase in fatty acyl unsaturation at lower temperatures is best known, it is not, by any means, the only change that can take place.

Table 4.11 Effect of fatty acyl groups on the temperature (T_c) of the gel to liquid-crystalline phase transition of phosphatidylcholines

Lipid species	T_c (°C)
(a) Saturated	
Dimyristoyl-PC (14:0, 14:0-PC)	24
Dipalmitoyl-PC (16:0, 16:0-PC)	41
Distearoyl-PC (18:0, 18:0-PC)	55
(b) Unsaturated	
Palmitoyl, oleoyl-PC (16:0, 18:1-PC)	–1
Dioleoyl-PC (18:1, 18:1-PC)	–19
Stearoyl, oleoyl-PC (18:0, 18:1-PC)	6
Stearoyl, linoleoyl-PC (18:0, 18:2-PC)	–16
Stearoyl, linolenoyl-PC (18:0, 18:3-PC)	–13
Stearoyl, arachidonyl-PC (18:0, 20:4-PC)	–13

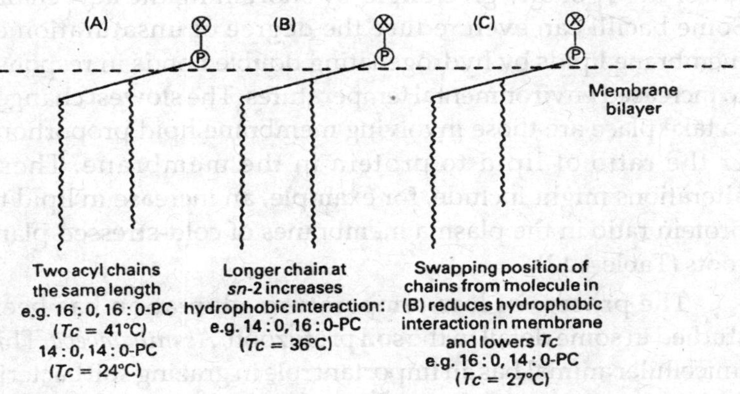

Fig. 4.16 The basis for re-tailoring of lipids during temperature adaptation.

The process of temperature adaptation has been studied in some detail in cyanobacteria by Murata's group and in the protozoon *Tetrahymena* and the alga *Dunaliella* by Thompson in Texas. A number of lipid changes has been observed by these workers and others (Table 4.12), but they occur on a different time-scale. A sudden change in growth temperature could immediately alter membrane fluidity with harmful consequences for the organism concerned. Murata has proposed that an emergency response involves re-tailoring (altering the position of fatty acid acylation) of molecular species. Such re-tailoring has been shown in vitro to lead to significant differences in the T_c of the lipid concerned. This is because the glycerol backbone of a membrane lipid is not exactly parallel to the surface of the bilayer and, hence, two acyl chains of equal length penetrate the hydrophobic membrane region to a different extent (Fig. 4.16). After the emergency response has occurred, other changes in lipid composition can take place on a longer time-scale. For organisms operating the aerobic pathway for desaturation an increase in *cis*-unsaturation is a relatively quick and easy modification to make. In contrast, bacteria such as E. *coli* could only increase their unsaturation by *de novo* synthesis of more palmitoleate and vaccenate. Such micro-organisms have been known, however, in some cases to lower the T_c of any given lipid by shortening the acyl chain. Some bacilli can even reduce the degree of unsaturation of membrane lipids by hydrogenating double bonds in response to increased environmental temperatures. The slowest changes to take place are those involving membrane lipid proportions or the ratio of lipid to protein in the membrane. These alterations might include, for example, an increase in lipid to protein ratio in the plasma membranes of cold-stressed plant roots (Table 4.12).

The process of low-temperature adaptation has been studied in some detail in the soil protozoon, *Acanthamoeba*. This unicellular animal has an important role in grazing soil bacteria and, hence, keeping their numbers roughly constant. It is also unusual (for an animal) in being able to synthesise linoleic acid. When *Acanthamoeba* is cooled, a succession of events occurs (Table 4.13) involving increased gene expression of the oleate desaturase that forms linoleate. In turn this leads to increased enzyme activity, membrane fluidity changes and a renewed ability to phagocytose.

Table 4.12 Changes in membrane composition caused by lowered growth temperatures in different poikilotherns

1. Re-tailoring of molecular species	Quick change involving swapping acyl chains on the glycerol backbone.
2. Increase of fatty acyl unsaturation	Relatively quick for organisms using the aerobic pathway of desaturation; involves .*in situ* modification of lipids.
3. Chain shortening	Relies on the lower melting temperatures of shorter acyl chains; relatively uncommon to see shortenintg of existing acyl chains; same net effect more commonly seen when *de novo* synthesis produces shorter chain products
4. Alteration of fatty acid type	Some micro-organism may alter amounts of branched-chain acids—this will also modify fluidity.
5. Lipid classes altered	Changes in membrane lipid classes often seen; where phospholipid interconversion is easy (e.g. PE— PC) then this could be adaptive because PC has a lower T_c than PE for the same molecular species.
6. Lipid-to-protein ratio changed	Usually the ratio is increased significantly at lower temperatures—due to increased net lipid synthesis.

An interesting application of membrane lipid biochemistry to a natural phenomenon has been made by Murata in Okasaki. He has examined the process of frost sensitivity in plants. It is well known that, whereas some plants can survive low or, indeed, freezing temperatures, many others drop their leaves and/or are killed by the first cool nights of autumn. It seemed logical to look for differences in the membrane composition of these two groups of plants to explain the difference. A superficial examination showed that both types of leaves contained the same classes of lipids with the usual exceptionally high content of linolenic acid. Murata noticed, however, that the phosphatidylglycerol fractions from frost-sensitive as opposed to frost-resistant plants were markedly different. Thus, frost-sensitive species had a much higher proportion (40-80%) of the 'disaturated' species, which had T_c values of about 40°C. (For this purpose the unusual *trans*-3-hexadecenoic acid, which is present in phosphatidylglycerol and which has a similar melting point to palmitic acid, is classed as saturated.) In contrast, frost-resistant leaves only had about 15% of the disaturated species. Murata, therefore, proposed that the basis for frost sensitivity lay in the phase separation of a significant quantity of phosphatidy lglycerol during exposure to low temperatures—this leading to membrane damage, leakage of intra-organellar contents and cell death. The experiment also emphasized how important it is to examine lipid composition in detail rather than merely taking average values.

Table 4.13 Sequence of events to restore membrane function on cooling *Acanthamoeba* (a poikilotherm)

1. Drop in environmental temperature causes loss of membrane fluidity and this is sensed by the animal (protozoon).
2. Induction of transcription of Δ12-fatty acid desaturase gene caused.
3. Activity of Δ12-desaturase causes oleate in membrane lipids to be converted to linoleate.
4. Membrane lipid fluidity is restored by increased unsaturation of acyl chains.
5. Phagocytosis recommences and *Acanthamoeba* can feed, grow and reproduce again.

More recently, Murata has engineered plants to become low-temperature tolerant. This has been done in two ways. Either the ability to form unsaturated fatty acids has been increased by cloning desaturase genes into the plants or, alternatively, the pathway of phosphatidylglycerol synthesis has been modified in order to increase the production of molecular species containing unsaturated fatty acids. In both cases, the transgenic plants are lowtemperature tolerant—thus showing that genetic modification of membrane lipids can be used to create useful new crop varieties.

In further experiments with Hungarian co-workers Horvath and Vigh, Murata suggested that the 'trigger' for increased gene expression when organisms are cooled can be the physical state of the membrane itself. Thus, reduction in the membrane 'fluidity' either by cooling or by chemical modification caused increased expression of various desaturase genes. Therefore, organisms have the ability to self-regulate their membrane fluidity.

LIPIDS AND MEMBRANE FUSION

Membrane fusion is a very common process in cells. In eukaryotic organisms, many different fusion events can occur at any given time, participating in such diverse functions as secretion, endocytosis, intracellular digestion in lysosomes, cell division and the adjustment of mitochondrial numbers to a cell's energy requirements. Just a short consideration of the overall events of fusion will make it obvious that, during the process, the normal lipid bilayer structure must disappear.

First of all, biochemists studied the process in model systems, employing lipid vesicles. It has been recognized for soine time that such vesicles can be induced to fuse when incubated in the region of their gel to liquid-crystalline transition temperature. Furthermore, it was also noted that Ca^{2+} addition would induce fusion in phosphatidylserine-containing systems, possibly by inducing phase separation of the lipid and, hence, a local crystalline point for fusion. However, it was also noted that in more complex mixtures, Ca^{2+} was not always able to induce fusion.

An important clue as to the mechanism of lipid membrane fusion has come from the work of de Kruijff and Cullis. They proposed that fusion proceeded because of the ability of lipids to undergo polymorphism (i.e. to adopt different structures). Three types of observation support the hypothesis. First, fusogens (such as monoacylglycerols) induce H_H (inverted micellar structures; Fig. 4.11) phase structures, consistent with a role for non-bilayer structures in fusion. Second, promotion of fusion of, for example, phosphatidylserine-containing systems by Ca^{2+} is accompanied by H_H structures. Third, a number of factors, such as pH changes or elevated temperatures that cause H_H formation, also promotes fusion of lipid vesicles.

Siegel and coworkers have examined fusion in more detail particularly in relation to two possible mechanisms – via inverted micelle intermediates or by stalk structures (Fig. 4.17). They produced evidence for either process being important although more recent experiments favour the stalk hypothesis. In either mechanism, however, non-lamellar lipid phases are initially involved. Furthermore, artificial membranes with inverted cone-shaped lipids (Fig. 4.11 supporting positive curvature) added to the cis-sided leaflets block stalk formation whereas their addition to *trans*-leaflets promotes pore opening. Cone-shaped lipids (Fig. 4.11) have the opposite effects. These results illustrate the potential for different-shaped lipids to influence membrane stability in various structures.

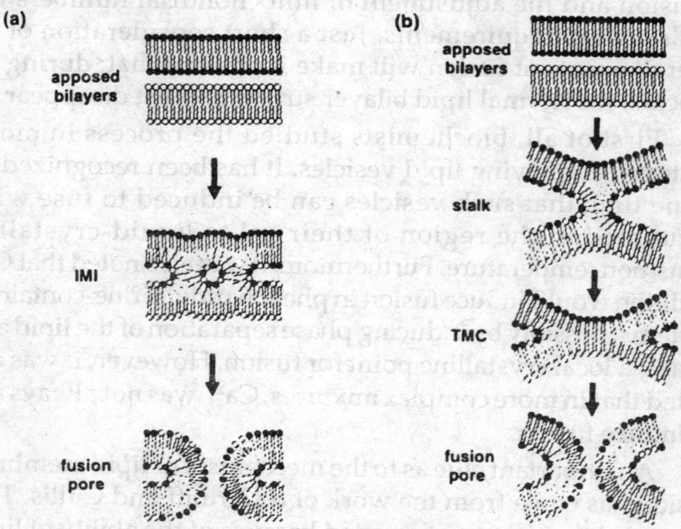

Fig. 4.17 Two possible mechanisms for membrane fusion. In method (a) inverted micellar intermediates (IMI) are involved and, in (b), the formation of a stalk.

Extension of the Cullis-de Kruijff hypothesis to natural membrane events has been difficult to prove. Nevertheless, confirmatory evidence has been obtained with a number of systems such as the release of chromatin granule contents during stimulation of the adrenal medulla. The exocytosis accompanying this event seems to depend on the ability of Ca^{2+} to promote non-bilayer structures.

Evidence from freeze-fracture microscopy has also implicated hexagonal II phases in two natural membrane systems where a kind of 'arrested fusion' exists. These are in the tight junctions between epithelial cells and in the contact sites between the inner and outer membranes of mitochondria or of Gram-negative bacteria like E. *coli*.

Fusion of biological membranes is now thought to require the action of specific fusion proteins. Of these, the viral fusion proteins that help the virus fuse with the host cell membrane during infection are the best understood. These proteins undergo dramatic and spontaneous conformational changes upon activation. For the influenza and the human immunodeficiency (HIV) viruses, the fusion peptide inserts itself into the membrane and then reorients itself thus forcing the fusion membranes together and allowing lipid mixing.

Fusion of intracellular eukaryotic membranes involves several protein families. These are termed SNAREs [soluble NSF (N-ethylmaleimide Sensitive Factor) receptors], SM proteins (Sec l/Munc 18 homologues where, for example, Sec derives from *see* mutants of yeast that have temperature-sensitive blocks in the secretory pathway) and Rab proteins (small GTPases, which in yeasts genetically interact with SNAREs and SM proteins). The latter probably function in the initial membrane contact connecting the fusion proteins, but are not involved in fusion itself. Although there is much genetic and structural information about these three components, which are essential for fusion (evidence from gene deletions or mutations), exactly what they do in the process is still unclear. Furthermore, although the mechanism of membrane fusion seems broadly similar in different situations, there must clearly be specificity to proteins such as in intracellular membrane traffic.

In the foregoing section we have suggested ways in which cells could initiate membrane fusion through allowing areas of hexagonal II phase. Conversely, there may be situations where the synthesis of large amounts of hexagonal II-forming lipids could destabilize the normal bilayer structure. Thus, when the anaerobic bacterium *Clostridium butyricum* is grown in the absence of biotin (when it is reliant on exogenous fatty acids) its membrane lipid composition can be made more unsaturated by supplying *cis*-monounsaturated acids rather than saturated ones. Since phosphatidylethanolamine and its plasmalogen are major membrane lipids in the bacterium, and their unsaturated species form the hexagonal II phase (Fig. 4.11) an increase in unsaturation might destabilize the membrane. *C. butyricum* reacts by reducing the proportion of phosphatidylethanolamine and increasing that of its glycerol acetal. Since the latter forms bilayers then this change restabilizes the membrane structure. Similarly, in *Acholeplasma laidlawii* where monoglucosyldiacylglycerol and diglucosyldiacylglycerol are major membrane lipids, the former is hexagonal II-forming while the latter gives bilayers (Fig. 4.11). Under conditions such as increased temperature or unsaturation, which would tend to favour the formation of non-bilayer structures, the monoglucosyl-lipid is converted to diglucosyldiacylglycerol. This change in lipid proportions would again be expected to preserve membrane bilayer stability.

Lipids and proteins interaction

As discussed (Fig. 4.11) the shape of complex lipid molecules can give rise to polymorphism. In addition, they may also contribute to curved regions of bilayers. Israelachvili suggested that, for regions of high curvature (such as at the ends of thylakoid stacks in chloroplasts) cone-shaped lipids would be needed in one leaflet and inverted cone-shaped lipids in the other leaflet (Fig. 4.11).

This is an attractive idea but examination of the transbilayer distribution of chloroplast lipids by various techniques has shown that the enrichment of monogalactosyldiacylglycerol (inverted cone) is the opposite of that required, i.e. it is more abundant in the outer leaflet. Nevertheless, because it is entirely possible that a small proportion of certain lipids may be concentrated laterally in certain membrane regions, it is still plausible that lipids alone may play a significant role in determining natural membrane shape. Experiments with whole membrane fractions would not pick up such a specific distribution, of course, but would merely provide an averaging of transbilayer composition. Indeed, some of the results with differently shaped lipids during membrane fusion experiments would support a role for different lipids in helping to stabilize regions of high membrane curvature.

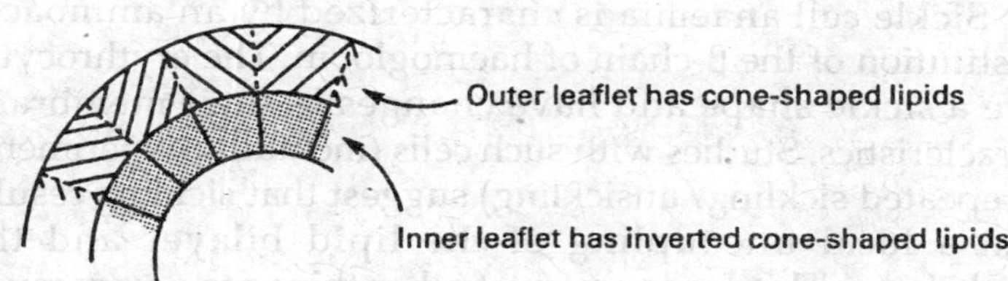

Fig. 4.18 How lipids could cause bilayer curves.

Much more likely is the co-operation of membrane lipids and proteins in controlling membrane shape and properties. For example, membranes containing very large integral protein complexes often contain high amounts of lipid able to form HII phases when purified. Thus, chloroplast thylakoids contain around 50% of the non-bilayer-forming monogalactosyldiacylglycerol while the photosynthetic membrane of purple non-sulphur bacteria is enriched in unsaturated phosphatidylethanolamine (also HII-forming). It has been suggested that these inverted cone-shaped lipids may help to package the integral proteins and ensure a good bilayer structure, which must of course be impermeable to ions and other molecules.

A situation where proteins and lipids appear to interact in determining membrane properties is in the erythrocyte. This simple cell can have its membrane composition altered in a non-destructive fashion by the use of phospholipid exchange proteins. The Dutch biochemists op den Kamp and van Deenen and their colleagues introduced specific phosphatidylcholine molecules into the outer leaflet by such means and showed that, depending on the molecular species substituted, the cells became more or less fragile and displayed susceptibility to osmotic shock. They could lose their flexibility and show an altered ion permeability. A surprising finding was that substitujion with certain species resulted in a changed cell shape. These results were explained by assuming that replacement of the natural slightly cone-shaped phosphatidylcholine molecules by the cylindrical dipalmitoyl species would cause the membrane to become convex while enrichment with the cone-shaped dilinoleoyl species would alter the shape to concave.

Sickle cell anaemia is characterized by an aminoacid substitution of the β-chain of haemoglobin. The erythrocytes have a sickle shape and have changes in their membrane characteristics. Studies with such cells (including experiments on repeated sickling/unsickling) suggest that sickling results from a local uncoupling of the lipid bilayer and the cytoskeleton. This leads amongst other things to a very much increased transbilayer movement (flip-flop) of lipids. In agreement with this suggestion it has been noted that membranes that do not contain a strictly organized cytoskeleton, such as the endoplasmic reticulum of rat liver or the plasma membrane of Gram-positive bacteria, also show relatively fast bilayer lipid movement.

Other experiments with erythrocytes also confirm the interaction of protein and lipids. In patients with hereditary pyropoikilocytosisthe spectrin dimers cannot associate into the normal tetramers and this gives rise to enhanced thermo sensitivity of the cytoskeleton. The cells also have an abnormal cell shape, increased osmotic fragility and an instability of the lipid bilayer manifesting itself in the temperature-sensitive nature of flip-flop movement. Also, in mutants lacking one or more of the membrane-spanning proteins that anchor the cytoskeleton in the bilayer, transbilayer mobility of lipids was enhanced. These observations emphasize that proteins play a role in determining lipid properties in membranes as well as vice versa.

Why are there so many membrane lipids

In biological membranes from different organisms, there is a bewildering array of different lipid molecules. Moreover, each lipid class also contains a variety of molecular species that differ in such aspects as their acyl moieties or sphingosine backbones. The fact that the lipid composition of a given membrane is generally maintained within strict limits points to a functional advantage. Perhaps this is most easily seen in the photosynthetic membranes of the oxygen-evolving organisms. Thus, in the millions of years since cyanobacteria entered into a symbiotic relationship as chloroplasts in simple algae and these developed to higher plants, the unique glycosylglyceride/ phosphatidylglycerol composition has been kept.

A clue as to why there are so many lipids may be developed from a realization that membrane lipids have many, diverse, functions:

(1) they act as permeability barriers for cells and organelles;

(2) they provide the matrix for assembly and function of a wide variety of catalytic processes;

(3) they directly participate in metabolism, e.g. as donors for the synthesis of molecules or as substrates for desaturation;

(4) they participate in a multitude of membrane fusion events;

(5) they actively influence the functional properties of membrane-associated processes.

Most experiments aimed at showing particular functions for membrane lipids use in vitro reconstituted systems. Unfortunately, although this approach yields important clues as to function, it is fraught with artefacts and may not reflect the in vivo situation at all. To look at the in vivo situation, genetics has been used. However, mutations cannot be made directly on the lipids themselves, but only on enzymes metabolizing them. Moreover, mutations in such genes may compromise many interacting functions or even the viability of the organism itself! The more complex the organism, the more difficult the experimental interpretations become. However, by careful manipulation of genes in *E. coli* it has recently been possible to develop secure arguments for the function of particular lipids in that bacterium.

In *E. coli* it has been demonstrated that the acidic phospholipids, phosphatidylglycerol and diphosphatidylglycerol (cardiolipin), which together make up less than 30% of the total phospholipid, had important functions for DNA-binding proteins (such as DNA topo-isomerase or polymerase). Phosphatidylglycerol was absolutely essential and null mutants were non-viable. By changing the expression of the phosphatidylglycerol synthase gene, the phospholipid could be varied from 3% to 20% and functions such as the initiation of replication and in the membrane translocation of proteins were revealed. Cardiolipin was found to function with the ADP/ATP carrier while phosphatidylethanolamine was needed for active transport events and the proper folding of proteins like lactose permease.

These experiments with mutants of *E. coli,* together with data from the glycosylglyceride-enriched *Acholeplasma laidlawii,* also showed clearly that organisms maintain the proportions of non-bilayer-forming versus bilayer-forming lipids rather carefully. Thus, for example, phosphatidylethanolamine mutants of *E. coli* regulate their diphosphatidylglycerol content according to the divalent cations present in the medium to ensure that normal membrane structures are maintained. In wild-type *E. coli* fatty acid composition rather than lipid class is usually changed while *Acholeplasma* will adjust its monoglycosyl-(non-bilayerforming) to diglycosyldiacylglycerol (bilayerforming) ratio in response to the addition of sterol, hydrocarbons or detergents to its growth medium.

Liposomes and drug delivery systems

We described how many membrane lipids, when isolated, could spontaneously form bilayers in aqueous systems. To prevent any contact of the hydrophobic acyl chains with the aqueous medium, such bilayers close to form vesicles. Usually, under the experimental conditions used to rehydrate lipid samples and with the commonly used membrane lipids (such as phosphatidylcholine) stable multilayered structures are formed. These structures, which have been compared to the multilayered appearance of onions, were termed"liposomes' by Bangham. He realized when studying such systems that, because of their non-toxic nature and impermeability to many solutes, they would make very good agents for drug delivery and other clinical applications. Because liposomes contain alternating layers of bilayer and aqueous space, hydrophobic solutes can be contained within the bilayer and water-soluble materials within the aqueous compartments. Both types of molecules are prevented from leaking by the hydrophilic or hydrophobic layers, respectively.

There are several fundamental problems to the use of liposomes for drug delivery (Table 4.14). First, the liposomes must be able to carry a sufficient quantity of trapped drug. Their capacity can be increased by making liposomes of very large volumes using special procedures such as the reversed-phase evaporation method. A recent technique uses transmembrane pH gradients across large unilamellar vesicles to encapsulate lipophilic, cationic drugs (e.g. the anticancer drugs doxorubicin or vincristine) with high (nearly 100%) efficiency and at high concentrations. Second, once in the blood circulation, serum factors such as

Table 4.14 Problems with the design of liposomes for drug delivery

1. Producing liposomes with a sufficient volume and concentration of trapped drug.
2. Preventing leakage of contents induced by serum constituents such as lipoproteins.
3. Reducing the preferential uptake of liposomes by macrophages of the reticulo-endothelial system.
4. Allowing access to extravascular tissue.
5. Specific targeting of liposomes to the required cells/tissues.

lipoproteins tend to interact with the liposomes and cause leakage. This leakage can be reduced by using more saturated phospholipids and increased cholesterol proportions in the liposomes. Third, because of the size of liposomes they tend to be taken up by the macrophages of the reticulo endothelial system. This means that the liposomal contents are almost always discharged in the liver or spleen, regardless of which organ site is really the target. Covalent linking of hydrophilic polymers (e.g. polyethyleneglycol) to the head group of certain lipids will reduce this. In addition, liposomes will not normally cross the endothelial barrier and, hence, gain access to extravascular tissue. Access, is a particular problem for the brain with its 'blood-brain barrier'. Finally, attempts are being made to encourage specific interaction with and uptake by certain cells or tissues through the covalent attachment of antibodies.

Despite these problems, the enormous potential for the use of liposomes in medicine ensures a very active research field. Some successes have been recorded already. Thus Fabry's disease, which mainly affects the liver, has been treated by enzyme-replacement therapy while parasites such as *Lieshmania,* which reside in macrophages, have been destroyed by drugs delivered in liposomes. Macrophages can also be activated by certain activating factors that can be liposome-delivered. Such treatment renders the macrophages remarkably active in recognizing and destroying diseased tissues including transformed cells.

A particularly exciting development of the use of liposomes has been the attempt to utilize the changed patterns of the surface glycoproteins and glycolipids in malignant cells. The aim is to incorporate a specific recognition molecule onto the surface of liposomes that carry highly toxic molecules such as the protein ricin. Treatment of patients should result in the specific uptake of liposomes of only by cancer cells, which would then be destroyed. In theory these techniques would enable very small numbers of cancer cells to be destroyed before they had a chance to form a sizeable malignancy and might be particularly useful for the treatment of secondaries or in following up surgical intervention. Certainly some success has been achieved by encapsulation of anticancer drugs (e.g. vincristine) when the toxic side-effects of the therapy seem to be reduced.

Liposomes have more recently beep used as a vehicle for DNA transfection. Ultimately this technique can be used for gene therapy. The method has been termed 'lipofection' and is

Table 4.15 Lipid anchors for proteins

Type	Attachment	Examples
1. Myristoylation	Co-translational; glycine at N-terminus.	cAMP protein kinase; G-proteins (Go, Gi); various viral proteins.
2. Palmitoylation	Post-translational; cysteine in body of protein.	Rhodopsin; fibronectin; Ca^{2+} - ATPase.
3. Prenylation	Post-translational; cysteine near C-terminus. Separate enzymes for famesyl or geranylgeranyl groups. Often carboxymethylated also.	*Ras* proteins; G-protein (γ-subunits); lamin B.
4. GPI (glycosylphosphatidylinositol)	Post-translational; at C-terminus; sensitive to phospholipase C.	Alkaline phosphatase; 5′-nucleotidase; acetylcholinesterase.

very dependent on the lipid mixture and dioleoyl-phosphatidylethanolamine seems to be rather effective. Unsaturated phosphatidylethanolamine readily forms non-bilayer structures so it is thought that, after uptake of the DNA-containing liposome by endocytosis, the phosphatidylethanolamine disrupts the endosome sufficiently for some DNA to escape before its destruction in the lysosome. The escaped DNA can then enter the nucleus where transcription can take place. Significant rates of transfection have been achieved and uses for human gene therapy are being investigated.

An interesting commercial application of liposomes has been the development of skin preparations. These types of moisturizing creams are lotions that are claimed to remove wrinkle-lines and are anti-ageing. The main constituents of these preparations are liposomes and the use of such vehicles might be expected to aid in the skin penetration of other components. Although the scientific basis for some of the claims made by the manufacturers are difficult to understand, the lotions do represent a novel and unexpected use of liposomes!

LIPID ANCHORS FOR PROTEINS

Proteins can interact or become embedded in the membrane in various ways. One type of interaction is via lipid anchors. There are four major classes of covalently bound lipids for membrane proteins (Table 4.15).

Acylation

Acylation of proteins can use myristoyl or palmitoyl groups. Myristoylated proteins (Fig. 4.19) come in many shapes, sizes and functions. The myfistate is attached co-translationally to a glycine at the N-terminus. A 7-10 amino-acid consensus sequence is usually present: $G^1N^2X^3X^4X^5X^6R^7R^8$. Position 2 is usually a small, uncharged residue such as Asp., GIn or Ser. Positions 3 and 4 are permissive and X^5 is usually Ser or Thr. Basic residues are preferred at positions 7 and 8.

N-Myistoylation confers upon a protein a significant tendency to associate with membranes but the interaction is not hydrophobic enough to maintain a long-term association. However, these properties are ideal for the function of many of the myristoylated proteins that have been discovered – such as protein tyrosine kinases or the ADP ribosylation factor.

Unlike myristoylation, the attachment of palmitate is a post-translational and reversible modification. A specific palmitoyl-CoA:protein S-palmitoyltransferase is involved, which in rat liver is located in the plasma membrane. At least two thioesterases are known to remove the fatty acid. Palmitoylation can occur on more than one cysteine residue for a given protein as in G-protein subunits or *Ras* proteins.

Because of its reversible nature, protein palmitoylation can be regarded as a method for regulating protein targeting in membranes. In some cases, palmitoylation enhances the loose association of previously myristoylated (or famesylated; see later) proteins with membranes. A good example is with the G-protein subunit. This is myristoylated and randomly and reversibly associates with different membranes until it encounters the plasma membrane where it binds to $G_{\beta\gamma}$. In this complex palmitoylation of G_{α} is facilitated and the doubly lipidated protein remains in stable association with the plasma membrane.

In animals and yeast cells acylated and particularly doubly acylated proteins are enriched in plasma membrane lipid 'rafts' and in caveolae-vesicular invaginations commonly found in animal cells.

Prenylation

A third type of modification involves prenylation. A covalent attachment is formed with either a C15 farnesyl or a C20 geranylgeranyl isoprenoid chain. These hydrophobic chains are bound via a stable thioether bound to a cysteine residue near the carboxyl terminus. Specific transferase enzymes are involved and the prenyl chains remain for the life of the protein.

As mentioned above, attachment of a farnesyl residue only makes a protein loosely associated with membranes. This association is considerably increased (10–20 x) by modification of the carboxyl terminus. Thus, the remaining amino acids are removed by a specific carboxy peptidase and the prenylated cysteine is carboxylmethylated.

The longer geranylgeranyl chain gives the protein a much higher membrane affinity (up to 45x) than a farnesylated protein. Thus, geranylgeranylated proteins are usually membrane located whereas farnesylated ones are cytosolic and only loosely associate with membranes. Reversible methylation/demethylation can be used to influence the strength of the latter association.

Prenylated proteins have been implicated in a wide variety of functions including signalling, cell division and stress responses.

GPI-anchors

A fourth type of lipid anchor is via glycosylphosphatidylinositol (GPI). This was actually the first type of lipid anchor to be discovered when it was noticed that exposure of plasma membranes to a phosphatidylinositol-specific phospholipase C released a number of hydrolytic enzymes. Further examination revealed that all were attached via a GPI-anchor . Well over 150 distinct proteins are now known to be linked to the outer surface of the plasma membrane via a highly conserved core structure (Fig. 4.19). Examples of proteins believed to be attached thus to membranes include alkaline phosphatase, acetylcholinesterase, 5′-nucleotidase and the variant surface glycoprotein (VSG) from *Trypanosoma brucei*. (The latter is responsible for the ability of the parasitic protozoon to evade the victim's immune defence system and so cause sleeping sickness.)

The core structure, first determined for the VSG protein, has been found in a wide variety of GPI-anchors from protozoa and plants through to mammalian brain tissue, although there may be specific additions (which can include saccharides, fatty acids or another phosphorylethanolamine) for individual GPI-proteins.

It will be noted that it is only the diacylglycerol domain of phosphatidylinositol that holds the membrane protein in place. However, such an anchor is at least as strong as a hydrophobic sequence of amino acids spanning the entire membrane.

At present we do not have a definite reason as to why some membrane proteins are attached through phosphatidylinositol. One possibility is that it may provide a means of selectively mobilizing these proteins. Indeed, there is some evidence that endogenous phospholipases may cause the release of VSC and alkaline phosphatase in soluble forms and other proteins such as acetylcholinesterase are also known to be released under certain physiological conditions.

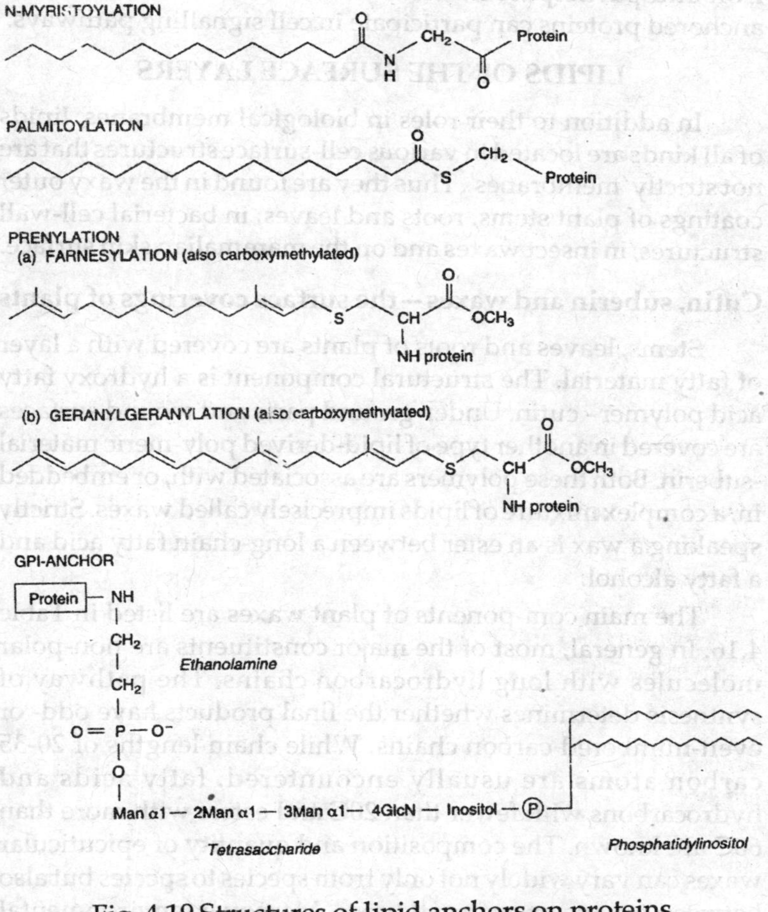

N-MYRISTOYLATION

PALMITOYLATION

PRENYLATION
(a) FARNESYLATION (also carboxymethylated)

(b) GERANYLGERANYLATION (also carboxymethylated)

GPI-ANCHOR

Protein — NH — CH_2 — CH_2 (Ethanolamine) — $O=P-O^-$ — O — Man α1 — 2Man α1 — 3Man α1 — 4GlcN — Inositol — (P) (Tetrasaccharide; Phosphatidylinositol)

Fig. 4.19 Structures of lipid anchors on proteins.

Another advantage is that the anchor itself targets the attached protein from the endoplasmic reticulum, where it is made, to the cell surface. This is achieved by association with plasma membrane destined lipid 'rafts'.

Once on the surface, CPI-anchored proteins (e.g. alkaline phosphatase, 5′-nucleotidase) can recruit nutrients for the cell by metabolizing non-utilizable substrates. Other CPI-anchored proteins may act as ligands to bind nutrients such as folate and iron and participate in their internalization. Further CPI-anchored proteins can participate in cell signalling pathways.

LIPIDS OF THE SURFACE LAYERS

In addition to their roles in biological membranes, lipids of all kinds are located in various cell-surface structures that are not strictly 'membranes'. Thus they are found in the waxy outer coatings of plant stems, roots and leaves; in bacterial cell-wall structures; in insect waxes and on the mammalian skin surface.

Cutin, suberin and waxes — the surface coverings of plants

Stems, leaves and roots of plants are covered with a layer of fatty material. The structural component is a hydroxy fatty acid polymer - cutin. Under-ground parts and wound surfaces are covered in another type of lipid-derived poly-meric material -suberin. Both these polymers are associated with, or embedded in, a complex mixture of lipids imprecisely called waxes. Strictly speaking a wax is an ester between a long-chain fatty acid and a fatty alcohol.

The main com-ponents of plant waxes are listed in Table 4.16. In general, most of the major constituents are non-polar molecules with long hydrocarbon chains. The pathway of synthesis determines whether the final products have odd- or even-numbered carbon chains. While chain lengths of 20-35 carbon atoms are usually encountered, fatty acids and hydrocarbons with fewer than 20C and esters with more than 60C are known. The composition and quantity of epicuticular waxes can vary widely not only from species to species but also between organs or even cell types. Moreover, environmental influences can change the composition.

Table 4.16 The major components of plant waxes

Compound	Structure	Occurrence
n-Alkanes	$CH_3(CH_2)_nCH_3$	Most plants; major components usually 29C or 31C.
iso-Alkanes	$CH_3CH(CH_3)R$	Not as widespread as *n*-alkanes; usually 27C, 29C, 31C and 33C.
Ketones	$R^1.CO.R^2$	Not as common as alkanes; usually 29C and 31C.
Secondary alcohols	$R^1.CH(OH)R^2$	As common as ketones; usually 29C and 31C alcohols.
	$CH_3CH(OH)R$	Uncommon in cuticular waxes; more common in suberin waxes; odd-chain length, 9-1S5C.
β-Diketones	$R^1.COCH_2CO R^2$	Usually minor, but in some species (e.g. barley) may be major components; mainly 29C, 31C and 33C.
Primary alcohols	$R.CH_2OH$	Most plants; even chains predominate, usually 26C and 28C.
Acids	R. COOH	Very common; even chains predominate, usually 24-28C.

Abbreviations: R, R^1 and R^2 are alkyl chains.

Whereas cutin contains dihydroxy fatty acids as major components, suberin has ω-hydroxy and dicarboxylic fatty acids. Phenolic constituents are present in both (Table 6.17). They are especially prevalent in suberin where they are believed to have an anti-microbial function. The exact composition of these polymers varies with the age and development of a given tissue.

Cell walls of fungi have significant amounts of aliphatic hydrocarbons, which are presumed to act as a desiccation barrier – similar to higher plants. In contrast, bacteria do not usually contain large quantities of waxy materials in their surface layers. However, as will be seen in subsequent sections, different bacterial classes contain other characteristic lipids.

Mycobacteria contain specialized cell-wall lipids

The mycobacterial cell wall contains three components. First, there is a skeleton consisting of arabinogalactan mycolate covalently linked through a phosphodiester bond to peptidoglycan. In addition, peptides that can be removed by proteolysis and free lipids are important components.

The skeleton is a branched polymer of D-arabinose and o-galactose in a 5:2 ratio. Every tenth arabinose has a mycolic acid esterified to the 5-hydroxyl. These mycolic acids are 60C-90C fatty acids that are 2-branched or 3-hydroxylated (Fig. 4.20). They may also contain other substituted groups such as cyclopropane rings or methoxy groups. The acids are often named after the type of bacteria in which they are found (e.g. nocardomycolic acid from *Nocardia* spp.).

Table 4.17 The main compositional differences between cutin and suberin

Monomer	Cutin	Suberin
Dicarboxylic acids	Minor	Major
In-chain-substituted acids	Minor	Minor(sometimes substantial)
Phenolics	Low	High
Very long chain (20-26C) acids	Rare and minor	Common and substantial
Very long chain alcohols	Rare and minor	Common and substantial

Name	Formula (examples of producing organism)
β-Mycolic acid	$CH_3(CH_2)_{17}CH(CH_3)\cdot C(=O)(CH_2)_{17}CH{-}CH(CH_2)_{19}CH(OH)\cdot CH((CH_2)_{23}CH_3)\cdot COOH$ (cyclopropane ring: CH_2 bridging CH—CH) (*Mycobacterium tuberculosis*)
α-Smegmamycolic acid	$CH_3(CH_2)_{17}CH{=}CH(CH_2)_{13}CH{=}CH\cdot CH(CH_3)(CH_2)_{17}CH(OH)\cdot CH((CH_2)_{21}CH_3)\cdot COOH$ (*Mycobacterium smegmatis*)

Fig. 4.20 Structure of two mycolic acids.

The extractable free lipids comprise a mixture of cord factors, mycosides, sulpholipids and wax D. Together they represent about 25-30% of the mycobacterial cell wall. Cord factors are esters of the disaccharide, trehalose, with two mycolic acids (Fig. 4.21). Cord factor is so-called because it is found in the waxy capsular material of virulent strains of tubercle and related bacteria. The factor causes the bacteria to string together in a long chain or cord. The compound is highly toxic and is somehow intimately associated with the virulence of the organism. It interacts strongly with host cell membranes thus impairing their function. Cord factor probably acts as a hapten and binds to albumin in plasma, thus forming an antigen. Like extracts of mycobacterial cell walls, cord factor acts as an immunostimulant. Killed mycobacteria (or their cell walls) are suspended in Freund's adjuvant, which is commonly injected with antigens to increase the titre of antibody.

The sulpholipids (Fig. 4.22) consist of trehalose, which is sulphated at the 2-position and acylated at several positions on both sugar residues. Instead of mycolic acids, the sulpholipids have a mixture of palmitic acid and very long chain (31C-46C) fatty acids with up to 10 methyl branches - known as phthioceranic acids.

Mycosides (Fig. 4.22) are again characteristic of mycobacteria. The basic structure for mycosides A and B is a long-chain, highly branched, hydroxylated hydrocarbon terminated in a phenol group. Acyl groups are esterified to hydroxy groups of the hydrocarbon chain. In contrast, mycoside C is a glycolipid peptide.

Lipopolysaccharide

In Gram-negative bacteria, the wall is far more complex than for Gram positives and contains glycopeptide, lipopolysaccharide, phospholipid and protein. Also, there is not such a clear distinction between the wall and the membrane as in Gram positives. Up to 20% of the wall contents may be lipids, but only a proportion of these are readily extractable by conventional solvent methods. This is because of the covalent nature of the lipopolysaccharide linkages.

The cytosol of Gram-negative bacteria is surrounded by a complex cell envelope consisting of at least three layers (Fig. 4.23).

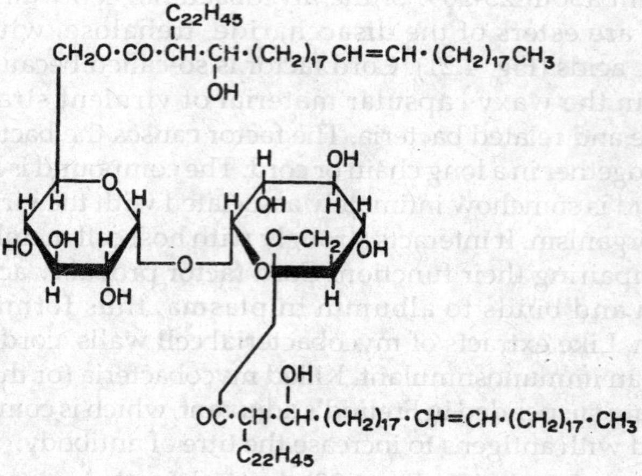

$$CH_2O \cdot CO \cdot CH(C_{22}H_{45}) \cdot CH(OH) \cdot (CH_2)_{17} CH{=}CH \cdot (CH_2)_{17} CH_3$$

$$OC \cdot CH(C_{22}H_{45}) \cdot CH(OH) \cdot (CH_2)_{17} \cdot CH{=}CH \cdot (CH_2)_{17} \cdot CH_3$$

Fig. 4.21 Cord factor.

The cytoplasmic membrane is composed of the familiar phospholipid bilayer with integral and peripheral proteins. Then comes a peptidoglycan layer, which is separated from the outer membrane by a periplasmic space. The outer membrane is extremely asymmetric, but contains lipopolysaccharide and enterobacterial-common antigen in its outer leaflet. Phospholipid molecules are missing from the outer membrane in many Gram-negative bacteria but are present in some species. In those bacteria containing phospholipids in the outer membrane, phosphatidylethan-olamine is, by far, the most common constituent (~85%).

The lipopolysaccharide is involved in several aspects of pathogenicity. It is a complex polymer in four parts (Fig. 4.24). Outermost is a carbohydrate chain of variable length (called the O-antigen), which is attached to a core polysaccharide. The core polysaccharide is divided into the outer core and the backbone. These two structures vary between bacteria. Finally the backbone is attached to a glycolipid called lipid A. The link between lipid A and the rest of the molecule is usually via a number of 3-deoxy-D-manno-octulosonic acid (KDO) molecules. The presence of KDO is often used as a marker for lipopolysaccharide (or outer melnbrane) even though it is not present in all bacterial lipopolysaccharides.

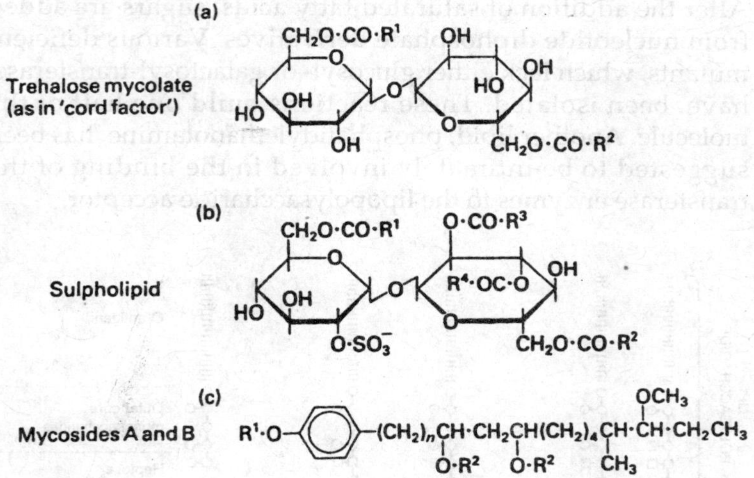

(a) Trehalose mycolate (as in 'cord factor')

(b) Sulpholipid

(c) Mycosides A and B

Fig. 4.22 Some mycobacterial lipids. Structures of mycobacterial lipieds: (a) trehalose mycolate, where R^1 and R^2 are mycolic acids; (b) sulpholipid, where R^1–R^2 are palmitic acid or very long chain branched fatty acids: (c) mycosides A and B, where R^1 is a mono- or trisaccharide and R^2 is a 12C–18C saturated fatty acid or a mycocerosic acid [*e.g.* $CH_3(CH_2)_{21}(CH(CH_3)CH_2)_2CH(CH_3)COOH$].

Lipid A is composed of a disaccharide of glycosamines (Fig. 4.25). The amino groups are substituted with 3-hydroxymyristate while the hydroxyl groups contain saturated (12C-16C) acids and 3-myristoxymyristate. Phosphate and KDO groups are also substituted. Unsaturated and cyclopropane fatty acids, which are common in other lipid types, are absent from lipopolysaccharide.

Studies on the synthesis of lipopolysaccharide have made extensive use of bacteria mutants that are deficient in one or more of the necessary reactions. Rothfield, Horecker and their colleagues in the USA initiated such experiments. It was discovered that formation took place in several stages.

Lipid A, which anchors lipopolysaccharide in the membrane, is made first. Hydroxy acids are added first to the disaccharide, followed by KDO and then saturated fatty acids. The hydroxy fatty acids come from acyl-CoA substrates whereas CMP-KDO is the source of the second addition units. After the addition of saturated fatty acids, sugars are added from nucleotide diphosphate derivatives. Various deficient mutants, which lack either glucosyl- or galactosyl-transferase, have. been isolated. These reactions build one half of the molecule. Another lipid, phosphatidylethanolamine, has been suggested to be intimately involved in the binding of the transferase enzymes to the lipopolysaccharide acceptor.

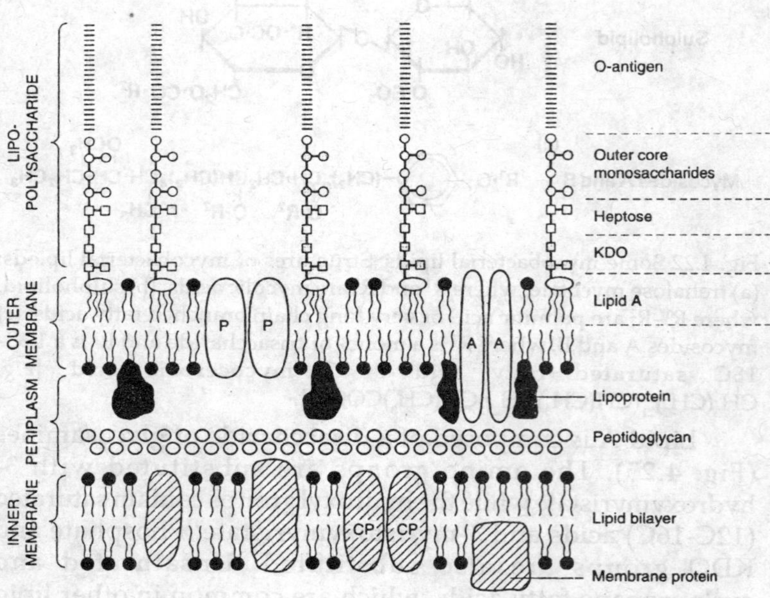

Fig. 4.23 Organization of the membrane system of a Gram-negative bacterium. P = porin; A = transmembrane peptidoglycan-associated protein; CP = carrier protein.

The *O*-antigen is made in three stages. The oligosaccharide units are transferred from nucleotide diphosphate carriers to a galactose attached to another lipid carrier; a 55C polyisoprenoid molecule. The oligosaccharide units are then polymerized and lipid carriers are released in the process. Finally, the complete *O*-antigen is transferred to the R-core, with the release of the final isoprenoid carrier.

Lipopolysaccharide synthesis occurs on the inner membrane from which the molecule must be transferred to the outer membrane. This occurs at specific membrane sites generally where the two membranes adhere to each other.

Gram-positive bacteria

The cell walls and membranes of most Gram-positive bacteria contain a series of highly anionic polymers. Most

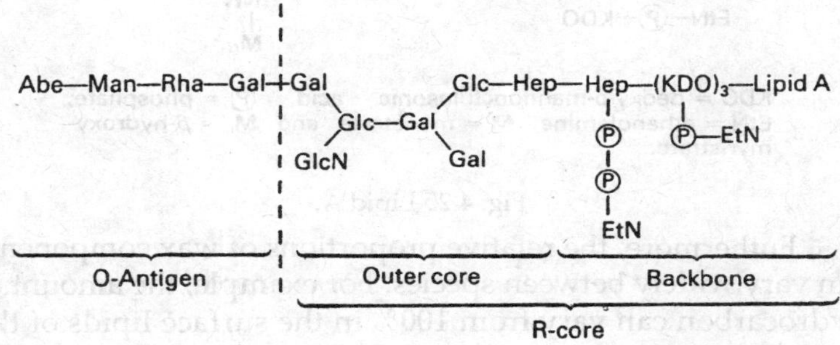

Fig. 4.24 Generalized structure of lipopolysaccharide from a Gram-negative bacterium.

important among these is teichoic acid, which is a polymer of glucose I-phosphate or ribitol phosphate. Membrane teichoic acids are based on glucose I-phosphate and a proportion may be linked to glycolipids to give lipoteichoic acids. The details of these structures and the amount of substitution of the teichoic acids vary with bacterial species.

Apart from the glycolipid linked to teichoic acid, the only significant lipid in most Gram-positive bacteria is that in the cytoplasmic (protoplast) membrane. Notable exceptions are the mycobacteria the special lipids of which have been dealt with above.

Insect waxes

A majority of studies on animal waxes have been done with arthropods, particularly insects. These waxes consist of a complex mixture that includes straight-chain saturated and unsaturated hydrocarbons, wax esters, sterol esters, ketones, alcohols, aldehydes and acids. The composition may be influenced by the plants on which the insect feeds.

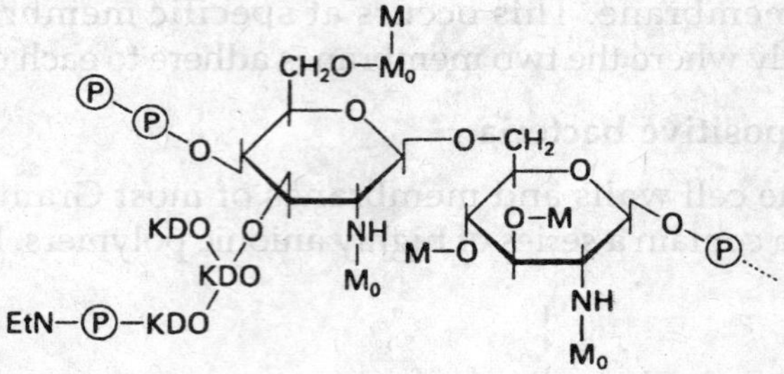

KDO = deoxy-D-mannooctulosonic acid, Ⓟ = phosphate, EtN = ethanolamine, M = myristrate and M_0 = β-hydroxy-myristrate.

Fig. 4.25 Lipid A.

Futhermore, the relative proportions of wax components can vary widely between species. For example, the amount of hydrocarbon can vary from 100% in the surface lipids of the field cricket *(Nemobius fasciatus)* to less than 1% in tobacco budworm pupae.

Not only do insect waxes prevent desiccation they may also have other diverse functions such as acting as pheromones or kairomones, helping in recognition and in thermoregulation.

Lipids of skin

The main function of the skin - one of the largest organs of the human body—is that of a protective barrier. The epidermis, which is separated from a predominantly acellular dermis by a basement membrane, is composed of four layers. The outer-most of these four layers, the stratum corneum, consists of impermeable cellular layers composed of flattened, densely packed, anucleate cells embedded in a waxy lamellar matrix of complex intercellular lipids. The latter are derived from the lipid-filled membrane-coating granules of the cells of the next layer, the stratum granulosum. In addition, the lipid mixture sebum is secreted by the sebaceous glands. Typical compositions of sebum and epidermal lipids are shown in Table 4.18.

Table 4.18 The main lipid classes of sebum and the skin's epidermis

Sebum lipids	Epidermal (*Stratum corneum*) lipids
40% Triacylglycerol	50% Ceramides
25% Wax esters	25% Fatty acids
15% Fatty acids	19% Cholesterol
12% Squalene	4% Cholesterol sulphate
8% Others	2% Others

The composition of human skin surface lipids varies over the body, but in sebaceous gland-enriched areas (such as the face) the secretion, sebum, may represent 95% of the total. When sebum is freshly formed it contains mainly triacylglycerols, wax esters and squalene. However, due to bacterial action, unesterified fatty acids are rapidly released. These acids are both an irritant and comedogenic (give rise to 'black-heads') and have been implicated in acne vulgaris—an extremely common complaint of puberty. It is often stated that the severity of acne correlates with the amount of seborrhoea and many treatments are designed primarily to remove such lipids from the skin surface.

More serious than acne, but less prevalent, are the skin diseases atopic eczema and psoriasis. Lipids have been suggested as being involved in both of these complaints although it must be stated that the evidence at present is equivocal. It has been noted in patients with atopic eczema that there is a disturbance in the normal tissue complement of polyunsaturated fatty acids. For convenience, blood samples have usually been analysed and these usually exhibit reduced levels of γ-linolenic (all *cis*-6,9, 12-octadecatrienoic) acid. γ-Linolenate is an intermediate in the normal conversion of the essential fatty acid, linoleate, to the eicosanoid precursor arachidonate:

$$c,c-9,12-18:2 \xrightarrow{\Delta 6\text{-desaturase}} c,c,c-6,9,12-18:3 \xrightarrow{\text{elongase}}$$
$$c,c,c-8,11,14-20:3 \rightarrow c,c,c,c-5,8,11,14-20:4$$

Moreover, symptoms of atopic eczema have been reported to be alleviated by γ-linolenate-rich lipids, such as the oil from evening primrose seeds. These results are interpreted to indicate a deficiency of the Δ6-desaturase responsible for linoleate metabolism, but they do not explain the beneficial effect of some other treatments.

Psoriasis is a severely debilitating skin disease affecting 2% of the population world-wide. It is responsible for more days spent in hospital than any other skin complaint and present treatments are rudimentary and rather unsatisfactory. Characteristically there is an impairment of normal cell differentiation and development, excessively high levels of some arachidonate metabolites and a potentiation of the symptoms in patients with lithium therapy. The symptoms have been drawn together recently with the suggestion that defective phosphoinositide metabolism is involved.

Finally a few remarks should be made about essential fatty acid (EFA) deficiency. Linoleic acid is the precursor of the eicosanoids, leukotrienes, prostaglandins, thromboxanes and related compounds. Until recently it was thought that the classic skin symptoms resulting from a deficiency were due to a lack of eicosanoids. However, the specific role of *n*-6 acids (such as linoleate) as opposed to *n*-3 acids (such as α-linolenate) in membrane fluidity and membrane function cannot explain their essentiality. For example, aspirin ingestion, which effectively prevents prostanoid formation, does not result in the appearance of EFA-deficiency symptoms. It is also generally accepted that at least some cell lines in culture do not have an EFA requirement—thus suggesting that EFAs are not essential for the formation and function of cellular membranes in general.

Recently, several linoleic acid-rich lipids such as acylglucosylceramide, acylceramide and a specific wax ester have been identified in human, pig and rat epidermis (Fig. 4.26). These lipids form part of the intercellular lipid-rich matrix serving as a water permeability barrier. In EFA deficiency the linoleate of these lipids is replaced by oleate, a change that is associated with a loss of the barrier function.

SUMMARY

A major characteristic of eukaryotic cells is the presence of cellular organelles. These organelles are bound by one or more membranes that serve as a barrier to the free passage of molecules – thus maintaining the unique composition and environment of individual organelles. Not only do individual organelles have a characteristic complement of enzymes (and, therefore, metabolic pathways) they also have membranes made up of a special blend of lipid molecules. In a few cases, such as the diphosphatidylglycerol of the inner mitochondrial membrane, these lipid distributions are so specific that they can be used to aid identification of subcellular fractions. In some instances, organelles can be induced or caused to undergo functional development or morphological change. By the use of such systems, the role of lipids in membrane biogenesis and function can be studied.

Most cellular membranes are composed predominantly of proteins and phosphoglycerides. Therefore, phosphoglycerides are significant components of prokaryotes and eukaryotes. In animals and non-photosynthetic cells of higher plants, phosphatidylcholine is the main phosphoglyceride. This molecule is not, however, a major component in bacteria. Phosphatidylethanolamine is a major component in most organisms while phosphatidylglycerol is common in prokaryotes and is the dominant (in many cases exclusive) phosphoglyceride of the photosynthetic membranes of oxygen-evolving species. As well as these molecules, phosphatidylinositol is quantitatively important forming a smaller but significant percentage of the membrane lipids in many cases. Phosphorylated derivatives of phosphatidy-linositol are found in the plasma membranes of animals and plants, while diphosphatidylglycerol is exclusively located in the inner mitochondrial membrane of eukaryotes. Together with phosphatidylethanolamine and phosphatidylglycerol, it is a major component of many bacteria.

In contrast to the quantitative importance of the phosphoglycerides, phosphonolipids (which contain a stable carbon-phosphorus linkage) have been found in significant quantities in only a few organisms such as sea urchins and protozoa. Similarly, ether-linked lipids also are less prevalent than the acyl phosphoglycerides, although phospholipids such as choline and ethanolamine plasmalogens are widespread but usually minor components in animals. The betaine ether-linked glycerolipids are major lipids in many lower plants.

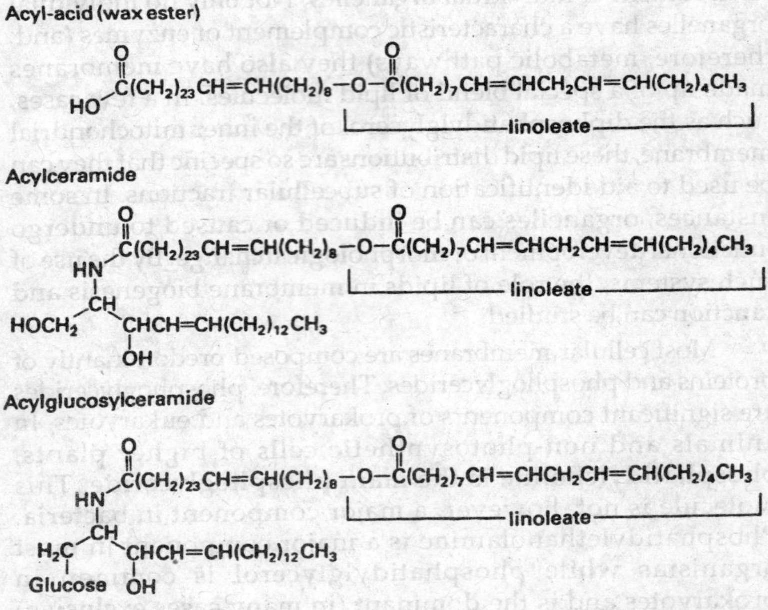

Acyl-acid (wax ester)

$$HO{-}\overset{O}{\overset{\|}{C}}(CH_2)_{23}CH{=}CH(CH_2)_8{-}O{-}\overset{O}{\overset{\|}{C}}(CH_2)_7CH{=}CHCH_2CH{=}CH(CH_2)_4CH_3$$

(linoleate: $\overset{O}{\overset{\|}{C}}(CH_2)_7CH{=}CHCH_2CH{=}CH(CH_2)_4CH_3$)

Acylceramide

$$HN{-}\overset{O}{\overset{\|}{C}}(CH_2)_{23}CH{=}CH(CH_2)_8{-}O{-}\overset{O}{\overset{\|}{C}}(CH_2)_7CH{=}CHCH_2CH{=}CH(CH_2)_4CH_3$$

(linoleate)

$$HOCH_2{-}CH(NH{-}){-}CH(OH)CH{=}CH(CH_2)_{12}CH_3$$

Acylglucosylceramide

$$HN{-}\overset{O}{\overset{\|}{C}}(CH_2)_{23}CH{=}CH(CH_2)_8{-}O{-}\overset{O}{\overset{\|}{C}}(CH_2)_7CH{=}CHCH_2CH{=}CH(CH_2)_4CH_3$$

(linoleate)

$$Glucose{-}H_2C{-}CH(NH{-}){-}CH(OH)CH{=}CH(CH_2)_{12}CH_3$$

Fig. 4.26 Structures of linoleic acid-rich skin epidermal lipids.

Glycosylglycerides are the major lipids of the photosynthetic membranes of higher plants, algae and cyanobacteria. Because of the preponderance of such membranes the main glycosylglyceride, monogalactosyldiacylglycerol, is actually the most abundant lipid in the world. The other glycosylglycerides found in plants and algae are digalactosyldiacylglycerol and the plant sulpholipid, sulphoquinovosyldiacylglycerol. Different lipids containing various glycosidic residues are found in micro-organisms such as *Pneumococcus* or *Mycoplasma* while the galactosylglycerides are minor constituents of mammalian brain.

The glycerolipids are distributed in individual membranes at characteristic ratios. There is also good evidence that biological membranes are asymmetric with regard to their lipid as well as their protein distribution. There may also be some lateral heterogeneity with particular regions of the membrane being enriched with certain lipids.

In contrast to the glycerolipids, sphingolipids are based on a sphingosine backbone. Sphingolipids are only found in any quantity in animals where they tend to be concentrated at the cell surface – in keeping with their function there. Sphingolipids have an acyl residue attached through the amino group of sphingosine while carbohydrates (or phosphorylcholine in the case of sphingomyelin) are linked via the alcohol residue. Depending on the nature of the sugar residues, ceramides, sulphatides and gangliosides are formed. Each cell contains a characteristic pattern of such molecules.

Sterols are based on the cyclopentanoperhydrophenathrene structure, which consists of a fused four-ring system. Three of the rings are six-membered while ring D has five carbons. A side-chain is attached to ring D and sterols contain an hydroxyl function at C3 of ring A. Cholesterol is, by far, the most common sterol in animals while different sterols occur in other organisms. Thus, ergosterol is important in yeasts while stigmasterol and β-sitosterol are common in plants.

Important steroids include the (pro-)vitamins D, bile acids and various steroid hormones such as testosterone, oestrogens, progesterones and glucocorticoids. Sterols may also be derivatized to give sterol esters, sterol glycosides and acylated sterol glycosides.

The basic structure of membranes is usually envisaged as the fluid-mosaic model of Singer and Nicholson. In this, amphiphilic lipids are arranged in a bimolecular layer with the hydrophobic moieties in the centre of the membrane and the hydrophilic head groups at the two surfaces. Stability is achieved by various ionic or hydrophobic interactions between the lipids and the membrane proteins. The latter can be embedded within the membrane (intrinsic proteins) or attached to the surface (extrinsic proteins). Many membrane proteins are large enough to actually span the membrane.

By a combination of enzyme digestion, chemical or antibody labelling or physical techniques such as NMR; it has been shown that all membranes thus examined are clearly asymmetric with regard to their lipids as well as their proteins. This asymmetry is maintained because, although lipids show rapid rotation and translational movement, they only move !rom one leaflet to the other at very slow speeds.

For a membrane to function effectively, the acyl chains of its lipids must be above their gel-liquid transition temperature. Organisms take care to maintain membrane fluidity by a variety of methods. Membrane fluidity contributes to important properties such as the operation of transport proteins and to the rapid diffusion of hydrophobic electron carriers like quinones.

Apart from a general function in providing the essential bimolecular layer, which is the foundation of biological membranes, lipids have more specific functions. One essential role is in membrane fusion—a process essential for secretion, endocytosis, intracellular digestion by lysosomes and cell division. The ability of membrane lipids to adopt non-bilayer phases is important for fusion. Moreover, lipids and proteins interact to determine membrane shape and properties and recent experiments with bacterial mutants have demonstrated the essential role of particular lipid components in specific membranes.

Artificial membrane systems have been used for many years to help scientists understand how membranes work. More recently, lipid vesicles (usually multilayered and called liposomes) have been employed as drug carriers for medical treatments. They have high capacities for both hydrophobic and hydrophilic drugs and often show better release and toxicity characteristics than other forms of treatment.

Lipids can also be used to modify proteins eovalently. Four types of lipid anchor are used—myristoylation, palmitoylation, prenylation and the glycosyl phosphatidylinositol anchor. Depending on the structure, they will cause proteins to be tightly or loosely associated with membranes.

The surface layers of different organisms often have lipids as significant components. The surface coverings of plants consist of waxes and cutin on the aerial parts while suberin covers the underground tissues or wound surfaces. Very long chain lipids are characteristic of these surface coverings.

Dependent on whether the surface layer is wax, cutin or suberin and also, to some extent, the plant tissue in question then the individual lipid composition will vary. Hydrocarbons very long chain acids and alcohols, their esters and long-chain aldehydes and ketones are all found with hydroxylated fatty acids prominent in suberin.

In contrast to plants, micro-organisms have rather specialized and, in some cases, unique surface coverings. Mycobacteria contain large amounts of cell-wall lipids such as mycosides while Gram-negative bacteria contain lipopolysaccharide. The latter is composed of four parts. On the outside of the cell is a polysaccharide of variable structure often called the *O*-antigen. This is attached to a core polysaccharide in two—parts—the backbone of which is linked to lipid A. In contrast, Gram-positive bacteria have a completely different surface structure. Lipids are less prevalent although a proportion of the membrane teichoic acids are linked to a glycolipid, such as diacyldiglycosylglycerol, to give lipoteichoic acid.

As in plants, waxes are important on the outer surface of insects. The composition of the waxy layer varies widely and is often influenced by the plant food that the insect eats.

The skin of mammals contains large amounts of lipids that act as a protective barrier. The epidermal lipids themselves contain a high proportion of ceramides, many of which contain linoleic acid as a major component. Cholesterol and its sulphate ester are also important. The lipid content is supplemented by the secretion of sebum from the sebaceous glands. Sebum is rich in triacylglycerols, wax esters and the terpenoid squalene and its secretion helps to maintain the skin in a waterproof and flexible condition. Several skin diseases, including acne, eczema and psoriasis, have lipids as major factors in their progenesis.

Chapter 5

Production of Flakes

The purpose of maneging a flanking roll is to improve the consistency of the flakes produced (improve extraction efficiency) and to eliminate or at least significantly reduce down time required for maintenance (reduce operating costs). To produce consistent flakes across the full face of a flaking roll requires control of the roll profile. Roll profile is dependent on roll shape, deflection, and temperature.

ROLL SHAPE

Roll shape is affected by the service a roll receives (full face and/or roll end grinding) and by wear. Due to the nature of material flow, the wear in the center of the roll will always be greater than at the roll ends. In order to maintain a true roll shape, periodic grinding of the roll ends will be necessary.

DEFLECTION OF FLAKING ROLL

Deflection of a flanking roll is primarily a function of the length to diameter ratio and the closing forces applied. Generally speaking, the roll length should be no more than about twice the diameter. Rarely in the oil extraction industry will roll closing forces be greater than roll with these dimension guide lines can tolerate.

Roll temperatures are influenced most significantly by the

uniformity of the feeding from end to end on the roll. Since roll ends lose heat more rapidly than the center (more radiating surface exposed at the ends) the feed may need to be slightly reduced in the center of the roll to maintain uniform temperatures from end to end. Roll shape also influences roll temperatures, as high areas on the roll face will do more work. More work translates directly into higher operating temperatures. As the roll temperature increases the diameter of the roll increases.

To manage a flaking roll, on the simplest level, requires gathering information regarding the operation of the flaking roll, analyze that information and take appropriate actions when and where required in order to achieve the desired results.

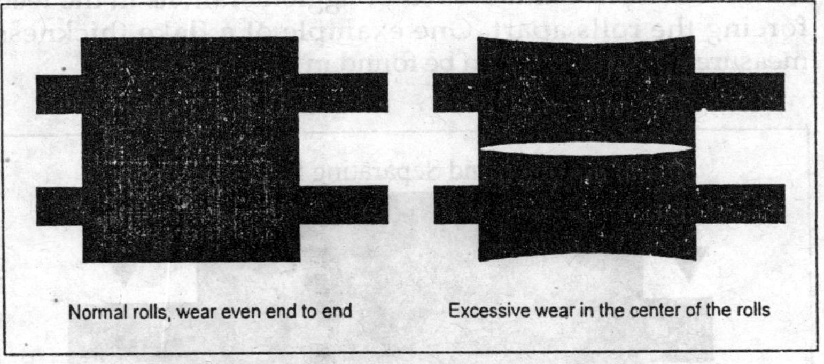

Fig. 5.1 Effect of Wear on Roll Profile

The single most essential piece of information required to manage a flaking roll is an accurate description of the flakes being produced. Since a flaking roll will produce approximately 15,000 flakes per revolution, it is not possible to measure every flake produced. However, a representative sample may be taken from each end and in the center of the roll and the results may be assumed to generally represent the total output from the machine. Flake thickness measurements must be made accurately and in a manner that can be easily repeated. While a conventional 'C' frame micrometer can be used to make flake thickness measurements, precise dial thickness gauges may also be applied. The flake thickness gauge has a distinct advant~ge in that the flake thickness is read directly from the dial, where a micrometer requires more skill to read and interpret. Ordinarily a number of flakes should be measured to determine the typical or average flake thickness. Frequent cleaning of the anvil faces of any measuring device will be required to ensure accurate results. When measuring thicker or curled flakes, it may be necessary to apply additional closing force of the measuring device to ensure the accuracy of the measurement. The greater the number of flakes measured, the more statistically correct the final readings will be. By taking 12 flake thickness measurements at each location (left end, center, right end) and rejecting the high and low in each series, a reasonable average can be determined. If significant discrepancies are noted, it is often worthwhile to investigate the possible cause, such as an improperly adjusted scraper blade allowing some flakes to wrap around the roll or agglomerations in the feed forcing the rolls apart. One example of a flake thickness measurement analysis can be found in Table 5.1.

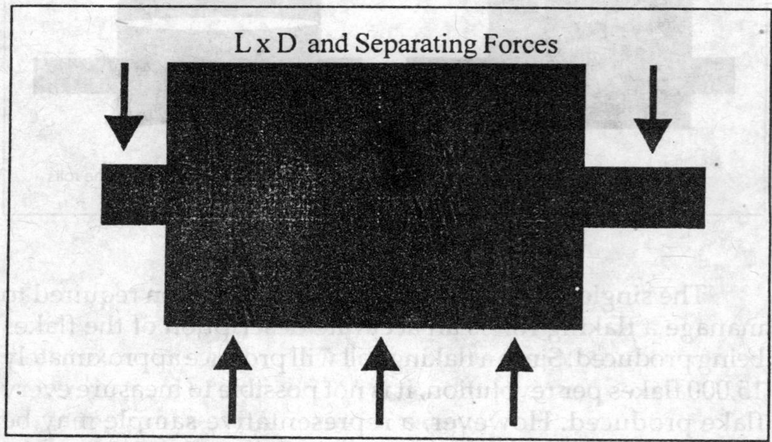

Fig. 5.2 Effect of Deflection on Roll Profile

The second most important piece of information required to allow an operator to control a flaking mill is an idea of the roll temperature across the full face of the roll. This can be quickly and easily determined through the use of a simple hand held infrared pyrometer. This 'Heat Gun' will allow the roll surface temperature to be read even when the flaking mill is in operation. (Caution: Proper safety practices must be followed when working around any moving parts or equipment). In order for the results to be accurate and repeatable, the calibration of the device must be correct and the procedure (distance and relative position to the roll) must be consistent from one measurement to the next. When some high oil materials are being processed (i.e. corn germ) the oil coating on the roll will affect the 'temperature' read by the infrared device, so the operating conditions must be similar for results to be comparable from one reading to the next. Normally, the best readings will be taken while the mill is actively processing material. While simply scanning the roll length and observing the temperature will yield valuable information, recording the temperatures across the roll face will provide a more complete and accurate tool that can be employed to facilitate operating and maintaining the flaking roll.

Table 5.1: Flake thickness reporting*

Left	Center	Right
14.0	13.5	12.0
13.0	14.0	14.5
12.5	14.5	14.0
13.5	14.5	14.5
14.0	14.0	13.0
12.5	13.5	14.0
13.5	14.0	14.5
14.0	14.5	13.5
14.5	13.5	14.5
13.0	14.5	14.0
12.5	13.5	14.0
14.0	14.0	13.0
13.4	14.0	13.9

* Flake thickness in .001"

From time to time, bits of agglomerated material by finding their way into the flaking mill feeder partially block the flow of material or lodge the in roll nip. Since one area of the roll is either being underfed (shows as a low temperature) or overfed (shows as a high temperature) the flakes produced in those areas will be affected. As the roll material cools, the diameter decreases and the flakes produced in that area will tend to be thicker. Alternatively, in an area that receives excessive feed, the roll temperature will be higher and the diameter will increase producing thinner flakes. Other factors that may affect product flow, such as material accumulating on magnets or on the pan of a vibrating feeder can also disrupt the product flows into the mill resulting in uneven heating of the roll.

Regular inspections made by operators in each shift are powerfully enhanced by a more thorough analysis of the flaking mill performance on a regular and timely basis. By observing the roll temperatures over a period of time, the condition of the roll and the need for routine service such as roll end grinding can be readily identified. Completing this type of evaluation one or two times per week, establishes a benchmark against which unusual operating conditions can be compared.

Effect of Temperature on Roll Profile

- As temperature increases, roll diameter increases
- Larger diameter rolls increase more per OF
- Cycling roll pressure of and on well allow rolls to back up, accommodating the increase in roll diameter

Roll Diameter	10°F	20°F	30°F
20"	0.0012"	0.0024"	0.0035"
24"	0.0014"	0.0028"	0.0042"
28"	0.0017"	0.0033"	0.0050"

COMMON OPERATIONAL PROBLEMS

Washboard/Corrugations/Flat Spots/Rattle Marks

Flaking rolls wear more in the center than at the ends, eventually leading to a condition where the roll end are larger in diameter than the center of the roll, also referred to as high ends. If this condition is left unchecked, the pressure exerted on the ends of the flaking rolls can become so great that a catastrophic failure occurs. Depending on the type of roll material and the flaking mill .installation, the failure inay be demonstrated as spalling, where material actually breaks out of the roll face or end. Alternatively the failure may become evident in the form of pressure bars forming on the face of the roll. This condition is known by various names in the industry, wash boarding, corrugations, flat spots, rattle marks, etc. The pressure bar itself is a band of work hardened material formed by metal to metal contact as the rolls slip and bounce against one. another. Adjacent to thfs work hardened matenal is an area of the roll not work hardened that continues to wear a normal rate resulting in the formation of the irregular working surface of the roll.

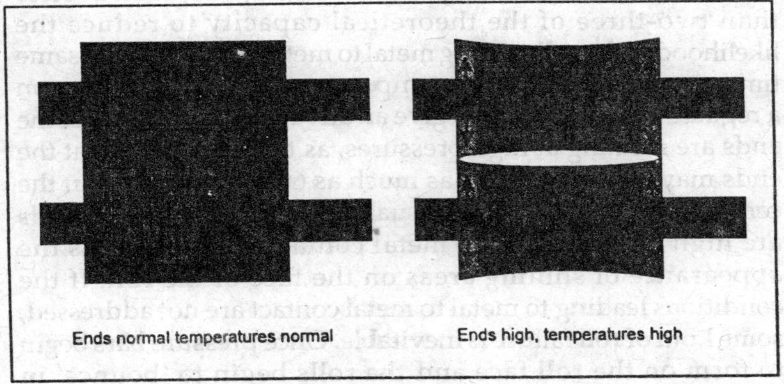

Fig. 5.3 Effect of Temperature on Roll Profile

Although the exact sequence of events and all contributing factors associated with pressure bars are not fully understood, metal to metal contact of the rolls over an extended period of time is clearly at the root. Machines with less rigid foundations and plants with less substantial mounting for the flaking mills seem to be more prone to this phenomena. Certain roll compounds available in the past ten years or so know as 'Indefinite Chill' rolls significantly reduce the possibility of spalling due to metal to metal contact, but are not able to reduce the tendency for these pressure bars to form when the rolls experience high loading such as when running with no feed or when the roll ends are 'high'. In fact, since the rolls are not likely to spall, there is a greater chance of roll ends running together unnoticed until these pressure bars begin to form.

While the process of developing pressure bars take some time to occur, the condition usually appears or is noticed 'overnight', and, if not treated immediately can quickly degrade to the point where rolls must be removed from the mill for proper treatment. Since this condition does not develop overnight and the warning signs are clear, it certainly is controllable. To reduce the possibility of pressure bars forming, every effort must be made to supply the flaking mill with an adequate stream of material across the full·face of the roll at all times. Additionally, flaking rolls should not be operated at less than two-three of the theoretical capacity to reduce the likelihood of the roll making metal to metal contact. At the same time, simply monitoring the temperatures across the roll face on a regular basis will usually give an advance indication that the ends are running at high pressures, as the temperature at the ends may be from 10°F to as much as 60°F higher than in the center of the rolls. Another visual indication that the roll ends are high or that metal to metal contact is occurring is the appearance of shining areas on the face of the roll. If the conditions leading to metal to metal contact are not addressed, some kind of roll failure is inevitable. Once pressure bars begin to form on the roll face and the rolls begin to 'bounce' in operation, the temperature differential may not longer be obvious as the rolls do not make continuous contact.

Q theoretical = (D x Lx d x RPM)/450

Q = Theoretical capacity, Metric Tons/Hour

D = Roll diameter in inches

L = Ro 11 length in inches

d = Flake thickness in inches

Shining areas or high temperatures at the roll ends indicate the need to relieve the roll ends. This roll end grinding or edge correction can be accomplished with a hand held 7" to 9" angle grinder or with a variety of grinding attachments. Many plants will design a simple fixture or bracket that is used to support and help guide a hand operated grinder to facilitate this roll end service. In most cases, a full face roll grind is not required to correct high roll ends, but it is important that both roll be ground at the same time. Many plants operate rolls for years without grinding the full face of the flaking rolls simply by maintaining roll ends on a regular basis. While a straight edge and flashlight will help guide the roll end grinding process, the only sure method of determining the proper amount to grind is operating experience. As an alternative to hand grinders, there is commercially available roll grinding attachments that mount to the flaking mill. These tend to be a bit cumbersome and difficult to set up and operate, but also reduce some of the risk of 'overcorrecting' with a hand held grinder. Generally speaking, devices that will grind rolls in the machine lack the rigidity to correct serious roll problems. According to one roll supplier, it takes approximately 8 hours in a cylindrical grinder designed for large flaking rolls to properly dress a roll suffering from pressure bars or with other significant surface defects. By the same roll manufacturers reckoning, an 'in place' grinder will take at least 5 times that long. If the damage cause by pressure bars is extensive (work hardened material may be as much at 020" deep) the correction process will be accelerated by first taking a cut off the roll face in a lathe.

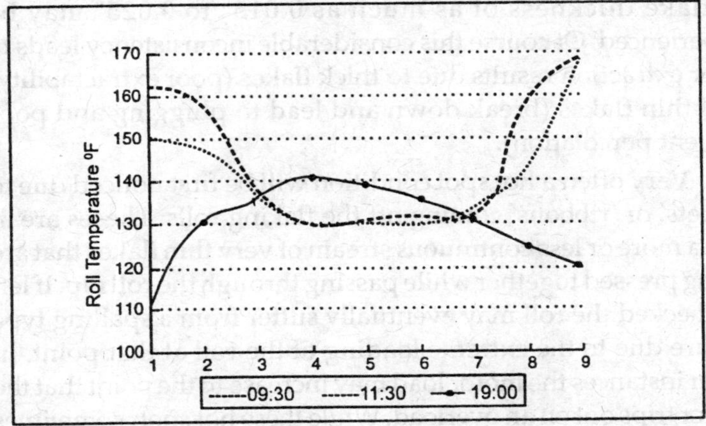

Fig. 5.4 Roll Temperature readings

Figure 4 reflects roll temperatures taken on a 28" diameter

x 62" long roll. Notice that at 09.30 the roll ends were 30°F higher than the center of the roll. The flakes produced from the roll ends were 0.012" to 0.014" while the flakes in the center were 0.018". The mill was stopped and the ends relieved with a hand grinder. The mill was restarted at 11.30. The roll temperature was taken again at 19:00. Notice how after grinding the ends were 20° to 25°F less than the center of the rolls. The flake thickness across the foll face varied from 0.012" on the left end, 0.011" in the center, and 0.013" on the right hand end of the mill.

Hot Spots

The second most common operational problem experienced with flaking rolls is a condition commonly referred to as hot spots. A hot spot is characterized by a localized heating of one area of the roll. This increase in temperature causes the roll to 'grow' (diameter increase to the increase in temperature) in one area reducing the gap between the rolls and causing more work to be concentrated in that areas. This lead to additional heating, more increase in the roll diameter, and thinner flakes. Since adjacent areas of the roll are not doing as much work, the roll temperature also decreases the thickness of the flakes produced and increases as the roll cools down. Inconsistencies in flake thickness of as much as 0.015" to 0.025" may be experienced. Of course this considerable inconsistency leads to poor extraction results due to thick flakes (poor extractability) and thin flakes (break down and lead to plugging and poor solvent percolation).

Very often a hot spot condition will be first noticed due to 'sheets' or 'ribbons' coming off the flaking rolls. Theses are in fact a more or less continuous stream of very thin flakes that are being pressed together while passing through the roll nip. If left unchecked the roll may eventually suffer from a spalling type failure due to the extreme loading of the roll at that point. In other instances the motor load may increase to the point that the motor trips out on an overload. While these hot spots sometimes appear, disappear or even 'move' for no obvious reason(s), the causes and preventions are in fact explainable and controllable.

Hot spots may be caused by :

Inconsistency in the feed overloading a portion of the roll face, i.e. a blockage in the feeder concentrating material in a small area of the roll nip

Lumps or agglomerations lodged in the roll nip blocking feed from and area of the roll and concentrating feed on adjacent areas

High roll ends

Failure to balance feed rate, roll pressures and motor loads

Roll diameter increasing due to warm up without 'relieving' the hydraulic closing mechanism

A flaking mill may be looked at in terms of a heat system. There are several sources of heat being added to the system; from the flaking mill motor, the conditioned material being put into the flaker and the frictional heating that occurs during the process, At the same time there are several avenues for heat to be dissipated form the flaking rolls. As the materials being processed are flaked, a tremendous increase in surface area is made.

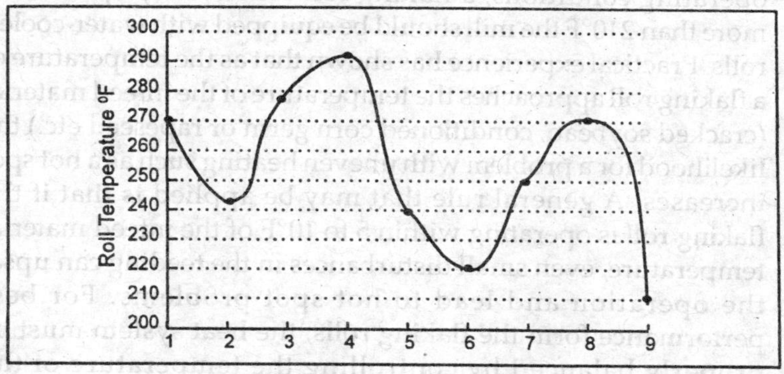

Fig. 5.5 Roll Temperature Readings

Because of the high temperatures involved in the flaking operation, some of the moisture in the products flashes off as it is lost to the atmosphere as steam. This changed state can be responsible for removing a significant proportion of the heat that is lost from flaking mill system. Additional losses occur to the atmosphere either by way of aspiration air being drawn through the mill and over the rolls or through the roll shafts and bearings and housings that act as heat sinks and radiators. Even the flakes themselves become a heat transfer medium when the rolls reach the steady operating condition, carrying much heat from the system and maintaining a general system balance.

Table 5.2 : Flaking Roll Temperature

Where does the heat come from?	Where does the heat go?
Flaking Mill Motor: 150 H.P. =.4 MBTU/Hour	Conversion of Moisture: Flashing .3% Moisture=.8 MBTU/Hr
Infeed Material: Cracked soybeans carry 0.8MBTU/H Frictional Heating	Atmosphere: Rolls, shafts, Bearings Radate 0.1 to .15 MBUT/Hr Flakes

According to roll manufacturers the maximum operating temperature of a flaking roll should be 210°F (100°C). If, due to operating conditions, a flaking roll consistently operates at more than 210°F the mill should be equipped with water-cooled rolls. Practical experience has shown that as the temperature of a flaking roll approaches the temperature of the infeed material (cracked soybean, conditioned corn germ or rapeseed etc.) the likelihood for a problem with uneven heating such as a hot spot increases. A general rule that may be applied is that if the flaking roll is operating within 5 to 10°F of the infeed material temperature, even small disturbances in the feeding can upset the operation and lead to hot spot problems. For best performance form the flaking rolls, the heat system must be properly balanced by controlling the temperature of the incoming materials, by insuring a uniform feed from end to end and by balancing the feed rate and roll pressure against the horsepower supplied.

Inconsistent Feeding

It is easy to see how a blockage in the feeder will create an uneven material flow into the nip of the flaking rolls.

Agglomerated materials from the conditioner or conveyors supplying the flaking mill occasionally become lodged in the feeder mechanism and disrupt the normal flow of product. Whether this blockage is in the feeder or in the nip of the rolls, the results are an increased flow of material on a small section of the roll face leading to increased roll temperatures and a decreased roll gap. In some cases an accumulation of materials on the magnets can also disrupt the normal flow of materials into the roll nip and initiate a hot spot. Obviously the corrective action for any of these conditions is to ensure a uniform flow of material in reaching the nip of the rolls from one end to the other.

Lumps or Agglomerations

The nature of the material being processed will have a significant influence on the likelihood of agglomerations and blockages occurring. High protein materials like soybeans are particularly prone to forming agglomerations and even more so when the cracked beans contain high levels of fines and moisture. A leaking steam tube in the conditioner is a common cause of agglomerated materials, and even more so in the case of soybeans that are cracked more finely. Other materials by their nature have a high percentage of trash, fines and or moisture that can lead to plugging and irregular material flow.

Obviously proper cleaning of the incoming material(s) and routine removal of any offending agglomerations is the only way to overcome these difficulties. In plants prone to generating agglomerations or passing trash to the flaking rolls, checking the feeder and flaking mill roll nip every hour may not be excessive. Every flaking should be checked at least one or two times per shift to make sure there is as uniform of feed stream as possible being presented to the flaking rolls.

High Roll Ends

High roll ends will also lead to high roll temperatures and thin flakes being produced at or near the roll ends. Regular checking of the roll temperature will allow the plant operations and maintenance personal to keep roll ends sufficiently relieved and eliminate this source of inconsistent performance. Since the failure to maintain roll ends has been thoroughly addressed in the preceding section, nothing more needs to be said concerning roll end maintenance at this point.

Balance of Feed Rate/Roll Pressure/Motor Load

Perhaps only 20 years ago flaking mills were limited by available technologies. Flaking rolls of approximately 20" diameter by 40" long (500mm x 1000 mm) operating at 150 to 200 RPM could process roughly 80 MTH of cracked and conditioned soybeans. This works out to be approximately 4.7 tons per day per square foot of roll working surface. With the advancement of roll manufacturing technologies and the availability of improved bearings and hydraulic loading mechanisms a 28" diameter by 52" long flaking roll operating at nearly 300 RPM today will process up to 280 MTD of cracked and conditioned soybeans. An additional benefit of the larger diameter roll was the ability to effectively process coarser material, reducing the need to produce very small pieces of cracked soybeans. The specific loading of these mills has increased to more than 8.8 tons per day per square foot of roll working surface. The need for significant improvements in roll management naturally accompanies this increased capacity.

While it may be possible to produce very thin flakes by running a low feed rate and high roll closing pressures, much of the applied horsepower will be transferred directly to the flaking rolls rather than to material being processed. As a result, in a short time the rolls will heat excessively and unevenly producing a hot spot which will prevent continued operation. Additionally, low feed rates run contrary to the rule of never operating at less than two-three of the theoretical capacity and may allow the rolls to run metal to metal further contributing to heating of the roll and potentially leading to damage of the roll surface.

The only way to produce consistent flakes is to balance the heat inputs to system (energy suppslied by the motor and heat in the incoming product) against the heat lost to the system (flakes, moisture loss, radiation to atmosphere). For every flaking mill there will be an operating range of feed rate/roll pressure/motor load within which a mill will operate successfully. Operating outside the range will result in the production of hot spots, inconsistent flakes and generally poor flaking mill performance. Generally speaking, flaking rolls will perform more consistently when operated at or near a full load since the flow of material through the rolls is greater, in creasing their ability to give up heat. Depending on the condition and size of the incoming materials, this may or may not be measured by a full motor load as the power required to produce flakes may be less than thê horsepower connected to the flaking mill. Suggested capacities for several popular sized flaking rolls are listed below.

While it is true that higher roll pressures (greater roll closing forces) will produce thinner flakes, there are limits to the rolls ability to give off the additional heat If the roll pressure is too high for the amount of material being processed; the rolls will begin to heat unevenly. Hot spots that develop as a result of low feed rates or excessive pressures normlally appear away from the roll ends and over a broad section of the roll initially. This is due primarily to the fact that the heat loss at the roll ends is greater than the heat loss from the middle of the roll. Left unchecked, the hot spot will tend to become more and more concentrated in a smaller and smaller area of the roll. Attempting to produce 'thinner flakes' by simply increasing the pressure on the flaking rolls without taking other operating characteristics into consideration will generally produce results far different than intended.

Table 5.3: Suggested capacities for popular sized Flaking Rolls

Flaking Roll Size		**Capacity - MTD**		
	H.P.	0.012"	0.016"	0.20"
20x40	75	65	80	100
24x48	125	210	270	330
28x52	125	280	340	400
28x62	150	330	390	450

While the motor load of a flaking mill is an important operating parameter, feed rate and roll pressures have a much more significant impact on the performance of a flaking mill. As the size and nature of the infeed materials change, the power required to flake those materials also change. As long as the feed rate (as determined by feed gate position or vibratory feeder controller setting) are correct and the roll pressure is sufficient to produce flakes of the required thickness, motor loading can be virtually ignored. Attempts to operate a flaking mill based on factors other than feed rate and roll pressure (i.e. motor load) while producing flakes of a given quality will be futile indeed. At the same time motor load information will offer a valuable insight as to the operation of the flaking mill. If the feed rate and roll pressures are within normal operating parameters a high amperage reading may indicate the presence of a hot spot. On the other hand, if feed rate and roll pressures are normal, low motor amperage may indicate a shortage of material to be processed.

Using the information concerning flake thickness measurements and roll temperatures will allow an operator to determine optimum flaking roll operating parameters and so achieve the best possible performance. High roll temperatures and thick flakes will normally be associated with high feed rates and corresponding high motor loads. High roll temperatures and thin flakes generally indicate insufficient feed either across the roll face, an uneven flow across the roll face due to some disruption of the feed stream, or an irregular face on the flaking roll, as in the case of high roll ends. As the flakes become thinner, capacity of the mill approaches the maximum theoretical limits and the temperature of the material being presented to the mill increases, the tendency for problems also increases. Even so, by understanding the relationship of feed rate, roll pressures, motor loads and flaking roll temperatures.

Roll Diameter Change

Accommodating Roll Diameter Change—All modern design flaking mill used in the oil seed extraction business today utilizes hydraulics to apply the roll closing forces. Hydraulic cylinders either close the rolls directly, or through a series of linkages and pivots. In any case, the controls to these hydraulic circuits must contain some system to prevent pressure spikes (high pressures developed instantaneously as an uncrushable object passed through the rolls, etc.) from being transmitted back through the pressure supply lines. This is particularly true of systems that employ one hydraulic power unit to supply several flaking mills and use black iron pipe to carry fluid under pressure. These pressure spikes can, ifnot isolated from the rest of the system cause failures of piping and fittings and in extreme cases could so over pressure a pump shaft as to cause a catastrophic failure. The use of check valves between the roll closing cylinders and controls in the hydraulic circuits effectively provides this isolation. At the same time, the check valves prevent the rolls from opening so long as the hydraulic pressure is applied. As flaking roll operate and the diameter changes, some provision must be made to accommodate this diameter change. In the case of rolls cooling down and decreasing in diameter due to reduced feed rates, lower pressures, or a drop in the temperature of the material being flaked the constant hydraulic pressure will continue to close the rolls and maintain the flake thickness.

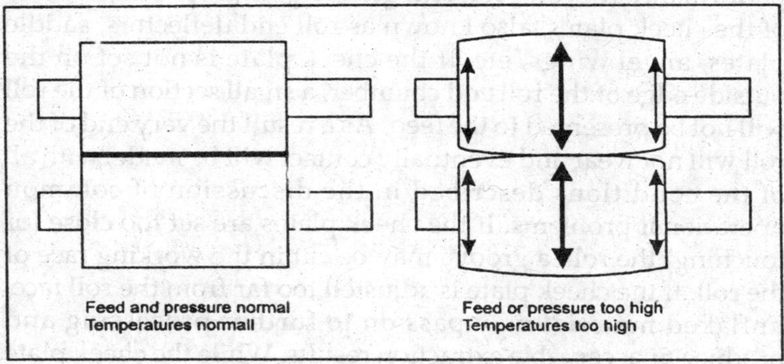

Fig. 5.6 Roll Profile and Temperature

In the case of roll heating up as when the temperature of the material being processed goes up or as the feed rate increases, the diameter of the rolls will increase. As noted above, the rolls cannot "back up" due to check valves in the hydraulic controls, so the roll gap becomes less and less. As a result, the flakes become thinner and thinner, the rolls get warmer and increase in diameter unless some provision is made to relieve the roll pressure and allow the rolls to back up. Simply cycling the roll pressure off and on again will in most cases dump the hydraulic pressure and bypass the check valves to allow the rolls to physically "back up". Reapplying the roll pressure then again establishes a roll gap (flake thickness) based on roll pressure and material passing the rolls, redistributing the load across the full face of the flaking roll and reducing the tendency for localized heating (hot spots) to occur. When roll pressures are decreased due to a reduction in the feed rate, it is generally beneficial to again cycle the roll pressure off and on again to allow the rolls to come to a normal operating position based on pressure and feed not being "held" in place by hydraulics in a closed system.

MISCELLANEOUS ITEMS

Cheek Plates

The most critical maintenance item that affects flaking mill performance (beyond roll end grinding) is proper adjustment of the cheek plates, also known as roll end deflectors, saddle plates, angel wings, etc. If the check plate is not set on the outside edge of the roll end chamber, a small section of the roll will not be presented to the feed. As a result the very end of the roll will not wear and eventually contact will be made, with all of the conditions described in the discussion of common operational problems. If the cheek plates are set too close (or touching) the roll, a groove may be cut in the working face of the roll. If the cheek plate is adjusted too far from the roll face, unflaked material may pass on to further processing and produce unacceptable extraction results. While the cheek plate itself is a fairly simple device and requires little routine maintenance, improper adjustment or service can have a dramatic impact on the performance of the flaking mill.

Scrapers

In many oil seed extraction operations, scraper blades are necessary to remove flakes that might otherwise stick to the roll face. Flakes sticking and wrapping around the roll can force the rolls apart, allowing unacceptable materials to pass on to the extraction process. Scraper blades are of particular importance when processing high oil or high moisture products that tend to stick more aggressively, such as corn germ, rapeseed and sunflower (meats). In some cases when flaking soybeans, scraper blades can be eliminated altogether. While scraper blades do perform an important function, they can also be a source of problems. The constant rubbing of the scraper against the roll may produce frictional heating. If any bit of mineral impurity lodges between the scraper blade and the roll, it is possible to cut grooves in the roll face. In the long run, scraper blades that perform best will be those that apply only as much force against the roll as required to keep the roll face clean.

Differential Speed Drives

Most flaking mills will be equipped with some type of drive between the rolls to maintain the desired differential roll speeds. The most common approach today is a series of double sided V -Belt and sheaves of varying diameters. The advantage of differential roll speeds are reduced roll pressures required to produce flakes, improved extraction of the flakes (more cell rupture), and promoting self-dressing of the rolls. If the differential speeds are not positively maintained not only will these benefits not be realize, but the maintenance requirement of the mill will be increased as well. For best long-term performance the belts must be aligned and tensioned, and every effort made to keep the operating environment clean.

Chapter 6

Vegetable Oil Processing

World oilseed production and processing is continuing to increase as the population grows and as societies become more urbanized. An aging, more health conscious, more educated, and more leisure-oriented society will have more time to discriminate between possible food choices. Technology development will be increasingly driven by health issues. This article will examine several recent technical trends that are occurring, largely in response to consumer demand but also due to rapid advances in technology.

MORE BYPRODUCT UTILIZATION

The last three years have seen an increasing recognition of the value of several molecules (Figure 6.1) that are wholly or partially removed from vegetable oil during the refining process. All of these molecules are sold in health food stores or drug stores in North America. Increasingly, it is likely that these molecules will be incorporated into functional foods or so-called designer foods. Table 6.1 summarizes six recent developments, which will likely increase the demand for vegetable oil by-products.

On April 7, 1998, the Food and Nutrition Board (FNB) of the National Academy of Sciences (USA) officially recognized choline as an important nutrient. The FNB sets Recommended

Table 6.1 Recent developments potentially affecting demand for byproducts of Vegetable Oil Refining

Recent Development	Date	Some New Products	Potential Consequence For Refiners
Establishment of dietary reference intake (DRI) for choline[1] (National Academy of Sciences, USA)	1998	• SMART Vitamins (C.E. Jamieson, Canada)	Increased demand for phosphatidyl Choline (lecithin)
First phosphatidylserine-containing products for memory enhancement[2, 3]			
a) Dietary supplement And	1995	• Leci-Ps™ (Lucas Meyer USA)	Increased demand for phosphatidylserine
b) Functional food product	1998	• Brain Gum™ (Gumm Tech International/Lucas Meyer, USA)	
First sterol-containing functional food product for cholesterol reduction[4, 5]	1995	• Benacol™ margarine (Raisio, Finland)	Increased demand for plant sterols
Recommendation to promote Vitamin E use for Alzheimer's disease[6] (American Psychiatric Association)	1997	–	Increased demand for Vitamin E

Continued

Study reporting that natural vitamin E is retained at least two times greater than the synthetic form[7]	1998	—	Increased demand for Natural Vitamin E
First tocotrienol-containing dietary supplement for cholesterol reduction[8]	1998	• NuTriene® Tocotrienol Concentrate (Eastman, , USA) • Tocomine (PORIM, Malaysia)	Increased demand for tocotrienols

Dietary Allowances (RDA's) and Dietary Reference Intake (DRI) for a variety of substances. Phosphatidylcholine or lecithin is the major source of choline in the diet. Lecithin is a vegetable oil by-product isolated from the gum fraction created by refining soybean. The FNB recommends 425 mg of choline daily for adult women and 550mg for adult men. Intakes of 450 and 550 mg for pregnant and lactating women, respectively, are recommended because of a high requirement for choline in growing fetuses and infants. Unfortunately, the richest dietary sources, of lecithin, such as eggs and organ meats, tend to be high in fat and cholesterol. The announcement of the new DRI for choline will likely increase the demand for lecithin.

In late 1995, Lucas Meyer Inc. introduced phosphatidylserine as a dietary supplement trademarked Leci-PS. Several human clinical trials including those by Cenacchi et al. (1993) and Crook et al. (1991) have proven the efficacy of phosphatidylserine in the improvement of age-associated memory impairment. Human clinical studies have also shown that phosphatidylserine improves the symptoms of patients with Alzheimer's disease. In 1998, Gumm Tech International launched Brain Gum, a gum that

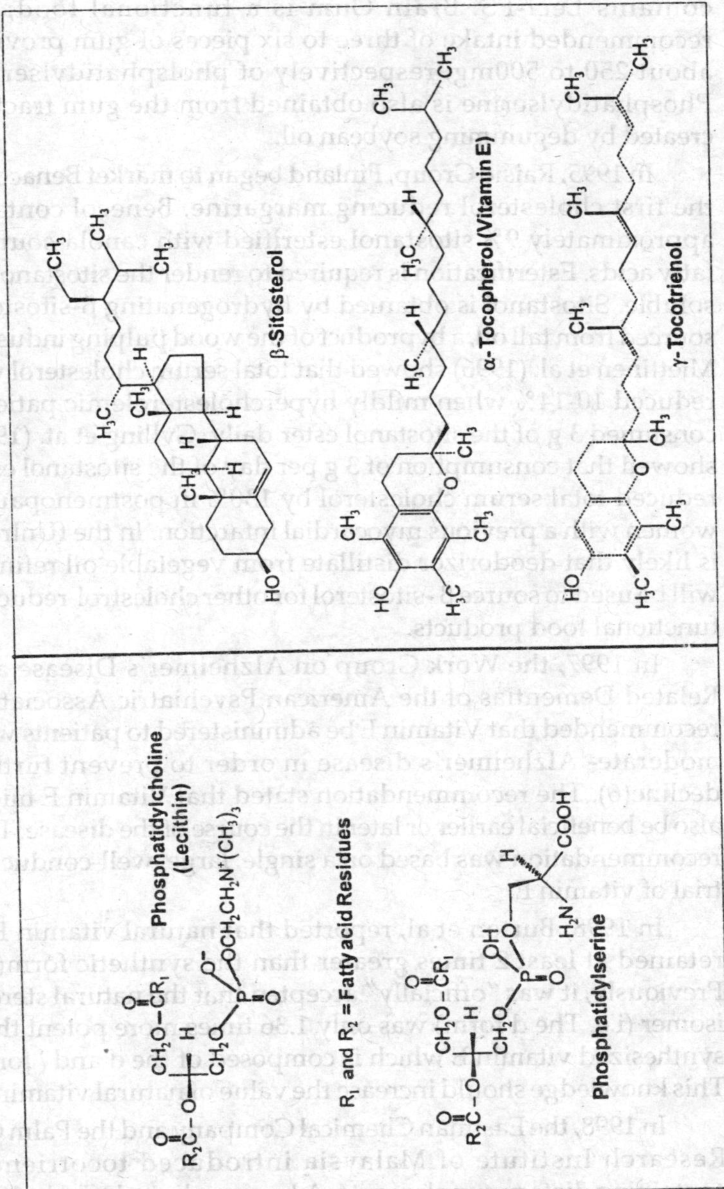

Fig. 6.1 Some Medicinal Compounds in Vegetable Oil Byproducts

contains Leci-PS. Brain Gum is a functional food; the recommended intake of three to six pieces of gum provides about 250 to 500mg, respectively of pholsphatidylserine. Phosphatidylserine is also obtained from the gum fraction created by degumming soybean oil.

In 1995, Raisio Group, Finland began to market Benacol™ the first cholesterol reducing margarine. Benecol contains approximately 9% sitostanol esterified with canola-sourced fatty acids. Esterification is required to render the sitostanol fat soluble. Sitostanol is obtanied by hydrogenating β-sitosterol sourced from tall oil, a byproduct of the wood pulping industry. Miettinen et al. (1996) showed that total serum cholesterol was reduced 10-14% when mildly hypercholesterolemic patients consumed 3 g of the sitostanol ester daily. Gylling et at. (1997) showed that consumption of 3 g per day of the sitostanol ester reduced total serum cholesterol by 130% in postmenopausal women with a previous myocardial infarction. In the fUnlre, it is likely that deodorizer distillate from vegetable oil refining will be used to source β -sitosterol for other cholestrol-reducing functional food products.

In 1997, the Work Group on Alzheimer's Disease and Related Dementias of the American Psychiatric Association recommended that Vitamin E be administered to patients with moderate- Alzheimer's disease in order to prevent further decline(6). The recommendation stated that Vitamin E might also be beneficial earlier or later in the course of the disease. This recommendation was based on a single, large, well-conducted trial of vitamin E.

In 1998, Burton et al, reported that natural vitamin E is retained at least 2 times greater than the synthetic form(7). Previously, it was "officially" accepted that the natural stereo-isomer (i.e. The d form) was only 1.36 times more potent than synthesized vitamin E which is composed of the d and *l* form. This knowledge should increase the value of natural vitamin E.

In 1998, the Eastman Chemical Company and the Palm Oil Research Institute of Malaysia introduced tocotrienol-containing dietary supplements. A human clinical trial in 1995 had shown that consumption of200220mg of tocotrienols for 4

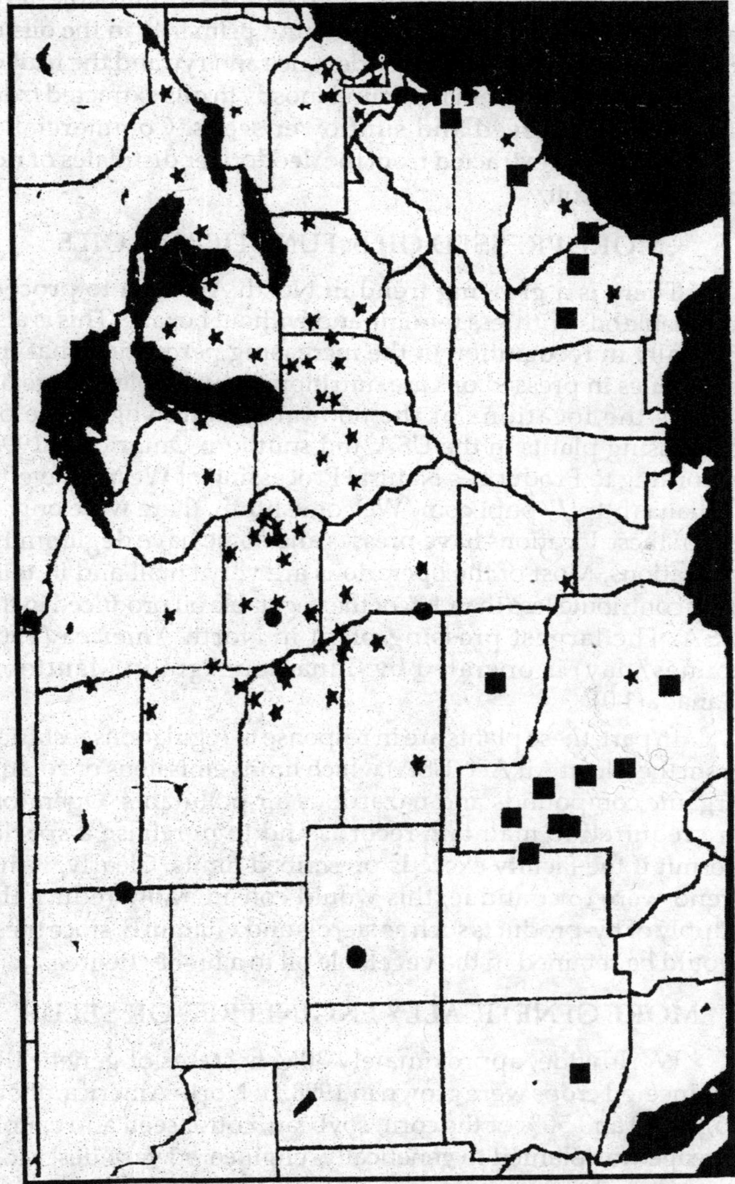

Fig. 6.2 Location [1,2] of Non-Hexane Vegetable Oil Processing Plants in the USA and Southern Ontario (⋆ soyabean; ▪ cottonseed; • other)

weeks in Hypercholesterolemic men reduces serum cholesterol by 10 to 13%(8). Tocotrienols are found primarily in the oils of cereal grains such as wheat, barley, rice and rye and the fruit of palm, whereas, vitamin E is found mostly in oils extracted from soybean, cottonseed and sunflower seeds. Commercially, tocotrienols are extracted from the deodorizer distillates of rice oil and palm oil.

MORE PRESSED OILS: FUNCTIONAL OILS

There is a growing trend in North America to process vegetable oils with less refining and without hexane. This is also partially in recognition of the increasing perception that the molecules in pressed oils are nutritionally valuable. Figure 6.2 shows the locations of the 86 non-hexane vegetable oil processing plants in the USA and southern Ontario. In 1995, according to Producer's Natural Processing of West Lafayette, Indiana (http://pnpi.com/Welcome.html), there were only 4. All of these locations have presses and some have degumming operations. Most of the operations are very small and in total they contribute less than 1 % of the vegetable oil produced in the USA. The largest pressing plant in North America (1500 tonnes/day) is operated by Canadian Agra in Manitoba, Canada(1 0).

In part, these plants are in response to regulations resulting from the Clean Air Act (USA) which limits emissions of volatile organic compounds and hazardous air pollutants. Operators are required to maintain records and to purchase a special permit if the facility exceeds prescribed limits. Clearly, if this trend were to continue, this would considerably reduce the supply of by-products such as sterols and vitamin E, since these would be retained in the vegetable oil to a higher degree.

MORE GENETICALLY ENGINEERED OILSEEDS

Worldwide, approximately 30M hectares of genetically engineered crops were grown in 1998. In North America, about 15, 30, 50 and 50% of the corn, soybean, cottonseed and canola acreage was planted to genetically- engineered varieties. Most engineered varieties have either been herbicide or insect tolerant. The beneficiary of this new technology to date, has largely been the producer and the chemical companies. This data clearly shows that genetic engineering has already had a profound effect on the production of commodity oilseed crops processed in conventional oilseed processing plants.

The adoption of genetically engineered crops in North America only took about three years; the rate of adoption of this technology was probably faster than any other in the history of agriculture. Farmers have adopted herbicide-tolerant crops because their crops have less dockage and their crop yields are higher.

Whereas, the first phase of genetic engineering in crops was centred on agronomic traits involving single genes, the next generation of genetically-engineered crops will be value-added varieties such as high laurate, high stearate, high omega fatty acids and high β-carotene vegetable oils, all of which are under development. The rapid introduction of the first wave of genetically engineered commodity crops is an indication of how quickly these niche, value-added crops could be introduced, if it is profitable to do so. Niche crops will likely be increasingly processed in smaller, pressing operations, possibly in agricultural processing complexes including chemical, cosmetic or other types of companies.

SUMMARY

Health foods containing vegetable oil by-products will be marketed, with limited claims, in retail grocery stores in the future. Consumers will favour obtaining their nutritional supplements as a part of their regular diet (i.e. functional foods) as opposed to the pill form. Secondly, at least in North America, there will be a continuing move away from questionable organic solvents in vegetable oil refining toward mechanical or other extraction methods. As consumers become more knowledgeable about the value of nutrients in pressed vegetable oils, there will be an increasing demand for this commodity. Thirdly, functional oils such as high β-carotene canola oil will begin to provide consumers some direct benefits from genetic engineering of crops. These niche, value-added crops will be segregated and then likely processed in smaller-scale pressing plants, perhaps located within multidisciplinary agricultural processing complexes.

Chapter 7

Sunflower Oil Processing

The processing of sunflower oil from a crude state to a consumer ready quality level is discussed in this chapter. The quality of the oil as it begins and leaves the refining process is discussed. Several systems of handling the oil at each stage to remove the non-quality materials are presented. A chart of the processes to be covered is shown in the figures.

CRUDE SUNFLOWER OIL

Sunflower oil is obtained from oil-type sunflower seed. The oil is extracted from the seed either mechanically (i.e., presses or expellers), with a solvent, or with a combination of mechanical and solvent extraction.

Crude sunflower oi l contains some lecithin, generally less than 0.5 percent, some color bodies, and other foreign material. The standards of the physical characteristics of sunflower oil are shown in Table 7.1.

DEGUMMING

Degumming is an optional process and is used to remove some of the lecithin and foreign materials which are present in the crude sunflower oil. Certain oil impurities such as phosphatides, protein or protein fragments, and gummy or mucilaginous substances, are soluble in oil only in the anhydrous form. The impurities can be precipitated and removed if they are hydrated.

Table 7.1 Crude Sunflower Oil Physical Characteristics

ITEM	VALUE
IODINE VALUE	120 - 142
SAPONIFICATION VALUE	188 - 194
UNSAPONIFIABLES (MG KOH/GM)	1.5 % MAX.
SPECIFIC GRAVITY 20°/20°C	0.918 -0.923
DENSITY @ 60°C	0.897
REFRACTIVE INDEX 40°C	1.467 - 1.469
TITER	16 - 20°C
FREE FATTY ACIDS	2.0 % MAX.
MOISTURE AND VOLATILE	0.5 % MAX.
COLOR - LOVIBOND (2"Cell)	Y =45, R=2, MAX.
FLAVOR AND ODOR	SWEET
FLASH POINT	> 121°C
INSOLUBLE IMPURITIES	0.3 % MAX.
PEROXIDE	10 MEG PEROXIDE/KG

Phosphatides and mucilaginous substances are emulsifying agents and in excessive quantity may increase the loss of neutral oil when the oil is caustic refined. If the oil is going to be physical or steam refined, the content of phosphatides and metallic prooxidants must be lowered by a degumming process before being refined. Water degumming generally removes about 90 per cent of the phospholipid material. There are, however, some non-hydratable phosphatides which do not react with water and are removed during caustic refining. Their presence tends to increase the neutral oil lost in the soapstock.

It is usually not economical to degum sunflower oil in a separate operation because the percentage of lecithin is so small. In addition to removing lecithin, degumming also removes some other foreign material. However, the quantity removed is usually so little that it can be easily removed during the alkali refining procedure. Additional equipment cost, labor expense and oil loss does not warrant degumming of sunflower oil if it is to be alkali refined.

NEUTRALIZATION OR REFINING

Neutralization removes nonglyceride impurities, which consist principally of free fatty acid (FFA), and significant quantities of mucilaginous substances, phosphatides, chlorophyll, and color bodies. The refining process removes the objectionable impurities in the oil with the least damage to either the glycerides, the tocopherols or other natural antioxidants.

Chemical (or caustic) refining converts the FFA to oil - insoluble soaps, removes all the substances, which become insoluble upon hydration, and removes some of the colour bodies present in the oil. Reactions are at a relatively low temperature, allowing for the refining equipment to be fabricated essentially from iron or mild steel.

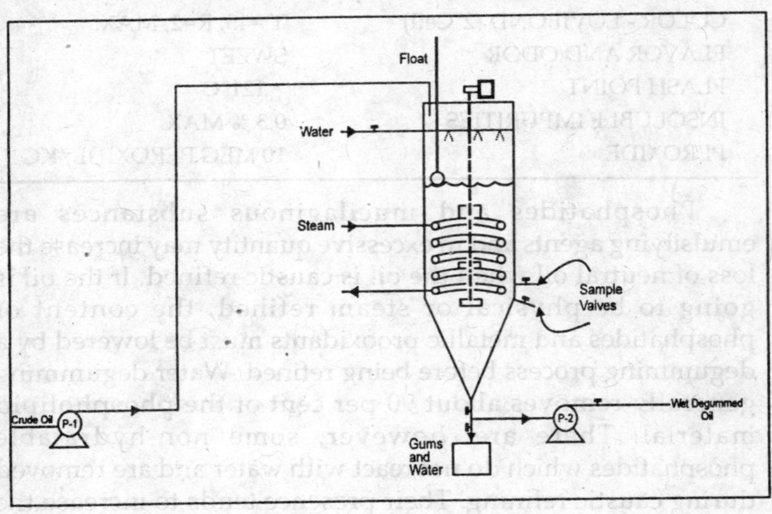

Fig. 7.1 Bath Degumming

The technology of caustic refining is concerned with: (a) the proper choice of alkali used, (b) the quantity of alkali used, and (c) the refining technique used to produce the desired purification (without excessive saponification of the neutral oil) and (d) the method used for efficient separation of the refined oil from the soapstock.

A disadvantage of the caustic refining method is the loss of a small portion of neutral oil, in addition to the FFA, as a result of saponification. This loss can be minimized, however, by utilizing good technology.

Sunflower oil can be alkali refined using either a batch method or a continuous method of refining. The batch method requires less capital investment, but; a) the yield of refined oil is considerably less than via the continuous method, b) the thruput rate is less and c) there is a higher labor cost.

Calculation of Caustic

Caustic calculation is the method of determining the quantity and concentration of the caustic solution required for alkali refining as it is applicable to both the batch and continuous methods.

The quantity of caustic required for neutralization can be determined in the following manner:

(1) Multiply the per cent FFA of the oil by 0.14286. This gives the per cent of solid sodium hydroxide required for neutralization.

(2) To this per cent, add a slight excess quantity, as the quantity calculated in Step 1 is the theoretical quantity. The excess usually ranges between 0.05 per cent and 0.20 percent, depending upon the per cent of FF A present in the oil and the quantity of non-oily constituents pre sent in the crude oil.

(3) Add the percent calculated in (1) to the desired excess established in (2) .

(4) The quantity of oil in pounds (or kilograms) is multiplied by the total per cent arrived at in (3). This gives the number of pounds (or kilograms) of solid sodium hydroxide required for the refining.

(5) The quantity of solid sodium hydroxide required is divided by the per cent of sodium hydroxide in the solution which is being used and this give the required quantity of caustic solution to be used in refining.

The per cent of excess caustic is very important as a large excess contributes to a high loss of refined oil. This is particularly true if the refining is being accomplished at a temperature of 71 °C (168°F) or higher. Some excess caustic is required however to obtain good soap formation and a minimum carryover of soap to the wash water centrifuge and the oil. It may be necessary to slightly increase the quantity of excess caustic in order to lower the soap content if the PPM (parts per million) of soap in the washed and dried oil is high.

Starting with an excess caustic of 0.10 - 0.15 per cent is recommended if the per cent FFA of the crude oil being refined is 0.7 - 1.2 per cent. The quantity of the excess caustic should be adjusted, if necessary, after the process has reached an equilibrium. If the per cent FFA in the crude oil is lower than 0.7 per cent, it may be possible to use an excess caustic of 0.05- 0.10 per cent. If the FFA are greater than 1.2 percent, the excess must be increased to between 0.15 per cent and 0.20 per cent. An excess of 0.25 percent is an extremely high quantity.

Caustic Calculation

As an example, assume a quantity of crude sunflower oil with FFA of 1.1 per cent and is refined at a rate of 15,000 Kg./hr. The following example shows how to calculate the required quantity of 16° Baume (Be′) caustic solution (11.06%):

$$\%\ \text{FFA} \times 0.14286 = \%\ \text{solid NaOH}$$

$$0.011 \times 0.14286 = 0.0015714$$

$$\text{Plus excess of} = \frac{0.0010000}{0.0025714}(0.1\%)$$

$$15{,}000\ \text{Kgs/hr.} \times .0025714 = \text{solid NaOH}$$

$$\frac{38.571\ \text{kgs. of solid NaOH}}{0.11.6(16^\circ\ \text{Be}')} = 384.74\ \text{Kgs. of } 16^\circ\ \text{Be}'\ \text{per hour}$$

It is suggested that a 16° Be′ caustic solution be used for refining sunflower oil. The concentration is not so high as to result in a significant amount of saponification of the neutral oil if the excess quantity was greater than the optimum. The Be′ of the caustic should be varied depending on various factors and conditions. If the per cent FFA of the oil to be refined is greater than 1.5 and the crude oil is relatively free from non-oily constituents such as gums and lecithin, it would be advisable to use a caustic solution having a Be' of 16.5-17.0. The reason for this is to avoid introducing an excessive amount of water which could inhibit the good separation.

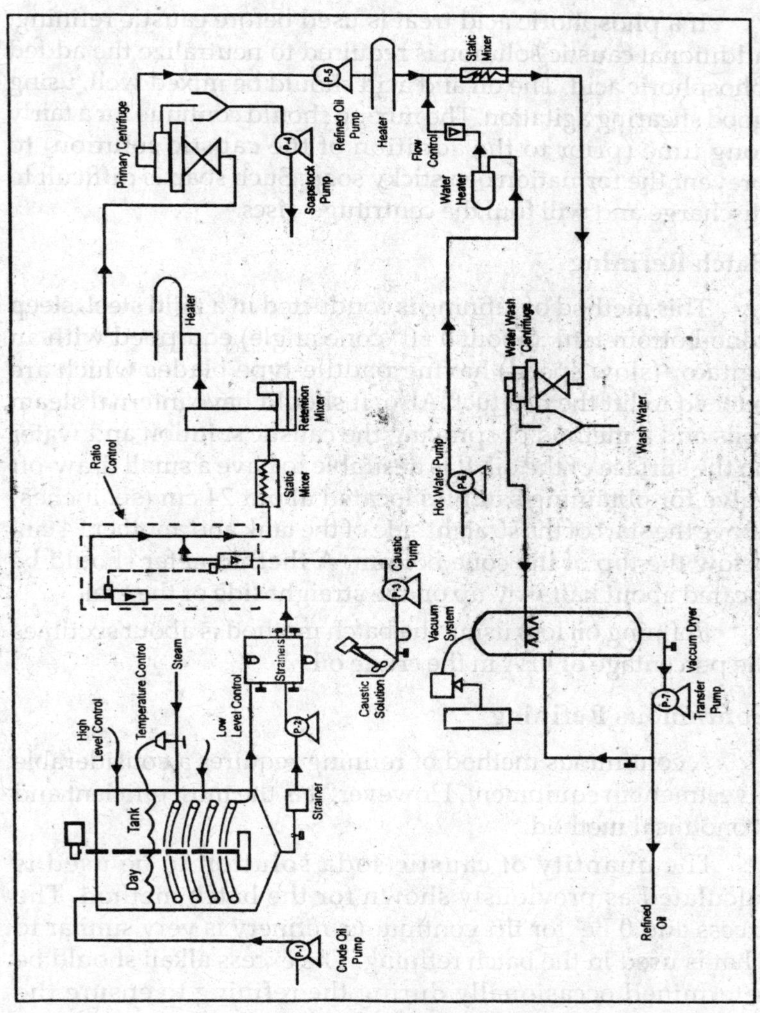

Fig. 7.2 Continuous Refining

If refining using the continuous method, adjust the quantity of caustic solution as the per cent FFA changes or the quantity of non-oily constituents in the feed stock change. In a continuous refinery a chart should be prepared which shows the quantity of caustic solution to be used as the per cent FFA changes during the run.

If a phosphoric acid treat is used before caustic refining, additional caustic solution is required to neutralize the added phosphoric acid. The oil and acid should be mixed well, using good shearing agitation. The mixing should continue for a fairly long time (prior to the addition of the caustic solution) to prevent the formation of a sticky soap. Such soap is difficult to discharge and will foul the centrifuge discs.

Batch Refining

This method of refining is conducted in a mild steel, steep cone-bottom tank (about a 60° cone angle) equipped with an agitator (slow speed) having paddle-type blades which are pitched to lift the product. Also, it should have internal steam coils and a method of spraying the caustic solution and water on the surface of the oil. It is desirable to have a small draw-off valve for obtaining samples located about 24 cm (six inches) above the start of the straight side of the tank and another 24 cm. below the top of the cone bottom. A thermometer should be located about half way up on the straight side of the tank.

Refining oil loss using the batch method is about six times the percentage of FFA in the crude oil.

Continuous Refining

A continuous method of refining requires a considerable investment in equipment. However, it is the most efficient and economical method.

The quantity of caustic soda solution to be used is calculated as previously shown for the batch method. The excess and 0 Be′ for the continuous refinery is very similar to what is used in the batch refining. The excess alkali should be determined occasionally during the refining to ensure the process is under control. It is desirable to determine the soap content after the water washing and vacuum drying. This is so small changes can be made in the refining conditions to minimize the soap level before further processing the oil.

Addition of a small amount of weak acid solution (e.g. citric acid) prior to vacuum drying may be desirable if the oil is going to be deodorized or further processed without decolorization with an acid activated clay. However, if the oil is to be bleached with bleaching clay, the acidity in the clay usually neutralizes the slight alkalinity of the oil.

The continuous method of refining permits varying the refining temperature considerably to obtain a good separation of the heavy phase using the centrifuge. A temperature as low as 43°C (110°F) has been used to produce a good quality refined oi l. A neutralization temperature of 57°C (135°F) gives both very good refined oil and good yield.

Refining Temperature

The caustic refining of vegetable oils is done at various temperatures. A temperature of 71°C (160°F) is quite often used. Some oils, however, are refined at temperature of 82°C (180°F). Usually, refining at a high temperature removes a little more of the colour bodies. However, a high temperature in conjunction with a high excess caustic causes too much oil loss due to saponification of the neutral oil.

The refining temperature is very critical in obtaining good yields. A compromise has to be made between removing more of the color bodies and obtaining a lower yield by using a higher refining temperature versus obtaining an improved yield using a lower temperature and achieving the desired color by decolorizing with an acid-activated clay.

Yields from continuous refining of relatively good grade of crude sunflower oil (no tank bottoms) at various temperatures have indicated that a refining temperature of 43°C(110°F) would produce a refining loss which would be approximately 1.42 times the per cent ofFFA. At a temperature of 57°C (135°F) the loss factor was 1.62, while at a temperature of 71 °C (160°F) the loss factor was 1.84. Thus, assuming an FFA content of 1.1 per cent, refining at 43°C versus 71°F saves 0.46 per cent oil.

Sunflower oil has been refined very successfully at a temperature of 57°C (135°F) using a continuous refining method. Also, high quality refined sunflower oil has been produced at a refining temperature as low as 43°C (110°F). The chlorophyll content of the oil refined at the 43°C (110°F) was somewhat higher than oil refined at 57°C (135°F) . After decolorization with acid activated clay, the chlorophyll was removed.

BLEACHING OR DECOLOURIZATION

After the oil has been refined it is usually bleached or decolorized to a specific colour. The bleaching process should be done under vacuum and at an approximate temperature of 77°C (170°F). The use of acid activated clay is recommended as the residual acid in the clay neutralizes the traces of soap that may be present. The chlorophyll removal is an important step as otherwise a green tint may appear in the oil after deodorization.

The oil should be dry before the addition of the clay if an acid activated clay is used. Any moisture present in the oil causes the acid to leach out of the clay, which causes an increase in the FFA content of the refined oil. If the filters are well blown, the oil loss in the bleaching process is about equal to the quantity of bleaching clay used in the process.

DEWAXING

The dewaxing process removes the waxy-type of the non-oily constituents that are not removed either by the alkali refining process or the degumming process. There are various methods, which can be used for dewaxing the oils such as a batch method or various types of continuous methods.

Batch Dewaxing Method

The batch method is relatively simple. It consists of an well-insulated tank of a given capacity, equipped with some means of agitation. A relatively slow speed mechanical agitator is preferred, however, it is possible to use a pump to circulate the oil from the bottom to the top. Under no circumstance should air be used for agitation of oil. Air bubbles will form in the oil which a) inhibit the heat transfer causing an extended processing time and b) increase the peroxide value of the oil.

The tank used for cooling the oil should also be equipped with a large number of relatively small diameter coils in order to give a large cooling surface. It would be preferable to have a vertical-type tank rather than horizontal, though either is

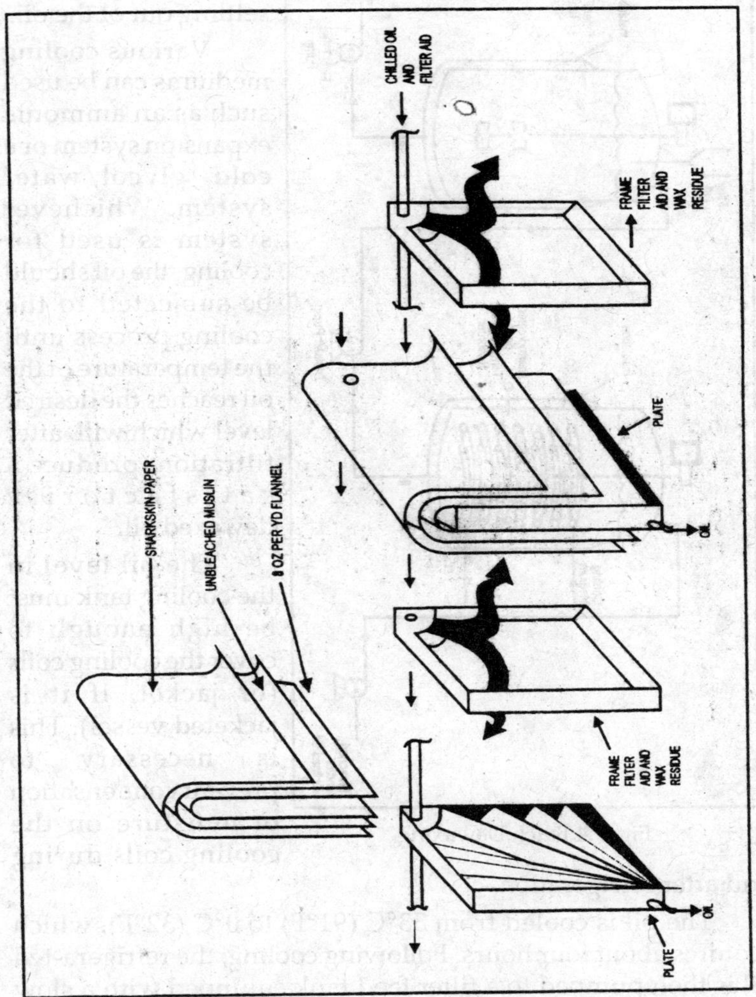

Fig. 7.3 Dressing the Filter Press- Dewaxing-

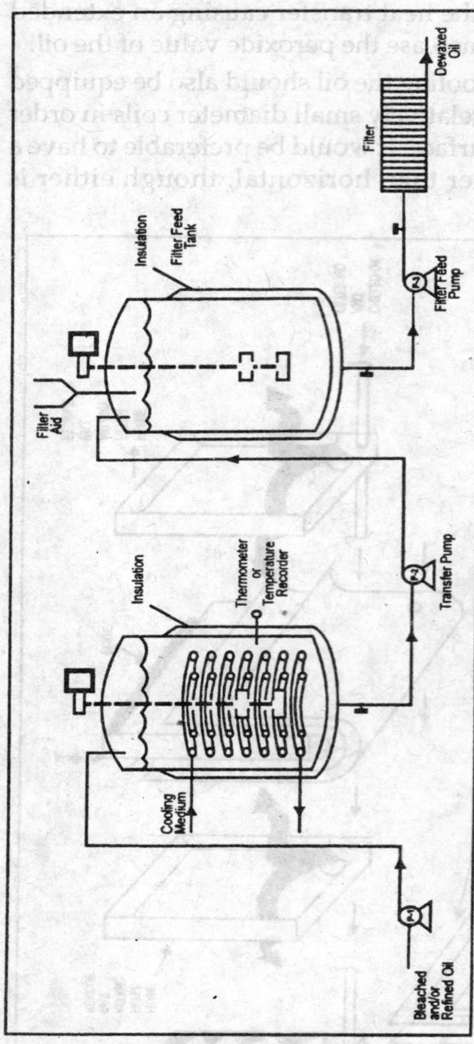

Fig. 7.4 Batch Dewaxing

acceptable. However, the agitation should move the oil rather slowly across the cooling coils, yet it should be enough to keep the waxes from settling out of the oil.

Various cooling mediums can be used such as an ammonia expansion system or a cold glycol/water system. Whichever system is used for cooling, the oil should be subjected to the cooling process until the temperature of the oil reaches the desired level which will, after filtration, produce a satisfactorily dewaxed oil.

The oil level in the cooling tank must be high enough to cover the cooling coils (or jacket, if it is jacketed vessel). This is necessary to prevent condensation of moisture on the cooling coils during and after refrigeration.

The oil is cooled from 33°C (91°F) to 0°C (32°F), which requires about four hours. Following cooling, the refrigera-ted oil is then pumped to a filter feed tank equipped with a slow speed agitator. There, about 1/4 - 1/2 per cent of filter aid base on the weight of oil is added to the oil. When the filter aid is thoroughly mixed with the oil, the mixture is pumped through a filter.

Various filters may be used, but the simplest that can be recommended is a plate and frame filter press. The press should

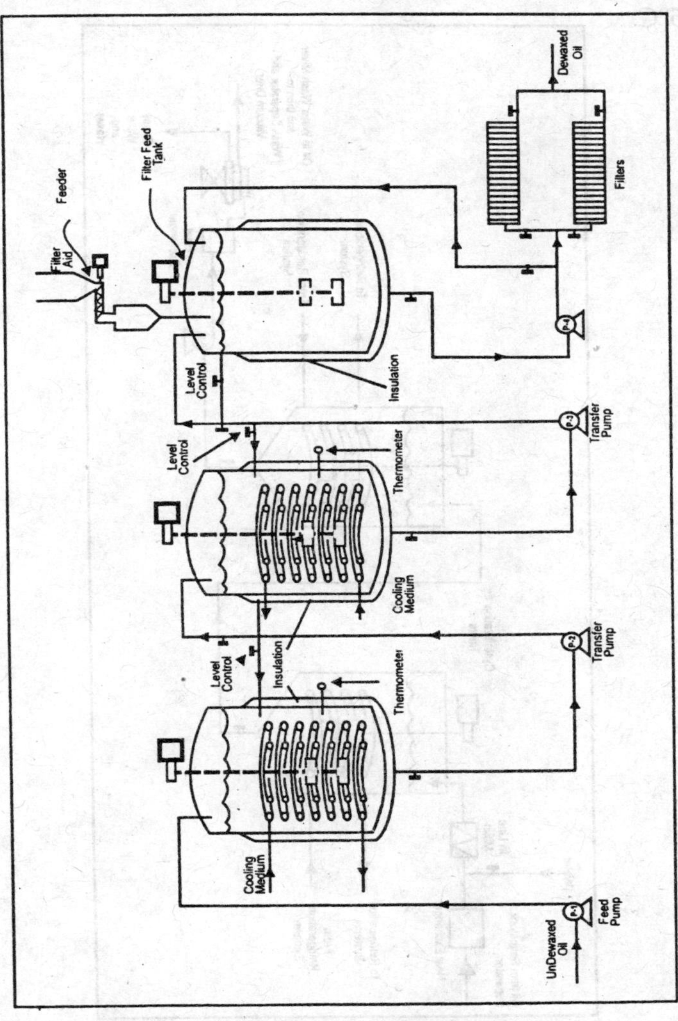

Fig. 7.5 Continuous Dewaxing

be dressed with flannel, muslin, and a sharkshin-type paper. If the press is so dressed the loss of refined oil during dewaxing is approximately equal to the quantity of filter aid used in the filtration.

The refined and dewaxed oil, using the above procedure, has a chill test of over 36 hours at 0°C and over 120 hours at 4.5°C.

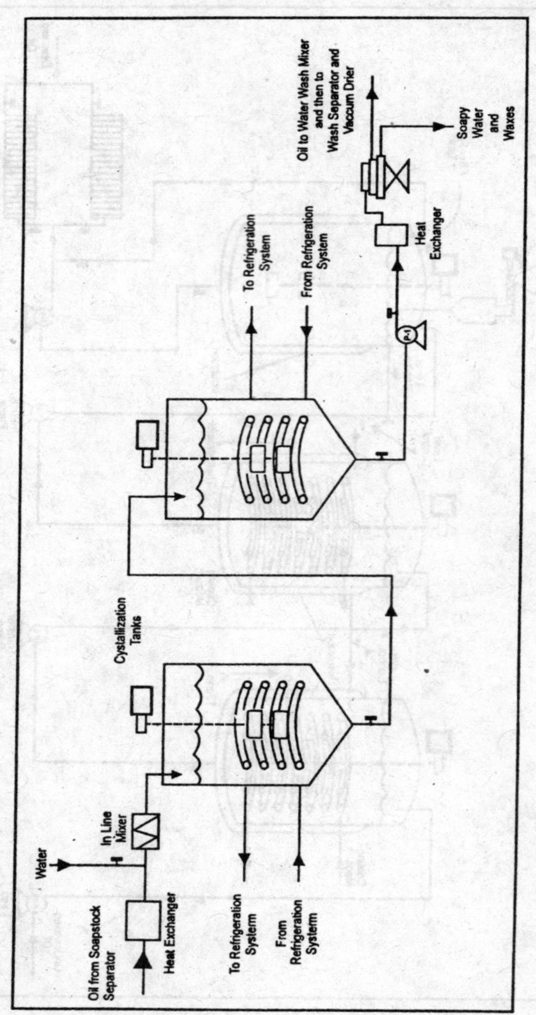

Fig. 7.6 Continuous Dewaxing (Using a centrifuge)

Continuous Dewaxing Method

A continuous system may use multiple tanks or a continuous chiller. The tanks should be insulated. The oil cools progressively as it is transferred from one tank to the next. The residence time in the system is long enough to permit reducing the temperature of the oil to the desired temperature before it is pumped from the final tank.

The chilled oil is then pumped to a filter feed tank where the filter aid is added and then pumped through a filter to remove the filter aid and waxes.

Alternatively, a continuous process may use a continuous chiller. With this process, the oil being chilled passes through a continuous pipe that has a cooling medium on the outside of a jacketed pipe. The flow rate is controlled so that the temperatume of the oil is at the desired temperature when discharged from the end of the continuous pipe. When this system is used, it is desirable to discharge the refrigerated oil to an insulated, agitated lank. This residence time in the tank permits agglomeration of the waxy material. Agglomeration of the material permits easier filtering. Filter aid may be added in the residence tank. The oil is then pumped through a filter press to remove the waxy material and the filter aid. Residence time should be about one hour before filtration.

Continuous Dewaxing Using a Centrifuge

Some caustic refining systems may incorporate a dewaxing process. The basic equipment for this process consists of heat exchangers, crystallization tanks, a cooling or refrigeration system, pumps, and a centrifuge. The degree of dewaxing possible by using this precess is questionable. However, if the requirement for the cold test of the oil is not stringent [stringent being 0°C(32°F) for 48 hours and 4.5°C (4°F) for 120 hours minimum], the continuous dewaxing process might be more advantageous than the lower temperature dewaxing of the refined oil. The low temperature crystallization, 0°C (32°F), and filtering of the refined oil will produce a better cold test oil and will meet the stringent requirements as stated above.

Chapter 8

Palm Oil Industry

COMPOSITION

Palm oil is obtained from the flesh surrounding the seed, simply by cooking, meshing and pressing. In this process the seeds are separated and after cracking and removing the shell, the kernel can be processed to yield palm kernel oil. Palm kernel oil forms about 10 percent of the total oil yield and has different composition and uses from palm oil.

As regards to fatty acid composition, palm oil occupies an intesmediate position among natural fats. Its saturated fatty acid content is 50 percent and the rest is monounsaturated and polyunsaturated. By virtue of the fatty acid composition, palm oil is semi solid containing mainly palmitic, oleic and linoleic acids, which are the most common components of all food fats. Palm kernel oil, on the other hand is rich in lauric acid and is interchangeable with coconut oil in most of its applications but is more flexible because it is more unsaturated compared to coconut oil.

Crude palm oil, besides its fatty acid composition also contains 1 percent minor components – A diagrammatic presen-tation of crude palm oil minor components is given in Fig. 8.1.

CAROTENOIDS AND TOCOLS

Carotenoids form a class of C40 polyunsaturated hydrocarbons (carotenes) and their oxygenated derivatives (xanthophylls). These compounds impart a rich orange-red colour to the oil.

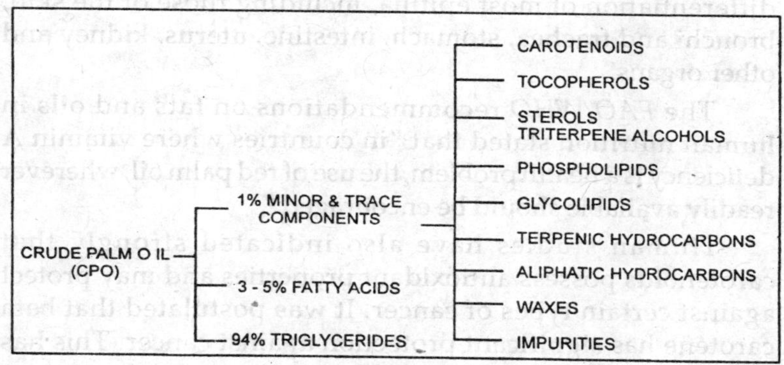

Fig. 8.1 Constituents of Crude Palm Oil

Crude palm oil is one of the richest natural plant sources of carotenes with the concentration of 500-700 ppm. It has 15 times more retinol equivalents than carrots and 300 times more than tomatoes. Analysis shows that alpha and beta carotenes constitute approximately 90% of the total carotenoid content and the rest are delta carotenes, gamma carotene, phytofluene, phytoene, zea-carotene, lycopene, neurosporene, alpha and beta zeacarotenes.

The production of palm oil from Malaysia is projected to reach over 8 million tonnes by the year 2000. Hence, the potential availability of carotenes in the year 2000 is about 4800 tonnes.

The tocopherols and tocotrienols are found in crude palm oil at 600-1000 ppm concentrations. The main constituents are 44% gamma tocotrienol, 22% alpha tocopherol, 12% delta tocotrienol, the rest being alpha and beta tocotrienols, gamma and delta tocopherols.

NUTRITIONAL BENEFITS

In human nutrition, carotenoids play important role as a

source of provitamin A. Beta-carotene, especially, is the most nutritionally active carotene as provitamin A. It is converted in the gastrointestinal tract into vitamin A. Vitamin A of course, plays an important role in regulating vision growth and reproduction. It is essential for the normal cellular differentiation of most epitilia, including those of the skin, bronchi and trachea, stomach, intestine, uterus, kidney and other organs.

The *FAO/WHO* recommendations on fats and oils in human nutrition stated that "in countries where vitamin A deficiency is a health problem, the use of red palm oil, wherever readily available should be encouraged."

Human studies have also indicated strongly that carotenoids possess antioxidant properties and may protect against certain types of cancer. It was postulated that beta carotene has sigpificant protection against cancer. This has sparked much interest and numerous reports have now appeared to shed new light on this question. An inverse association between serum beta carotene and the risk of squamous cell carcinoma of the lung has been found. It was also reported that beta carotene inhibited oral tumors when applied topically or given as a dietary supplement.

The antitumor effects of palm carotenes had been studied on the developments of chemically-induced skin cancer and viral-induced lymph cancer in *HRS/J* female hairless mice. Palm carotene was found to possess antitumorigenic effects and the inhibition of subcutaneous lymphoma was concentration dependent: 0:3% palm carotenes (PC>>.1 % PC>0.05% PC. All the palm carotenes group inhibited. epithelial pappilloma formation as well.

Vitamin E functions primarily as an antioxidant, especially to prevent oxidation and peroxidation of polyunsaturated fatty acid units of membrance phospholipid (within and on the plasma membrane of cells). This works against injury to cell membranes as in red blood cell fragility in man. It is also implicated in preventing photooxidative damage to the lens and thus in preventing cataract formation. The antioxidant properties of vitamin E have been implicated to prevent/ delay the onset of various degenerative diseases such as aging, cancer, arthritis, damage to cells caused by air pollution and also in preventing coronary heart diseases.

The nutritional benefits of palm vitamin E (palm vitee) has been studied on human as well as animal models. The hypocholesterolemic effect of palmvitee has been studied by various researchers. It was found in one study that the concentrations of serum total cholesterol (-15%), LDL cholesterol (-8%), apoprotein B(-10%) and thromboxane (–25%) decreased significantly in 15 hypercholesterolemic subjects fed 200mg palmvitee capsules/ day. Another study on 9 healthy volunteers taking one capsule of palmvitee perday also showed a reduction of serum cholesterol (reduction rage -5% to -35.9%) and LDL cholesterol (reduction range 0.9% to 37%).

The effect of palm oil vitamin E on isolated Langendorff perfused rat hearts has also been studied and results indicated that palmvitee eontaining a mixtue of tocotrienols and tocopherols may be more efficient than alpha tocopherol alone in the protection of the heart against oxidative stress induced by ischemia-reperfusion.

Large population studies have analysed the relationship between antioxidant nutrients such as carotenoids and vitamin E and degenerative diseases. A continuing study among 22,000 male physicians in the US yielded an interim finding that those taking a supplement of beta carotene reduced the occurrence of major cardiovascular events by almost 50%. It was reported that an inverse correlation between plasma vitamin E levels and mortality from ischemic heart disease in a cross cultural epidemiology.

CAROTENOIDS

Sources of Carotenoids

Various analytical methods have been used for the determination of the carotenoid profile of crude palm oil. Column chromatography has been used in earlier studies using different adsorbents and reverse phase *HPLC* has been shown to have several advantages for the separation of carotenoids in the oil.

Table 8.1 Carotenoid contents of Palm Oil Extracts

Palm Oil Extract	Carotenoid Content (ppm)
Crude palm oil	630-700
Crude palm olein	680-760
Crude palm stearin	380-540
Second pressed oil	1800-2400
Residual oil from fiber	4000-6000
Total carotenoids estimated at 446 nm	

Table 1 shows the carotenoid contents of various palm oil extracts. A carotenoid rich oil can be obtained from palm pressed fiber, a palm oil by product which is presently burnt as a fuel in palm oil mills. The pressed fiber was found to contain about 5-6% of residual oil with a carotenoid concentration of 4000-6000 ppm, and the carotenoid content of the residual oil in the pressed fiber of the hybrid oil palm fruit is even higher, at 6000-7000 ppm.

Carotenoids from the commercial crude palm oil are concentrated during the extraction and fractionation process. A system of palm oil extraction based on a double pressing technique has been implemented by several mills in Malaysia. The expected advantages of double pressing over the conventional single-stage pressing are lower oil loss in fiber, higher kernel extraction rate, less wear on the screw worm and cages, and reduction of contamination of the kernel oil with crude palm oil. More interestingly, the oil produced from the second pressing has a higher concentration of carotenoids. This could be due to the fact that the first pressing in a double pressing process is carried out at a lower pressure (to avoid cracking of the nut) and oil relatively higher in carotenoids from the second pressing is extracted. After removal of the nuts the fiber is then subjected to a higher pressure and more carotenoids are extracted out of the mesocarp, together with the residual oil from the first pressing.

Carotenoids are also being concentrated in the industrial process called fractionation. Palm oil is a semi-solid fat at ordinary room temperature. Fractionation is carried out to extend the uses of palm oil and the products obtained are the liquid oil (olein, 70-80%) and the solid fat (stearin, 20-30%). The carotenoid content in the palm olein (lower melting) fraction is enriched by 10-20%, as shown in Table 8.1.

Recovery and Extraction of Carotenoids

The carotenes on the market at present are chemically synthetic beta carotene, carotene extracted from carrots and carotene extracted from algae Dunaliella. The carotenoid composition of the various carotenes are shown in Table 8.2. The demand for natural carotenes is such that customers are willing to pay a premium for these, which they value for their perceived safety. Fig. 8.2 shows the major commercial uses of beta carotene in the global marketin which they are widely used as a food colorant.

Table 8.2 Carotenoid compositions of various Carotenes

	alpha-C	beta-C	Others
Carrot carotene	30-50%	40-60%	10-15%
Synthetic beta-carotene		100%	
Dunaliella carotene	6-10%	84-90%	
Palm oil carotene	30-35%	60-65%	5-10%

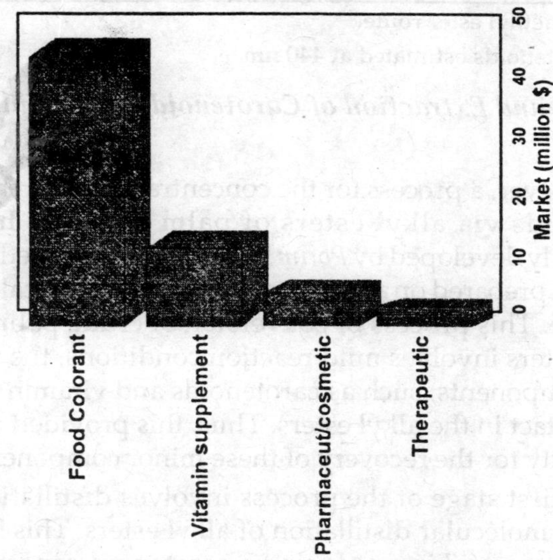

Fig. 8.2 Major commercial uses of beta carotene.

Because carotenoids are likely to grow in importance and value, the recovery of carotenoids from palm oil and its by-products is important. In addition, some studies are also being carried out to recover and concentrate palm-based carotenoids for pharmaceutical purposes.

Numerous extraction methods have been developed to recover the carotenoids from crude palm oil (Table 8.3). Most of the reported methods are difficult, costly or inefficient.

Table 8.3 Carotenoid concentrates from various methods

Method	Carotenoid content (ppm)	Recovery (%)
Vacuum distillation*	>20,784	<46
Molecular distillation*	>80,000	>80
Adsorption*		
C18	8000-9000	>90
Carbon	5000-7000	<50
Molecular distillation of crude palm oil	1290-1990	
Adsorption from crude palm oil (activated carbon)	3700-5600	<80

* through methyl ester route

Total carotenoids estimated at 440 nm

Recovery and Extraction of Carotenoids via Alkyl Esters Route

However, a process for the concentration and recovery of carotenoids via alkyl esters of palm oil route has been successfully developed by *Porim*. Volatile palm oil methyl esters have been prepared on a large scale as an oleochemical or diesel substitute. This process of conversion of crude palm oil into methyl esters involves mild reaction conditions, the valuable minor components, such as carotenoids and vitamin E would remain intact in the alkyl esters. Thus, this provided a unique opportunity for the recovery of these minor components.

The first stage of the process involves distillation steps including molecular distillation of alkyl esters. This has been proven successful in producing a carotenoid concentrate of

>80,000 ppm from both the laboratory and pilot plant scale. The major carotenes in the concentrate are alpha and beta carotene. The second stage involves column chromatography with an absorbent to give a carotene fraction. The concentration of this carotene fraction ranged from 90-100% (8.4 & 8.5). The carotene fraction can then be diluted to produce the desired commercial carotene concentrate in various concentrations (1-30%) for various applications.

Table 8.4 Carotene Content

Palm oil	600 ppm
Alkyl Esters	600 ppm
Carotene Concentrate	10,000-80,000 ppm
Carotenes	>90%

Lion Corporation of Japan has made efforts to recover carotene from palm oil since the 1970s. They have developed a process which is also via the methyl ester route (Fig. 8.3). Crude palm oil is converted into methyl esters by transesterification with methanol at low temperature which results in little loss of carotenes, if any. In the second step methyl esters are mixed in an organic solvent, and then the mixture is separated into a carotene rich layer and a decolourizsed methyl ester layer. The methyl ester is used for oleochemical purposes while the carotene may be further concentrated from the carotene rich layer by ordinary means. Finally, the carotene is purified by liquid chromatography with hexane. The palm oil carotens contains more than 95% of carotenoids (60-70% of beta carotenes, 30-35% alpha carotenes and small amounts of gamma and delta carotenes).

Carotech has also developed an integrated process for recovery of carotenoids and tocotrienols from crude palm oil, again via the alkyl esters route. Fatty acids in a vegetable oil are subjected to alcoholic esterification to form an ester-rich layer including fatty acid alkyl esters, carotenoids and tocotrienols. The ester rich layer is exposed to solvolytic micellization to form a carotenoid rich layer. The ester rich layer is separated from the carotenoid-rich layer. The carotenoids in the carotenoid-rich layer are concentrated and can be adsorptively separated from the carotenoid-rich layer. Fatty acid alkyl esters are separated from the ester-rich layer to form a tocotrienol-rich layer.

Table 8.5 Carotene composition (%)

Carotene	Concentrate	Crude Palm Oil
Phytoene	1.5	1.3
Phytofluene	0.3	0.1
Cis- β -Carotene	0.9	0.7
β - Carotene	49.9	56.0
α - Carotene	33.3	35.1
Cis- α -Carotene	5.5	2.5
γ -Carotene	1.7	0.7
ζ -Carotene	1.3	0.3
δ -Carotene	0.6	0.8
N eurosporene	0.1	0.3
β -Zeacarotene	1.3	0.7
α -Zeacarotene	0.4	0.2
Lycopene	3.4	103
Total (ppm)	80560	673

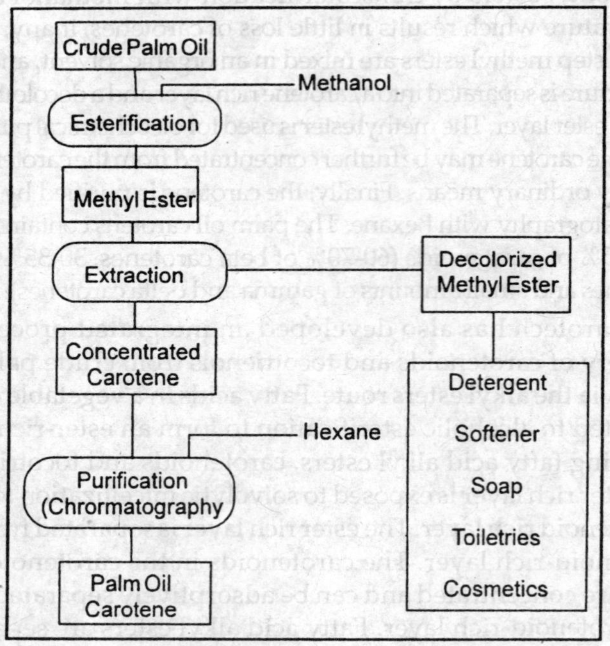

Fig. 8.3 Lion Procedure for Producing Palm Oil Carotene

Red Palm Oil

A process has also been developed by *Porim* to produce deacidified and deodorized red palm oil from the degummed palm oil with more than 80% of the original carotenes still intact as well as the vitamin E. This red palm oil is of similar excellent quality to the refined, bleached and deodorized (RBD) palm oil that is normally traded but contains no carotenes.

Table 8.6 Carotene composition (%) of deacidified and deodorized Red Palm Oil and Crude Palm Oil

Carotene	Red Palm Oil	Crude Palm Oil
Phytoene	2.0	1.3
Phytofluene	1.2	0.1
Cis- β -Carotene	0.8	0.7
β - Carotene	47.4	56.0
α - Carotene	37.0	35.1
Cis- α -Carotene	6.9	2.5
Carotene	1.3	0.7
ζ -Carotene	0.5	0.3
γ -Carotene	0.6	0.8
N eurosporene	trace	0.3
β -Zeacarotene	0.5	0.7
α -Zeacarotene	0.3	0.2
Lycopene	1.5	1.3
Total (ppm)	545	673

The process involves two stages i.e. pretreatment of the crude palm oil followed by deodorization and deacidification by molecular distillation. The pretreatment is carried out in a conventional manner using phosphoric acid followed by bleaching earth. This allows impurities and oxidative products in the crude oil to be removed. The deodorization and deacidification is carried out using molecular distillation unit at temperatures <165°C and a pressure of 20- 35 X 10–3 Torr.

Analysis by HPCL shows that the profile of carotenes are not destroyed during the process (Table 8.6) and the quality of red palm oil has also been found to be good (Table 8.7). Sensory evaluation carried out on red palm oil showed that it is of very evaluation carried out on red palm oil showed that it is of very good quality and comparable to RBD palm oil.

Table 8.7 Quality Parameters of Red Palm Oil

Carotenes	>80% intact
Tocopherols & Tocotrienols	>80% intact
Free fatty acids	<0.1%
Peroxide value	<0.2
Phosphorus content	< 2 ppm
Moisture & impurities	<0.1%

The commercial production of red palm oil involves the capital investment in a molecular distillation unit, which includes a vacuum system. The other costs involved are the power supply, water supply for cooling and storage tanks. The cost of production, based on a 5-year investment plan is given below:

Plant capacity (T/yr)	10,000	30,000
Pretreatment (RM/T)	17.00	17.00
Molecular distillation (RM/T)	99.00	96.30
Total cost of Red Palm Oil (RM/kg) (includes cost of crude palm olein)	1.2529	1.2374

Unitata has also developed a process to produce carotene enriched red palm oil. This is a chemical process which involves chemical neutralisation of the free fatty acids, followed by pretreatment, then deodorization using conventional method to remove the odour. The quality of the product obtaind from this process is also similar to the RBD palm oil, containing not less than 250 ppm of carotenes and 800 ppm of vitamin E. The oil produced has been marketed which bears the tradename of Nutrolein.

TOCOLS

Source of Tocols

Table 8.8 shows the various sources of tocopherols from palm oil and its products. Tocopherols accumulated in the palm fatty acid distillate *(Pfad)* after physical refining can be as high as 0.8% or greater so that the *Pfad* has been looked upon as a good source of natural tocopherols and tocotrienols. In fractionation or fractional crystallizations tocopherols tend to remain in the liquid or olein fractions thus rendering the solid fractions deficient in antioxidants.

Table 8.8 Tocopherols in Palm Oil (ppm)

Crude Palm Oil	600-1000
Crude Palm Olein	800-1000
Crude Palm Stearin	250-350
PFAD	150-8500
RBD Palm Oil	356-630
RBD Palm Olein	468-673
Palm Kernel Oil	(80-100)*

The vitamin E content of palm oil and other edible oils is shown in Table 8.9. Palm oil has an abundance of tocotrienols as compared to other oils.

Palm Vitamin E Isolation

The palm fatty acid distillate *(Pfad)* contains 0.4%-0.8% tocopherols and tocotrienols with the tocotrienols present as a major component. The bulk of the *Pfad* is free fatty acids.

The concept of isolating the tocopherols and tocotrienols (palm vitamin E) is based on the conversion of free fatty acids, partial glycerides and triglycerides into methyl esters after removal of which by distillation a concentrate of palmvitee is obtained. Purification of the concentrate will yield tocopherols and tocotrienols.

The purification of the crude mixture is effected by crystallization during which sterols are separated, followed by ion exchange chromatography, in which the tocopherols and tocotrienols are separated from squalene. Palmvitee is then washed dried and purified by molecular distillation. Finally it is deodorized before being dissolved in super olein prior to encapsulation.

The process flow of isolation of palmvitee is given in Fig.

Table 8.9: Total Tocopherol and Tocotrienol of Oils and Fats

	Tocopherol					Tocotrienol					
Oil and fats	α-T (ppm)	β-T (ppm)	γ-T (ppm)	δ-T (ppm)	%T	α-T3 (ppm)	β-T3 (ppm)	γ-T3 (ppm)	δ-T3 (ppm)	%T3	Total T+T3 (ppm)
Palm Oil	256	-	316	70	55%	143	32	286	69	45%	1172
Palm Oil	279	-	61	-	31%	274	-	38	69	69%	1081
Palm Oil	152	-	-	-	17%	205	-	439	94	83%	890
Soyabean Oil	101	-	593	264	100%	-	-	-	-	0%	958
Safflower Oil	387	-	174	240	100%	-	-	-	-	0%	801
Cosm Oil	112	50	602	18	100%	-	-	-	-	0%	782
Cottonseed Oil	389	-	387	-	100%	-	-	-	-	0%	776
Sunflower Oil	487	-	51	8	1 00%	-	-	-	-	0%	546
Peanut Oil	130	-	216	21	100%	-	-	-	-	0%	367
Cocoa Butter	11	-	170	17	99%	2	-	-	-	1%	200
Olive Oil	51	-	-	-	100%	-	-	-	-	0%	51
Coconut oil	5	-	-	6	31%	5	1	19	-	69%	36
Palm Kernel Oil	13	-	-	-	38%	21	-	-	-	62%	34
Lard	12	-	7	-	73%	7	-	-	-	27%	26

8.5. The concentration of vitamin E at different stages of the process is given in Fig. 8.5.

This pilot plant is capable of producing vitamin E with the following results:

purity	95-99%
overall yield	65%
production	5-6 kg/batch

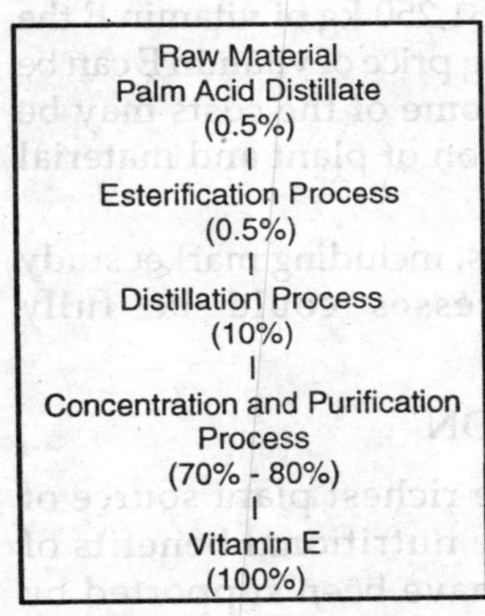

Fig. 8.4 Concentration of Vitamin E at Different Stages

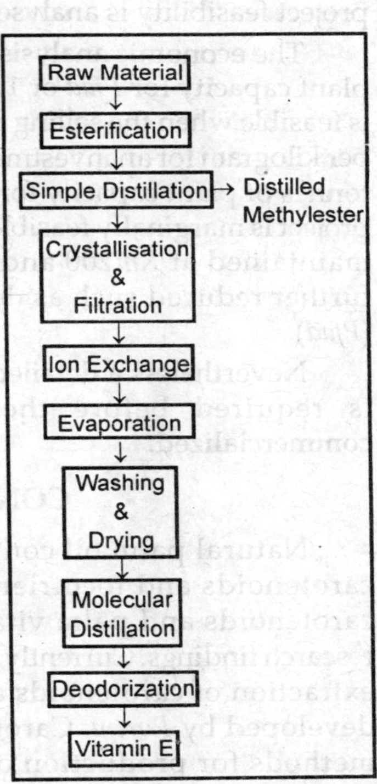

Fig. 8.5 Vitamin E pilot plant production process

ECONOMIC VIABILITY EVALUATION

It has been shown that it is technically feasible to produce tocopherols and tocotrienols from *Pfad* and preliminary economic assessment on the process has been attempted.

Vitamin E extraction is a concentration process from the

0.5% vitamin E content of the feed material, *Pfad* to 99% purity. There are 8 processing steps, and based on two capacities of 25,000 tonnes and 50,000 tonnes *Pfad* per year, the plant investment amounts to *Rm* 49.86 and *Rm* 75.50 million respectively (1989 estimates).

The material, utility and labour costs are estimated on the pilot plant experience. The effect of variation of selling prices on project feasibility is analysed.

The economic analysis indicated that for 50,000 tonnes of plant capacity for *Pfad* or 1,62,500 kg of vitamin E, the project is feasible when the selling price is maintained at about RM200 per kilogram for an investment of75.5 million ringgit. For 25,000 tonnes of plant capacity for *Pfad* or 81,250 kg of vitamin E the project is marginally feasible if selling price of vitamin E can be maintained at *Rm*200 and more, some of the costs may be further reduced such as depreciation of plant and material (*Pfad*).

Nevertheless a detailed analysis, including market study is required before these processes could be fully commercialized.

CONCLUSION

Natural palm oil contains the richest plant source of carotenoids and tocotrienols. The nutritional benefits of carotenoids and palm vitamin E have been supported by research findings. Currently, three commercial processes for the extraction of carotenoids are available, that is the process developed by *Porim*, Carotech and Lion Corporation. Two methods for production of carotene-enriched palm oil is available, one by *Porim* and the other by Unitata. The technology for isolation of palm vitee (the tocopherols and tocotrienols) from *Pfad* is ready for commercialisation.

Chapter 9

Deodorization of Oils and Fats

Deodorization is the final processing step in the refining of edible oils and fats. The process was developed to meet the growing demand of odorless and tasteless oils and fats as ingredients for the manufacture of margarine and to remove the unpleasant flavor from alkali refined cottonseed oil. Henry Eckstein was the first to introduce steam deodorization process in the United States. However, David Wesson brought the current deodorization process using the combination of high vacuum and high temperature.

NATURE OF THE PROCESS

Deodorization is primarily a high temperature, high vacuum steam distillation process to remove volatile, odoriferous materials present in the edible oils and fats. These odoriferous compounds are stripped from the relatively nonvolatile oil because of the differences in the volatility between triglycerides and the flavor compounds. The thermal treatment occurring during the deodorization process also heat bleaches the oil by destruction of carotenoids at elevated temperatures. The flavor and oxidative stability of the oil increases by removal of odor and flavor compounds and by thermal decomposition of peroxides.

Deodorization is essentially a steam distillation process and steam is used as carrier due to following reasons:

- It is cheap.
- It can be easily obtained.

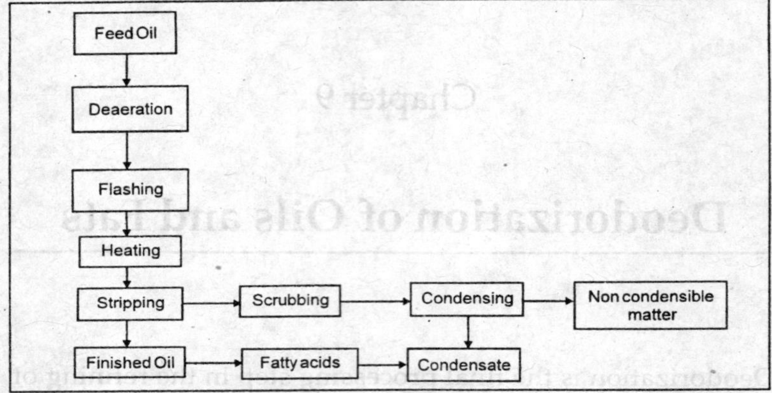

Fig. 9.1 Different steps involved in the deodorization process

- It can be easily condensed.
- It is rather less reactive with the oil.

The total deodorization process comprises a number of processing steps as given below.

- Deaeration of the oil.
- Heating of the oil to deodorization temperature.
- Steam stripping of the oil.
- Cooling of the oil.

All the above steps are carried out under vacuum. The separation steps comprise one or two individual deaeration and stripping step. A series of operations carried out for creating the vacuum and handling and recovery of the deodorizer distillate. The stripping operation is basically a purification of the liquid oil phase by mixing a stripping gas with the oil, which facilitates mass transfer of the volatile compound to the gas phase. The gas phase is continuously removed from the system thus preventing the volatiles from reentering the liquid. Fig. 9.1 illustrates the different steps involved in the deodorization process.

FLAVOR AND ODOR COMPOUNDS

The substances that impart taste and smell to the oils and fats are called flavor and odor compounds. These compounds are strongly soluble and tenaciously retained in the oils and fats and hence require separate process for their removal. The concentration of flavor and odor compounds rarely exceeds 0.1 % the weight of the oil. These flavor and odoriferous substances may belong to following two groups:

1. Substances, which are responsible for the characteristic odor and taste of the oils and fats and are therefore, present in the fresh fat from undamaged raw material.
2. Substances arising from the deterioration of the oils and fats through improper storage, transport and processing.

Flavor substances belong to above groups are aldehydes, ketones, alcohols, hydrocarbons and various other compounds formed by the heat decomposition of peroxides and pigments. In addition to above, oils and fats also contain some flavorless substances such as sterols and tocopherols. Part of these substances is also removed during the deodorization process.

Most of the odoriferous compounds have vapor pressures of similar order as those of the fatty acids and are therefore removed during the distillation process along with fatty acids.

THEORY OF STEAM DISTILLATION

Mass transfer from the liquid to the gas phase of a certain volatile component is driven by the difference between the equilibrium partial pressure and the actual partial pressure present at the spot of transfer. The equilibrium partial pressure is a function of the concentration of the volatile component in the oil phase. The actual partial pressure is determined by the resistance to mass transfer occurring in the system. The stripping component must be immiscible with the oil liquid and be present as gas at the deodorizing conditions in order to get maximum effect of stripping gas. These conditioys are ideally met by water vapor i.e., steam.

When a liquid is heated, the most volatile components tend to distill first, leaving behind the less volatile components. If the system is put under vacuum, the order of removal is the same but distillation occurs at a lower temperature because vapor pressure is proportional to temperature and less heat is needed to bring the vapor pressure upto the pressure in the system. The introduction of stripping steam increases vaporization and further reduces the temperature for deodorization.

The process of steam distillation by stripping steam under vacuum is governed by the laws ofRaoult and Dalton. By applying these laws, we obtain following basic formula for distillation.

$$S = \frac{PO}{EP_V A} \ln \frac{V_1}{V_2}$$

S = Moles of Steam

O = Moles of oil

P = Total system pressure at which distillation takes place

P_v = Vapor pressure of volatile components

E = Vaporization efficiency

A = Activity coefficient

V_1 = Initial concentration of volatile component in the oil

V_2 = Final concentration of volatile component in the oil

The vaporization efficiency is expressed as the ratio of actual partial pressure to the equilibrium vapor pressure of the volatile component. In practice, coefficient E represents the degree of saturation of steam with the volatile component during its passage through the oil layer. It is evident that higher the vaporization efficiency, the lower the quantity of steam required for reaching the same distillation effect.

The equation (I) simply states that amount of steam required for deodorization (or the time required at a definite steaming rate) is

- directly proportional to the size of the oil batch

- directly proportional to the total system pressure
- inversely proportional to the vapor pressure of the volatile components
- inversely proportional to the vaporization efficiency
- directly proportional to the logarithm of the ratio of initial to final concentrations of volatile components.

Table 9.1 Deodorization Time vs Temperature

Temperature	Time (minutes)						
[°C]	120	60	30	15	8	4	2
210	O	O					
230		O	O				
250		O	O				
270				O			
290					X		
310						X	
330							X

For continuous deodorization, Gavin has rewritten equation (I) using time dependent variables as

$$\frac{V_1}{V_2} = 1 + (K / P).p_v (S / O)$$

where V_1, V_2 represent moles/hour of FFA in the oil entering and leaving the deodorizer. K is an Experimental constant, and P, p_v S and O are as shown previously.

INFLUENCE OF DIFFERENT OPERATING VARIABLES

Influence of Temperature

The temperature of deodorization is the most important physical variable to be chosen with utmost care. From thermodynamic point of view, the logarithm of the vapor pressure of a volatile component is approximately proportional to its absolute temperature. Thus if a given increment in temperature will double the volatility of flavor substances, a further equivalent increment will quadruple their volatility and so on. This in turn reduce the amount of stripping steam required for deodorization. For example, the vapor pressure of linoleic acid increases from 32 mm Hg at 250°C to 67 mm Hg at 270°C. This would double its rate of removal during deodorization and for similar substances, the deodorization time would be halved for each increase in temperature of about 20oe. At high deodorization temperature, thermal decompositionof some constituents, naturally present in the oil such as peroxides, pigments and complexes, takes place and we obtain a light colored product. The work of Moser showed that there is a definite relationship between temperature and time of deodorization as shown in Table 9.1 where each 'O'. is from data and each 'X' is extrapolation.

However, operation at very high temperature may initiate undesirable side reactions like isomerization and polymerization etc. in the oils and fats and may convert natural cis-isomer into trans-isomer to some extent. These undesirable side reactions are related to time and temperature *of* deodorization but independent of composition of oil to be deodorized. Moreover, the natural antioxidants such as tocopherols and tocotrianois present in the oils and fats are destroyed/reduced by high deodorization temperature thereby reducing the oxidative stability of the finished product.

Influence of Vacuum

Another important physical variable to be optimized is the absolute pressure or vacuum under which the deodorization is carried out. The absolute pressure must be sufficiently low to make the vapor pressure of the volatile compounds sufficiently high to be distilled off from the oils & fats. Theoretically, the amount of stripping steam is reduced by lowering the absolute pressure in the deodorizer. A high vacuum in deodorizer leads to economy in operation, more rapid deodorization, low FFA content in the product and improved oxidative stability of the oil. For deodorization process to be economical, the absolute pressure should be in the range of 4 - 6 mm Hg. Table -2 shows boiling point of different fatty acids at different absolute pressures.

Table 9.2 Boiling Point of Fatty Acids

Absolute pressure (mmHg)	Palmitic acid (°C)	Stearic acid (°C)	Oleic acid (°C)
100	270	290	285
40	244	263	255
10	210	228	220
4	192	209	205
2	179	193	190
1	167	183	175
0.5	154	170	162
0.1	132	148	140

OIL LOSSES IN DEODORIZATION

The amount of odoriferous compounds to be distilled in deodorization is small and in the range of 0.1% of the weight of the oil. However, total loss of the material is considerably high due to unavoidable loss of other materials. The main source of oil loss in deodorization process is attributed to following three factors:

1. Compounds distilled from the oil. It include fatty acids, flavor compounds, small amounts of the oil itself, sterols, tocopherols and other unsaponifiable matter.
2. Oil droplets entrained in the exit vapors. This mechanical entrainment of oil by the stripping steam is the main source of oil loss in many deodorizer installations.
3. Losses due to the hydrolysis of oil to fatty acids by reaction with stripping steam.

The loss of material depends upon the time, steaming rate, pressure and temperature of deodorization as well as the amount of FFA and unsaponifiable material in the oil. The deodorization loss is more in batch process as compared to semicontinuous and continuous operation. It may vary from 0.30% to as high as 1.0% depending upon the processing conditions employed in the process and the design of the deodorizer.

METHODS OF DEODORIZATION

Three methods of deodorization are currently in commercial use:

1. Batch process
2. Semicontinuous process
3. Continuous process

Batch Deodorization

Batch deodorization is used mainly in plants of low capacity where small quantities of oil must be processed separately without intermixing or where the cost of continuous or semicontinuous deodorization does not justify its use. The conventional batch deodorizer is a vertical cylindrical vessel of welded construction with dished or conical ends and a height two to three times the diameter. The stripping steam is injected into the bottom of the vessel to bring the steam into close contact with the oil, through a distributor which usually consists of a flat spider of perforated pipes radiating from a central steam delivery pipe. While the steam passes up through the oil, volatile molecules are transferred across the surface area into the gas bubble, thereby reducing its content in the oil. Stripping is continued until the product meets the specifications. Deodorizer must be well insulated to avoid loss of heat by radiation. With equipment operating at optimum temperature and a 6-12 mm Hg absolute pressure, about 8-10 hrs is usually required for the complete cycle of charging, heating, deodorizing, cooling and discharging. Heating and cooling of the oil is generally accomplished by internal pipe coils. The principal advantage of batch deodorization system is its simplicity. It can be operated for as long as short a period as desired. However, the cost of utilities for the operation of a batch deodorizer is much higher than the cost of utilities for a semicontinuous deodorizer or continuous deodorizer and hence limits its use for deodorization purposes in refining of edible oils.

Semi Continuous Deodorizer

Semi continuous deodorizers operate on the basis of handling finite batches of oil in a timed sequence of deaerating-heating, holding-steam stripping and cooling such that each quantum of oil is completely subjected to each condition before processing to the next step. These plants are normally automated and controlled from a central panel with time cycle controller and interlocks such that the sequence steps are interrupted in the event of insufficient batch size, improper drop valve opening or closing or the oil not reaching the present heating or cooling temperatures in the allotted time. The main advantages of the semicontinuous deodorization are:

1. Better quality of the product due to identical treatment for all of the oil.
2. Frequent stock changes are possible.
3. No intermixing of oil between stock changeover.
4. Minimum loss in production in stock changeover.

The main disadvantages of the system are:

1. Less heat recovery.
2. High operational cost.
3. Due to frequent stock changeover, the temperature gradient in some parts of unit can generate stresses on welded parts and it could lead to failure from metal fatigue.

The first semicontinuous deodorizer was developed by the Votator company in 1948. It consists of a tall cylindrical shell of carbon steel. The shell consists of five trays of stainless steel construction. Fig. 9.2 shows flowsheet for a Votator semicontinuous deodorizer with heat recovery. Oil to be deodorized is pumped to a measuring tank. With the help of an automatic valve, a batch of measured quantity is charged into the top deodorizer tray where it is heated and deaerated under vacuum. After giving predetermined residence time, internal drop valves open in sequence and oil is subjected to deodorization temperature, holding, steam stripping, heat recovery and cooling operations in second, third, fourth and fifth deodorizing trays respectively. The cooled deodorized oil is discharged from fifth tray into a drop tank situated beneath the deodorizer. From drop tank, final oil is pumped through a polishing filter to next operation or storage.

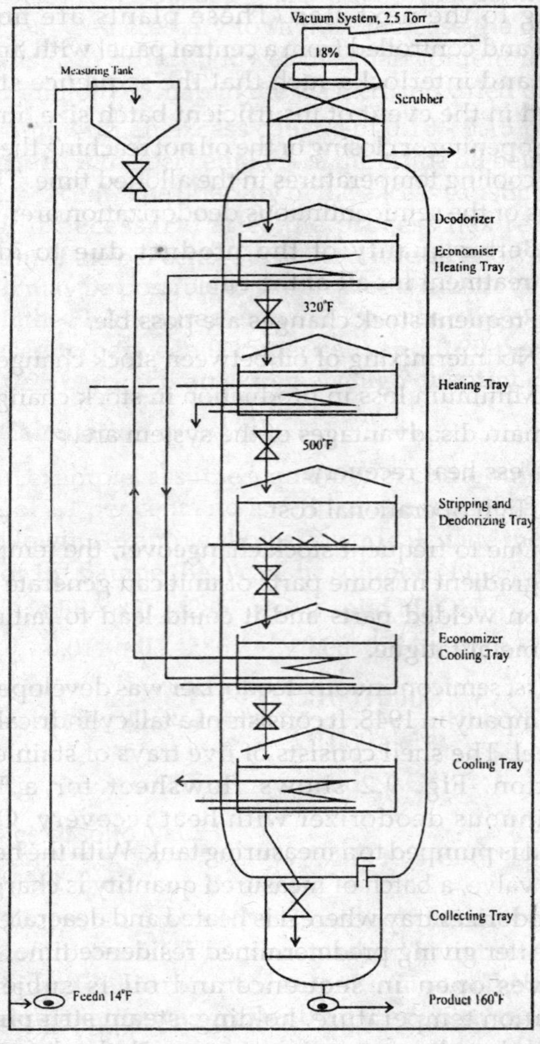

Fig. 9.2 Semicontinuous deodorizer

Continuous Deodorizer

Continuous deodorizers operate on the principle that oil flows continuously through the various stages of deaeration, heating of oil to deodorization temperature, holding, steam stripping, heat recovery and cooling during the deodorization process. The advantages of a continuous deodorizer are:

1. Simple to operate
2. Uniform utility consumption
3. Better heat recovery (about 80-85%)
4. Good quality of the product.

However, the main disadvantages of continuous deodorization are:

1. Any breakdown/stoppage will result in higher production loss
2. Whole plant has to be stopped for maintenance purposes
3. Higher initial capital cost of the system

All deodorizer designs are characterized by a very short contact time between the stripping gas and liquid, compared to the total stripping time, resulting in practically unchanged composition of the liquid phase during its contact with a specific gas molecule. In commercial deodorization, two principles of flow pattern are relevant: cross flow and countercurrent flow. In cross flow, flow directions of the phases cross each other wherein countercurrent pattern, liquid and vapor phases flow in opposite direction. In principle, it is true for all the designs that lower the pressure and higher the deodorization temperature, the more successful will be the removal of volatile compounds from the oil. Continuous deodorizers vary in their design and operations depending upon the way of contact between the vapor and liquid phases. Broadly, continuous deodorizers are of two types:

- Tray type deodorizers
- Thin film deodorizers

Tray deodorizers are based on a series of steam agitated trays or compartments usually vertically stacked in a cylindrical shell. Deaeration, heating of oil and stripping of free fatty acids and other volatile compounds are carried out simultaneously under vacuum. The retention time usually varies from 10 to 30 minutes. Liquid levels of typically 0.3 to 0.8 meter are maintained by overflow pipes or weirs. The trays are drained by separate discharge valves. Thin film deodorizers require less energy for driving the vacuum system. The term 'thin film' refers to the way the oil is distributed throughout the column over a large surface area. This large surface area is provided by a sort of packing or just by the wall of a pipe. The oil typically flows downward by gravity and the steam is injected at the bottom of the column.

Further, deodorizers may be of single or double shell design. In single shell deodorizers, the outside shell surfaces are exposed to atmosphere whereas in double shell deodorizers, it is common to hang multiple open trays inside a pressure shell and subject the entire assembly to an vacuum.

The main advantages of double shell type deodorizers are:

- Better quality of the product due to lower risk of exposing the hot oil directly to air in case of a leakage in the shell.
- Better economy of the operation due to lower heat losses or insulation requirements.
- Comparative lower capital cost as trays can be made from expensive high grade stainless steel whereas shell can be made from carbon steel.

The disadvantages of double shell design are:

- In double shell deodorizers, it is difficult to observe, inspect and service the internals as compared with single shell design.
- For high FFA feedstock, the cost of the system will be high as internal parts and shell both are fabricated with expensive high grade steel.

The single shell deodorizers have certain advantages over double shell design which are given below:

- The most important advantage of single shell deodorizer is its easy serviceability.

- Better control of the process as things can be seen through sight glasses mounted on critical positions.
- Ease in cleaning and maintenance of the equipment.

However, the main disadvantages of single shell design are:

- Greater risk of oxidation of the oil in case of leakage through the shell.
- Less economy in operation due to high cost of insulation to avoid heat loss in the atmosphere.

A number of continuous deodorizers have been developed over a period of time. They are extensively used for deodorizing edible oils. Some of the important designs are described here.

EMI Double Shell Continuous Deodorizer

Fig. 9.3 is a schematic diagram of EMI double shell deodorizer. Feedstock is pumped into the external deaeration vessel maintained under vacuum. The deaenited feedstock flow is regulated by a liquid level controller and pumped into an internal heat recovery tank where it is preheated in counter current fashion by the outgoing hot oil. About 50% heat recovery is achieved in this tank. The oil then flows by gravity to the heating tray. The oil is heated here to deodorization temperature by a thermal heater using Dowtherm or other fluid as heating media. The tray is continuously sparged with steam to avoid localised heating of the oil. After reaching the deodorization temperature, the oil flows by gravity to a stripping tower. Here the oil is deodorized under vacuum by passing the oil in a thin film over a series of stripping trays counter current to the flow of stripping steam which is injected into bottom of the section. After prestripping, the oil flows continuously down to the holding tank in order to provide required retention time for the thermal treatment of the oil. This tank is also steam sparged continuously. In final stripping, the oil from the holding tank is again deodorized in another series of stripping trays to remove any flavor compounds released during the holding period. The oil then flows by gravity into the heat recovery tank where it preheats the incoming oil for maximum heat recovery. From here, oil flows by gravity to a cooling tank where the product is cooled by flowing water in the coils. The cooled deodorized oil is then pumped for storage or next operation.

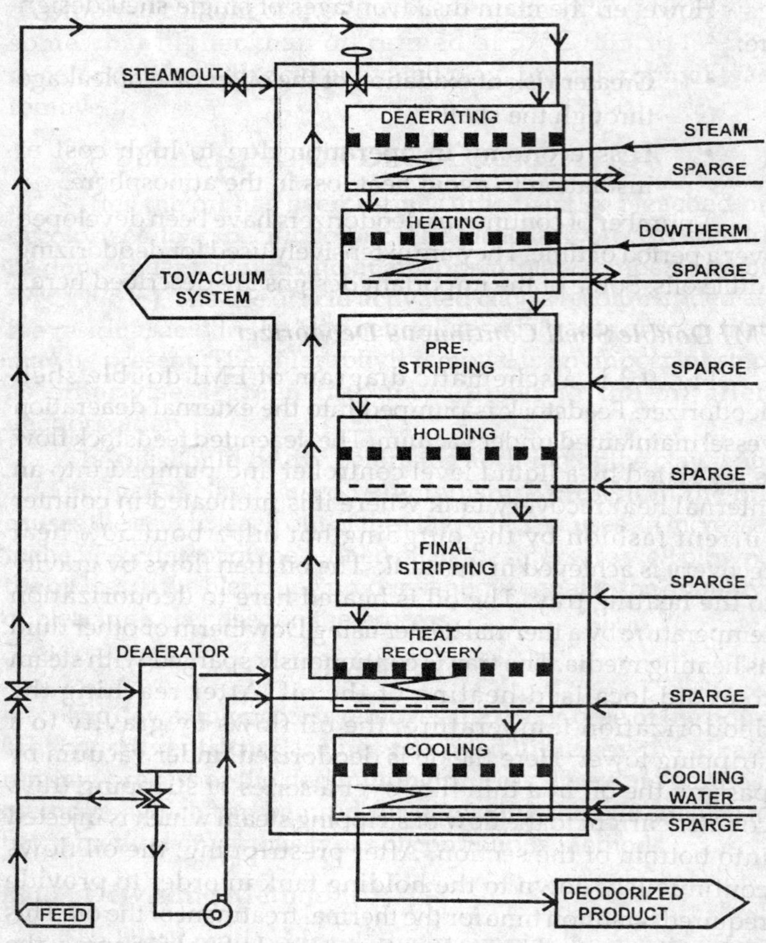

Fig. 9.3 EMI double shell deodorizer

The steam used in sparging and stripping alongwith the volatile impurities passes from each vessel into the outer annular shell space and is removed through a single connection to the vacuum system. Wire mesh type entrainment separators are provided in the covers of the tanks and towers to remove entrained oil from the vapors leaving them. Any entrained oil not removed together with any volatile materials that condense on the outer shell surfaces, drains to the bottom of the outer shell and is removed periodically.

The double shell deodorizing system is available in capacities of 6MT -30MT/hr. However for lower capacities, the single shell deodorizer is more economical tosabricate.

Campro Continuous Deodorizer

Campro deodorizer is a continuous unit operating at 4-6 torr absolute pressure and at temperatures upto 250°C. The main deodorizer consists of a horizontal steel shell enclosing a single rectangular stainless steel tray. The shell is fabricated in two sections, joined together by a flanged connection. The shorter section is fixed in position. Process and service piping enters through this section. The larger section slides away on beams to expose the tray for maintenance or cleaning. The tray contains three sections; the thermal oil heating section, the deodorizing section and the heat recovery section. The deodorizing section is composed of a series of unique thin film distillation stages.

The oil is pumped to the deaerator through prepolish filter, heat exchanger and the preheater where oil is deaerated and dried at 90°C under vacuum. The deaerated oil is pumped into the deodorizer through the exchanger, drop tank coils and the heat recovery panels to recover the heat from the outgoing product. The oil is heated to deodorization temperature by thermal oil panels. Here oil flows in a serpentine manner around the heating panels and is agitated by a small flow of steam to improve the heat transfer performance.

The hot oil then overflows the heating section baffle and is propelled through the stripping stages by the sparging steam. The final stripping stage empties the oil into the heat recovery section. Here also oil flows in a serpentine manner around the heat exchange panels where it preheats the incoming oil. Final oil overflows the heat recovery section baffle and drains out of the deodorizer to the drop tank by gravity.

From the drop tank oil is pumped to a dozing tank through a oil to oil heat exchanger. In the dozing tank, antioxidant is added to increase the shelf life of the product. From the dozing tank, oil is pumped through polish filter to storage tank after cooling it to 45°C.

A small portion of the fatty acids which are stripped from the oil within the deodorizer, may condense in the shell and drain to the scrubber system while the majority pass to the scrubber alongwith the steam. The vapor is cooled with a small quantity of sprayed water before entering the scrubber. In the scrubber, cooled fatty acid is recirculated to scrub the fatty acids from the vapor steam. The condensed fatty acids are collected in the fatty acid tank.

Vacuum is drawn on the deodorizer by means of a four stage condensing vacuum system. Vacuum is drawil on the deaerator by means of a two stage condensing vacuum system.

Packed Bed Continuous Deodorizer

A new deodorizer design known as Soft Column based on countercurrent thin film technology has recently been introduced by Alfa Laval. In this design, a large surface area is obtained by distributing the oil over a large surface e.g., a structured packing. Here, the oil flows by gravity over the packing and meets sparging steam in counter current fashion. The steam passes upward through the deodorizer at minimum pressure drop. The residence time of the oil in the structured packing is approximately five minutes. A three tray retention section at the bottom of the column is provided for giving desired thermal treatment to the oil. The retention time may vary from 10 to 90 minutes.

The packed column technology is superior to conventional tray technology as it requires less than one third the stripping steam needed by a tray deodorizer. The reduction in stripping steam consumption is due to the large surface to volume ratio, a shorter transfer path for volatiles, a countercurrent flow of oil and steam and there is no need for steam agitation. Also, the finished product has lighter colour, higher stability due to retention of tocopherols and tocotrienols and much better taste.

However, the large surface area increases the pressure drop across the packing thus reducing the driving force at the bottom of the column. Packings having a specific surface area in the range of 200-600 sq.m per cubic meter are being used commercially. Moreover, deposits of particles etc., tend to break the thin oil film thus reducing the effective contact area. Furthermore, settling of material may foul the packing and thus increase the pressure drop and reduce efficiency. The two most important reasons for fouling are probably formation of condensation reaction products of phospholipids and oxygen dependent polymerization of triglycerides. However, these shortcomings may be overcome by better input oil quality, proper design of the equipment, complete deaeration and avoidance of leakage during operation.

IMPORTANT/PARAMETERS FOR DEODORIZATION

The key factors in the design of a deodorization system are:

- Energy Consumption
- Product Quality
- Performance Reliability
- Investment Cost
- Environment and Pollution Concerns

The relative importance of these parameters varies greatly for different applications and locations. In practice, plant designs are a compromise between different design options available in the market.

Table 9.3 shows normal commercial deodorization conditions. Good quality can be produced with batch, continuous or semicontinuous deodorizers. Continuous and semicontinuous deodorizers are more energy efficient than batch since they use less steam and are more adaptable to heat recovery.

Several design features are essential in any deodorizer for obtaining a good quality product. Some of them are given below:

1. All surfaces in contact with hot oil must be stainless steel construction to avoid the prooxidant effect of iron.

2. Air contact with hot oil must be avoided by design and construction.
3. Oil should be deaerated before being heated.
4. Polish filters are provided before and after deodorizer to remove oxidized oil particles and other filterable impurities.
5. Design should eliminate the possibility of condensed distillate dripping back into the oil.
6. Provision should be made to seal the deodorizer so that air does not enter the hot deodorizer if the vacuum fails due to a loss of steam pressure.
7. Oil should be cooled before being exposed to air. A temperature of 60 degC is recommended for hydrogenated oils. For unhydrogenated oil, cooling to 32 degC is desirable.
8. Addition of a minimum of 50 ppm citric acid to the deodorized oil is recommended. This is most effectively done as a 20% solution in water added to the cooling section where the oil is still under vacuum and at a temperature below 70 degC. The addition of citric acid before deodorization is an acceptable practice, but it is not a substitute for the later addition, because the citric acid is decomposed by the high deodorization temperatures.
9. Sparging nitrogen gas into the oil exisiting from the deodorizer provides additional protection against oxidation. 10% by volume of nitrogen will saturate the oil and provide a small excess.

Periodic cleaning of deodorizers to remove oxidized oil buildup is essential. A good practice is to schedule this every three to six months. If cleaning is delayed too long there will be a decline in flavor quality and stability of the oil. Cleaning is done by circulation of hot detergent or caustic soda solution. This is followed by thorough rinsing with a final citric acid rinse to neutralise any alkaline materials.

Table 9.3 Commercial Deodorization Conditions

PARAMETERS	BATCH	SEMICON-TINUOUS	CONTINUOUS
Absolute pressure, mm Hg.	6-8	3-6	3-6
Deodorization temperature, 0C	180-190	210-260	210-260
Holding time at elevated temperature,	3-8 hrs.	15-120 minutes	15-120 minutes
Stripping steam: wt % of oil, %	5-15	1-5	1-5
Free Fatty Acids			
Feed, %	0.05-0.15	0.05-0.50	0.05-0.50
Deodorized oil, %	0.05-0.10	0.02-0.05	0.02-0.05

Once deodorized, oil should be carefully handled to avoid oxidation. Mild steel storage tanks are suitable if they are kept clean and not rusty. Stainless steel is preferred if the added cost is acceptable. Oils should be kept at the lowest possible temperature. Hydrogenated oils should be above the melting point and unhydrogenated soybean oil can normally be held at ambient temperature. It is desirable to have agitators in all tanks to provide uniform temperature and prevent precipitation of stearine on cold surfaces. Nitrogen blanketing is highly effective in preventing oxidation. Care should be taken that no contaminants are introduced. Oil should be checked daily for FFA, color, and peroxide value. If an appreciable increase is noted, tank should be emptied and thoroughly cleaned.

Chapter 10

Designer Lipids

The terms "Functional Foods" and "Designer Foods" are commonly used for foods that have been designed to offer specific health benefits (Fig. 10.1), although the food legislation in most countries does not specifically approve such designation unless the potential health benefits are scientifically proven. In Japan functional foods with scientifically proven beneficial effects on health can be designated as Foshu (Foods for Specific Health Use) according to the food legislation. Food products supplemented with substances with known or potential health benefits are also termed Nutraceuticals or Nutraceuticals, e.g. products supplemented with polyunsaturated fatty acids (PUFA) or phytosteryl esters (Fig. 10.1).

Foshu (Foods for Specific Health Use) - officially permitted by food legislation in Japan for products with scientifically proven health benefits

Nutraceuticals e.g.

Products supplemented with:

- ω3 and ω6 Polyunsaturated fatty acids

  (Pufa):

  - Eicosapentaenoic acid (EPA),
  - Docosahexaenoic (DHA),

- γ-Linolenic Acid (GLA),
- Arachidonic acid (ARA),
- Conjugated linoleic acids (CLA),
- Phytosteryl esters, etc.

Fig. 10.1 Designer Foods

(Foods designed to offer specific health benefits)

Designer lipids are in the broad sense fats and other lipids of which the triacylglycerol structure and composition have been altered to achieve specific physical, chemical, biochemical and nutritional properties. Designer lipids or structured lipids include structured triglycerides and phospholipids. Fig. 10.2 outlines a few major objective of lipid designing. Development of specialty products, such as cocoa butter substitutes with well defined physical properties from natural fats was in the past a major objective of lipid designing. Calorie reduction of dietary fats for the prevention of obesity and cardiovascular diseases is currently an important target of a lipid designer. Moreover, structured triglycerides for the nutrition of infants and patients have been developed with the aim to modulate the absorption and metabolism of definite fatty acids by locating them in specific positions of triacylglycerols.

Structured triglycerides are produced by conventional processing, e.g. hydrogenation and interesterification, and by biotechnology, e.g. by genetic modification and enzymatic reactions, such as lipase-catalyzed interesterification.

- **Specialty products, e.g. cocoa butter substitutes, with specific physical properties**
- **Calorie reduction in obesity and cardiovascular diseases**
- **Modulation of the absorption and metabolism of specific fatty acids by locating them in specific positions of triglycerides (Structured Triglycerides)**
  - **use in infant nutrition and in specific conditions of health and disease, e.g. dietetics, clinical nutrition, parenteral nutrition**

Fig. 10.2 Objectives of Lipid Designing

Strictly speaking, even chemically interesterified (randomized) natural oils and fats are also structured triglycerides, however, essentially no calorie reduction occurs through randomization of the acyl moieties in the sn-l, sn-2- and sn-3-positions of the glycerol backbone (Fig. 10.3). Recent studies on humans have revealed no difference between natural fats and chemically interesterified fats with regard to their effects on blood lipids and coronary risk factors.

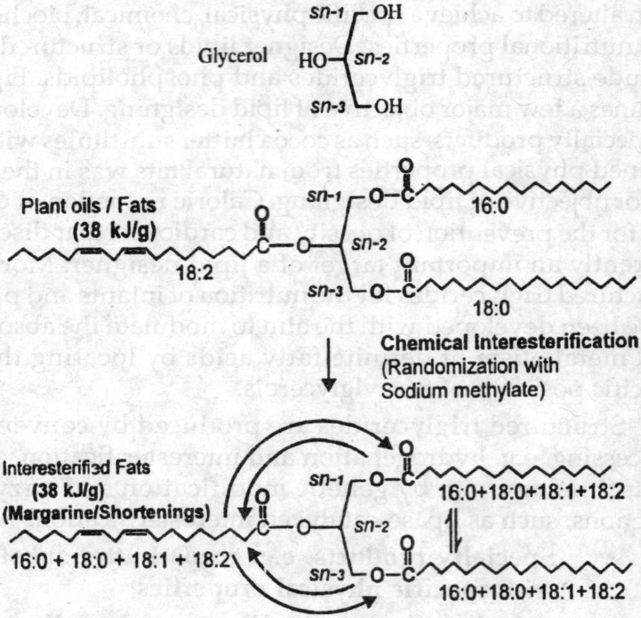

Fig. 10.3 Triglyceride structure of plant oils and interseterified fats prepared by chemical radomization

For the designing of structured triglycerides one has to take into account that the absorption of triacylglycerols in the gastrointestinal tract depends, not only on the chain length of the acyl moieties, i.e. long-chain fatty acids (LCFA) vs. medium-chain fatty acids (MCFA), but also on the location of an acyl moiety on the glycerol backbone.

Long-chain triglycerides (LCTG) are hydrolyzed in the intestinal lumen at the sn-1 and sn-3-positions by the action of pancreatic and gastric lipases and the resulting mono glycerides (2-acylglycerols) and the fatty acids are reesterified via coenzyme A esters during their passage through intestinal macosa (Fig. 10.4). The resulting triglycerides enter the circulation via the lymphatics as chylomicrons and are subsequently hydrolyzed to fatty acids that are partly catabolized in the liver or transported to adipose tissues and stored as triacylglycerols.

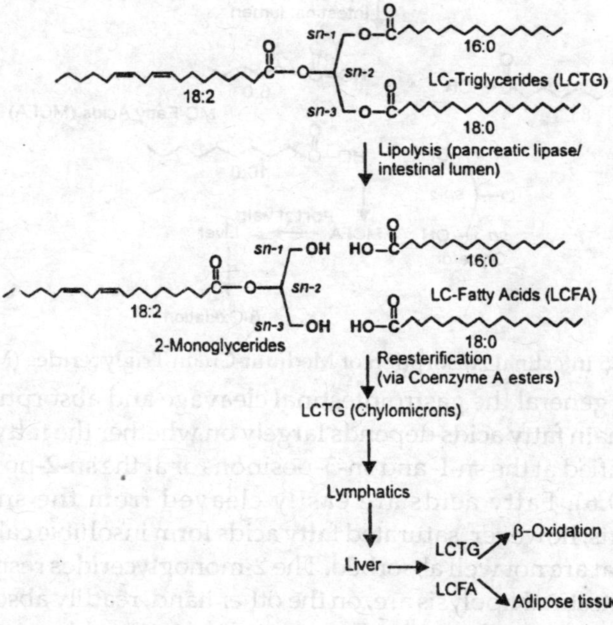

Fig. 10.4 Intestinal Absorption of Long-Chain Triglycerides (LCTG)

Medium-chain triglycerides (MCT) on the other hand are completely hydrolyzed in the intestinal lumen by the pancreatic lipase and the resulting medium-chain fatty acids are directly transported via the portal vein to liver where they are rapidly catabolized (Fig. 10.5).

Even when gastric or pancreatic lipase are deficient, the MCT can be directly absorbed through the intestinal mucosa, hydrolyzed to MCFA by intestinal lipase and transported to liver. Obviously, MCT are readily available source of energy as compared to LCTG.

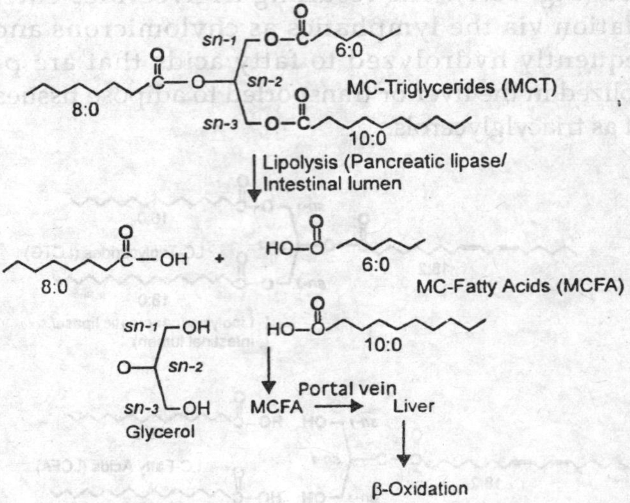

Fig. 10.5 Intestinal Absorption of Medium-Chain Triglycerides (MCT)

In general the gastrointestinal cleavage and absorption of long-chain fatty acids depends largely on whether the fatty acid is esterified at the sn-l- and sn-3-positions or at the sn-2-position (Fig.10.6). Fatty acids are easily cleaved from the sn-1,3-positions, however, saturated fatty acids form insoluble calcium salts that are not well absorbed. The 2-monoglycerides resulting from intestinal lipolysis are, on the other hand, readily absorbed through intestinal mucosa. Even saturated fatty acids esterified at the sn-2-position of dietary triacylglycerols are thus easily absorbed as 2-monoglycerides.

The above biochemical and physiological pathways have been taken into consideration in designing the following types of designer lipids.

DESIGNER LIPIDS BY CONVENTIONAL TECHNOLOGY

Medium-chain Triglycerides (MCT)

Esterification of fatty acid fractions from coconut oil or

palm kernel oil with glycerol yields MCT containing 25 to 35% capric acid and 65 to 75% caprylic acid (Fig. 10.7). Alternatively, MCT can be prepared by fractionation of coconut oil or palm kernel oil or by transesterification of alkyl esters of medium-chain fatty acids with glycerol. Such products are available

sn-1, *sn*-2, *sn*-3; R_1, R_2, R_3

$HO-C(=O)-R_1$ -Fatty acids easily cleaved from *sn*-1,3-positions, however,

$HO-C(=O)-R_3$ -saturated fatty acids form insoluble Ca-salts that are not absorbed

-2-monoglycerides are readily absorbed (i.e. fatty acids from the *sn*-2-position, even saturated fatty acids are well absorbed)

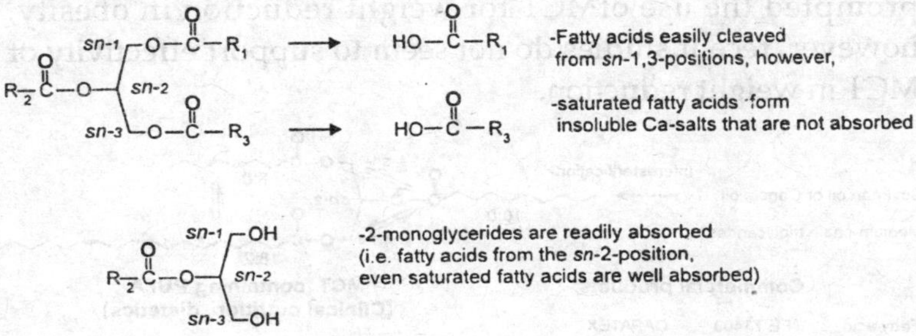

Fig. 10.6 Gastrointestinal cleavage and absorption of fatty acids esterified at the sn-1-3,- or sn-2-positions of dietary triglycerides

DESIGNER LIPIDS BY CONVENTIONAL TECHNOLOGY

Medium-chain Triglycerides (MCT)

Esterification of fatty acid fractions from coconut oil or palm kernel oil with glycerol yields MCT containing 25 to 35% capric acid and 65 to 75% caprylic acid (Fig. 10.7). Alternatively, MCT can be prepared by fractionation of coconut oil or palm kernel oil or by transesterification of alkyl esters of medium-chain fatty acids with glycerol. Such products are available

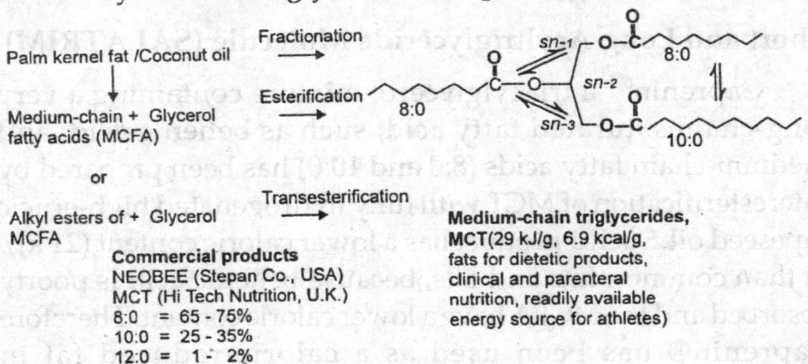

Fig. 10.7 Chemically Produced Medium-Chain Triglycerides

commercially as liquid oils or margarines. MCT as a readily available source of energy are used by athletes and as dietetic products in clinical and parenteral nutrition. The lower caloric content ofMCT (29 kJ/g) as compared to long-chain triglycerides contained in common fats and oils (38 kJ/g) has prompted the use ofMCT for weight reduction in obesity, however, recent studies do not seem to support effectivity of MCT in weight reduction.

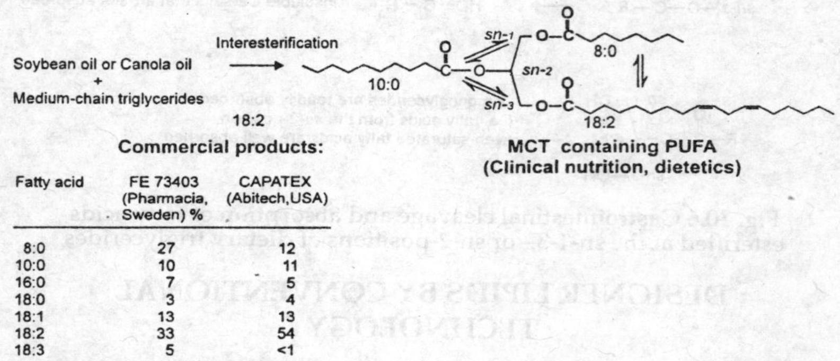

Fatty acid	FE 73403 (Pharmacia, Sweden) %	CAPATEX (Abitech,USA)
8:0	27	12
10:0	10	11
16:0	7	5
18:0	3	4
18:1	13	13
18:2	33	54
18:3	5	<1

Fig. 10.8 Chemically Produced Medium-chain Triglycerides Containing Polyunsaturated Fatty Acids (PUFA)

Medium-chain triglycerides containing polyunsaturated fatty acids (PUFA) are prepared commercially by interesterification of soybean or Canola-type rapeseed oil with MCT (Fig. 10.8). Such products are also used in clinical nutrition and dietetics.

Short and Long Acyltriglyceride Molecule (SALATRIM)

Caprenin® -a triacylglycerol mixture containing a very long-chain saturated fatty acid, such as behenic acid, and medium-chain fatty acids (8:0 and 10:0) has been prepared by interesterification of MCT with fully hydrogenated high-erucic rapeseed oil. Such a product has a lower caloric content (21 kJ/g) than common fats and oils, because behenic acid is poorly absorbed and the MCFA have a lower caloric content. Therefore Caprenin® has been used as a calorie-reduced fat in confectionery products, however, the inability ofCaprenin® to reduce postprandial lipemia, as compared to other dietary fats, led to withdrawal of this product from the market.

Low calorie (21 kJ/g) asymmetric triacylglycerols containing short- and long-chain acyl moieties (Short and Long Acyltriglyceride Molecule, *Salatrim*) are prepared by interesterification of fully hydrogenated soybean or Canola-type rapeseed oils with short-chain triacylglycerols, such as tributyrin(Fig.). The low calorie content of such products is due to poor absorption of stearic acid, especially from the sn-1,3-positions, and due to the butyroyl or propionoyl moieties. *Salatrim* products, such as BENEFAT 1 has received the *Gras* (Generally Regarded as Safe) status and are being used in many countries in confectionery as cocoa butter substitutes and for frying chips.

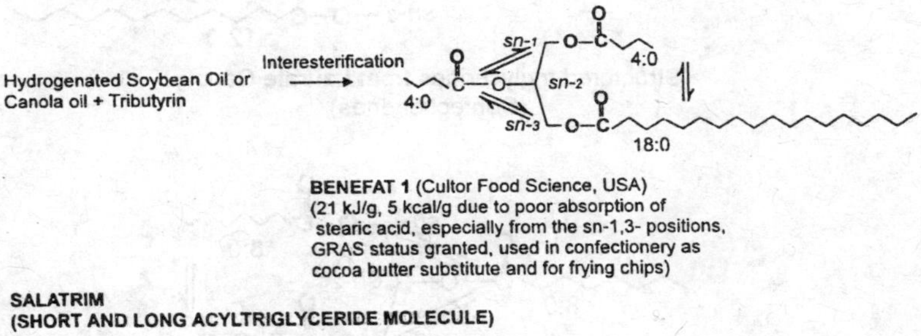

Hydrogenated Soybean Oil or Canola oil + Tributyrin —Interesterification→ [structure: sn-1, sn-2, sn-3; 4:0, 4:0, 18:0]

BENEFAT 1 (Cultor Food Science, USA)
(21 kJ/g, 5 kcal/g due to poor absorption of stearic acid, especially from the sn-1,3- positions, GRAS status granted, used in confectionery as cocoa butter substitute and for frying chips)

SALATRIM
(SHORT AND LONG ACYLTRIGLYCERIDE MOLECULE)

Fig. 10.9 Chemically Produced Structured Triglycerides Containing Short- and Long-Chain Fatty Acids

DESIGNER LIPIDS BY BIOTECHNOLOGY

Biotechnological approaches for the preparation of designer lipids includes genetic modification of oil plants, microbial production of oils and lipase-catalyzed transformation of fats and oils.

Structured Triglycerides by Genetic Modification

Genetic modification of oil plants has yielded oilseeds producing a wide diversity of designer oils. Especially, the development of transgenic rape (Laurate Canola) with a high level of lauric acid in the seed oil and its cultivation in the USA in large acerages are impressive.

The triacylglycerol composition of the designer oil in Laurate Canola (Fig. 10.10) shows that the lauric acid is esterified exclusively at the *sn*-1, 3positions and the C18- unsaturated acyl moieties are located at the sn-2-position (11). In comparison, MCT from palm kernel oil or coconut oil contain the medium chain acyl moieties randomly distributed in the *sn*-1-, 2- and -3-positions (Fig. 10.10). The unique triacylglycerol structure of the designer oil from Laurate Canola can be utilized as such or after hydrogenation for the preparation of special fats for confectionery products.

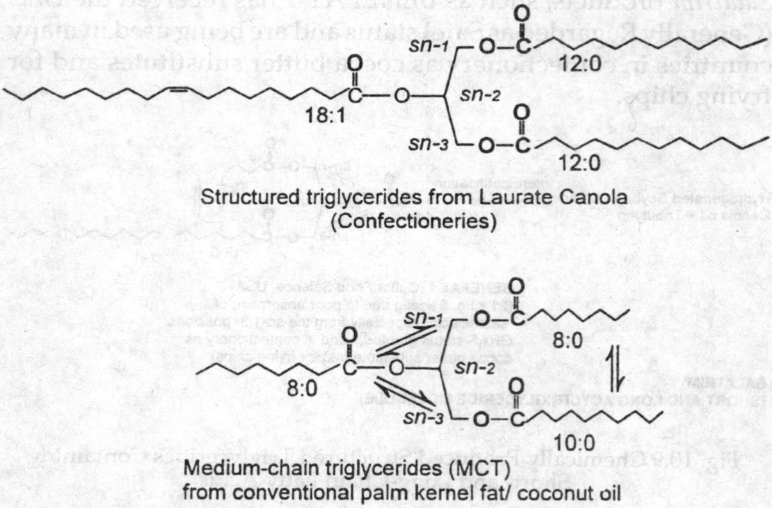

Structured triglycerides from Laurate Canola
(Confectioneries)

Medium-chain triglycerides (MCT)
from conventional palm kernel fat/ coconut oil

Fig. 10.10 Structured Triglycerides in Genetically Modified Oils as Coimpared to MCT from Palm Kernel Fat or Coconut Oil

Triglycerides and Other Lipids Containing ω3 and ω6 Fatty Acids

Nutritional significance of essential ω3 and ω6 fatty acids is well-documented. The ω6 fatty acids., e.g. linoleic acid, yield via desaturation and chain elongation arachidonic acid (ARA), a precursor of one set of eicosanoids, i.e. prostanoids of series-2 and leukotrienes of series-4, whereas the ω3 fatty acids, e.g. (α-linolenic acid yield correspondingly docosahexaenoic (DHA) and eicosapentaenoic acid (EPA), a precorsor of another set of

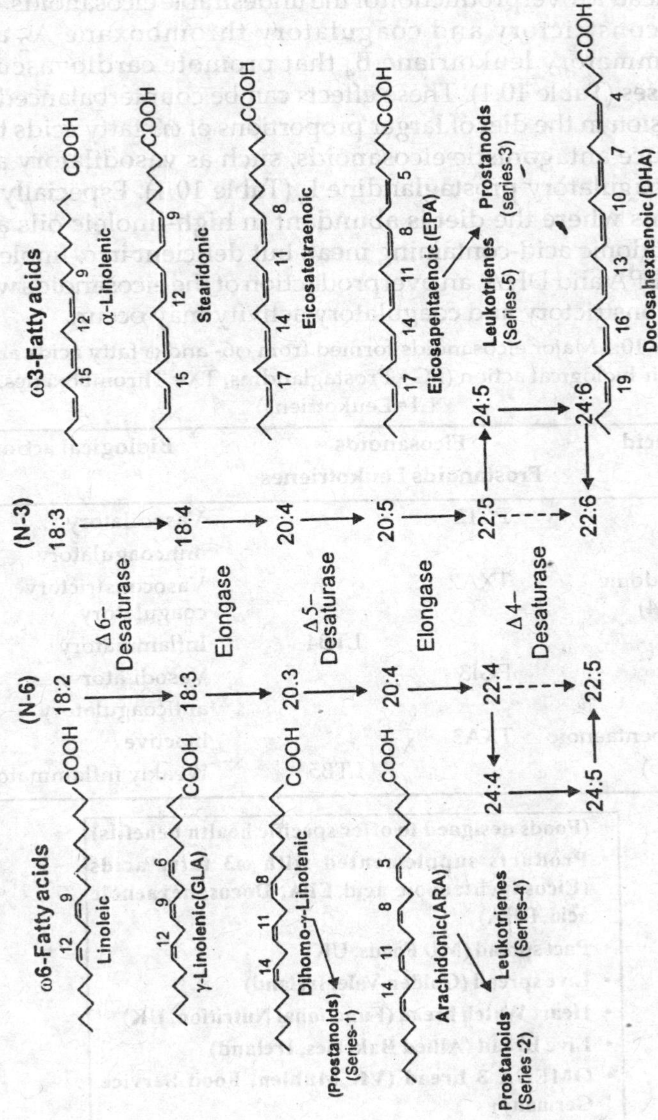

Fig. 10.11 Metabolism of Long-Chain Polyunsaturated Fatty Acids

eicosanoids, i.e. leukotrienes of series-5 and prostanoids of series-3 (Fig. 10.10). The abundance of ω6 fatty acids in the diet may lead to overproduction of the undesirable eicosanoids, e.g. vasoconstrictory and coagulatory thromboxane A_2 and inflammatory leukotriene B_4 that promote cardiovascular diseases (Table 10.1). These effects can be counterbalanced by inclusion in the diet of larger proportions of ω3 fatty acids that produce antagonistic eicosanoids, such as vasodilatory and anticoagulatory prostaglandine I_3 (Table 10.1). Especially, in regions where the diet is abundant in high-linoleic oils and arachidonic acid-containing meat, but deficient in α-linolenic acid, EPA and DHA, an overproduction of the eicosanoids with vasoconstrictory and coagulatory activity may occur.

Table 10.1 Major eicosanoids formed from ω6- and ω fatty acids and their biological action (PG= Prostaglandins, TX= Thromboxanes, LT=Leukotriens)

Fatty acid	Eicosanoids		Biological action
	Prostanoids	**Leukotrienes**	
	PGI2		Vasodilatory, anticoagulatory
Arachidonic (ω6 20:4)	TXA2		Vasoconstrictory, coagulatory
		LTB4	Inflammatory
	PGI3		Vasodilatory, anticoagulatory
Eicosapentaenoic (ω3 20:5)	TXA3		Inactive
		LTB5	Weakly inflammatory

(Foods designed to offer specific health benefits)

- **Products supplemented with ω3 fatty acids (Eicosapentaenoic acid EPA, Docosahexaenoic acid, DHA)**
- **Pact spread (MD Foods, UK)**
- **Live spread (Golden Vale, Ireland)**
- **Heart Watch bread (Functional Nutrition, UK)**
- **Live Bread (Allied Bakeries, Ireland)**
- **OMEGA 3 bread (VK Muhlen, Food Service, Germany)**
- **OMEGA-DHA eggs (Omega DHA Food, Germany)**
- **Mikata with DHA (Coca-Cola, Japan)**

Fig. 10.12 Functional Foods/Designer Foods

In many Western countries there seems to be a deficiency of ω3 fatty acids in the diet that should be counterbalanced by inclusion of ω3 fatty acid-rich fish or marine oils in the diet. Therefore, several food manufacturers have started fortifying foods and beverages with ω3 fatty acids as shown for a few selected items in Fig. 10.12.

Carbon source
(Carbohydrates / hydrocarbons / oils and fats)

⇓

Biomass production using whole microbial cells
(algae, yeasts, fungi, bacteria)

⇓

Downstream processing, Extraction, Refining etc.

⇓

Microbial oils (Single-cell oils, SCO) for food, nutraceutical and non-food applications

Fig. 10.13 Biotechnology of Microbial Oils

In addition to plant oils with substantial levels of α-linolenic acid, such as Canola-type rapeseed oil, a common source of ω3 fatty acids, e.g. EPA and DHA, is generally marine oils that are, however, objectionable to vegetarians.

Alternative sources of designer oils containing ω3 fatty acids are lipid-producing microorganisms, such yeasts, molds and microalgae that can be grown on a wide variety of carbon sources to yield biomasses from which microbial oils (Single Cell Oils, SCO) are obtained by extraction and refining (Fig. 10.13).

Table 10.2 lists selected species of microlagae and fungi that are known to produce specific ω3 and ω6 polyunsaturated fatty acids. The microalga Crypthecodinium cohnii and the fungus Mortierella are used for the commercial production of

SCO containing docosahexaenoic (DHA) and arachidonic (ARA) acid, respectively. Fatty acid composition of these oils that are used for fortification of infant food formulations is given in Table 10.3.

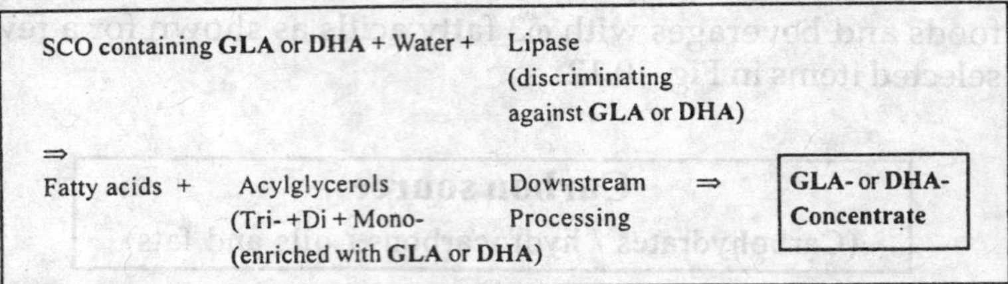

Fig. 10.14 Enrichment of γ-Linolenic Acid (GLA) or Docosahexaenoic Acid (DHA) from Single-Cell Oils (SCO)

Table 10.2 Polyunsaturated Fatty Acids in Biomass of Lipid Accumulating Microorganisms

Organism	**Major poly-unsaturated Fatty acid**	**(% w/w of total fatty acids)**
Microalgae		
Spirulina platensis	18:3 ω6	21
Dunaliella tertiolecta	18:3 ω6	32
Porphyridium Cruentum	20:4 ω6	60
Chlorella minutissima	20:5 ω3	45
Navicula saprophilla	20:5 ω3	22
Crypthecodinium Cohnii	22:6 ω3	40
Fungi		
Mortierella isabellina	18:3 ω6	8
Mortierella Rammaniana	18:3 ω6	26
Mortierella elongata	20:4 ω6	15
Mortierella alpina	20:4 ω6	43-66
Pythium irregulare	20:5 ω3	25
Thraustochytrium Aureum	22:6 ω3	40-50

Table 10.3 Composition of Commercially Available Single Cell Oils Containing Polyunsaturated Fatty Acids (ARA, DHA, GLA)

Organism	Carbon/ Nitrogen Source	Trade name/ Manufacturer, Supplier	Fatty acid (% w/w of total) 14:0	16:0	18:0	18:1	18:2ω6	18:3ω6	20:4ω6	22:6ω3
Mortierella alpina[b]	Dextrose/yeast extract vegetable protein hydrolyzate	**ARASCO** Martek Bioscience, USA		8	10	21	7		40 (ARA)	
Mortierella alpina	Dextrose/ yeast extract	**SUN- TGA25/** Suntory Co., Japan		14	6	14	26		25 (ARA)	
Crypthecodi- 40 DHA) nium cohnii	Dextrose/yeast extract	**DHASCO/** Martek Bio-science	15	16	<1	21	1			
Mortierella isabellina	Sugars	**Fungal oil/** Sigma, USA		21	4	37	22	10 (GLA)		

Commercially available SCO contain at the most 40% DHA or ARA. Ability of lipases to differentiate between specific polyunsaturated fatty acids from others has been utilized to enrich definite fatty acids from SCO by lipase-catalyzed selective hydrolysis. Thus lipa1se-catalyzed selective hydrolysis of SCO containing γ-linolenic acid (GLA) or DHA leads to preferential cleavage of fatty acids other than GLA or DHA and enrichment of GLA or DHA in tri-, di- and monoacylglycerols; downstream processing yields acylglycerols enriched with GLA or DHA (Fig. 10.14). Similarly, lipase-catalyzed selective esterification of fatty acids of SCO containing GLA or DHA with n-butanol leads to preferential esterification of fatty acids other than GLA or DHA and enrichment of GLA and' DHA in the unesterified fatty acids; downstream processing yields fatty acids enriched with GLA or DHA (Fig. 10.15).

Structured Triglycerides by Lipase-Catalyzed Reactions

The ability of regiospecific lipases to interchage, via interesterification, the fatty acylmoieties almost exclusively located at the sn-I,3-positions of triacylglycerols (Fig. 10.16) has been utilized to prepare several structured triglycerides shown in Fig. 10.17.

Fatty acids containing **GLA** or **DHA** + *n*-Butanol + Lipase (discriminating against GLA or DHA)
(from SCO)

⇒

n-Butyl esters (enriched with other fatty acids) + Fatty acids (enriched with **GLA** or **DHA**) ⇒ Downstream processing ⇒ **GLA**- or **DHA**-Concentrate

Fig. 10.15 Enrichment of g-Linolenic Acid (GLA) or Docosahexaenoic Acid (DHA) from Single-Cell Oils (SCO)

Cocoa butter substitutes containing palmitic and stearic acids esterified at *sn*-1 ,3-positions of the glycerol backbone and oleic acid esterified at the sn-2-position are prepared commercially by interesterification, catalyzed by sn-1 ,3-specific lipases, as shown in Fig. 10.18. Palm oil mid-fraction containing,palmitic acid esterified at the sn-l,3-Positions and oleIc acid at the sn-2-position is enzymatically interesterified with either stearic acid or ethyl stearate to yield the targetted cocoa butter substitutes.

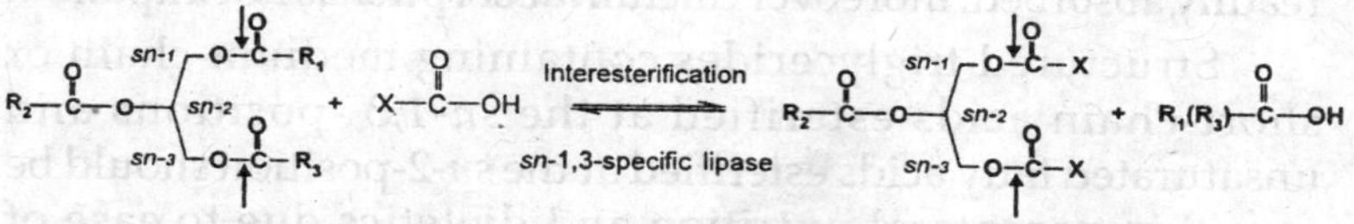

Fig. 10.16 Enzymatic preparation of structured triglycerides: Interesterification (Acidolysis) of a fat with a fatty acid X- COOH using a sn-1, 3-specific lipase (R1, R2, R3=fatty acyl moieties; ↑↓ = point of attack by the lipase)

Human milk fat substitutes containing palmitoyl moieties predom in antly at the sn-2-position of the glycerol back bone and unsaturated fatty acids esterified at the *sn*-1, 3-positions are produced commercially by acidolysis of tripalmitin with unsaturated fatty acids, catalyzed by sn-l,3-specific lipases as shown in Fig. 10.19. Such triacylglycerols are well absorbed by

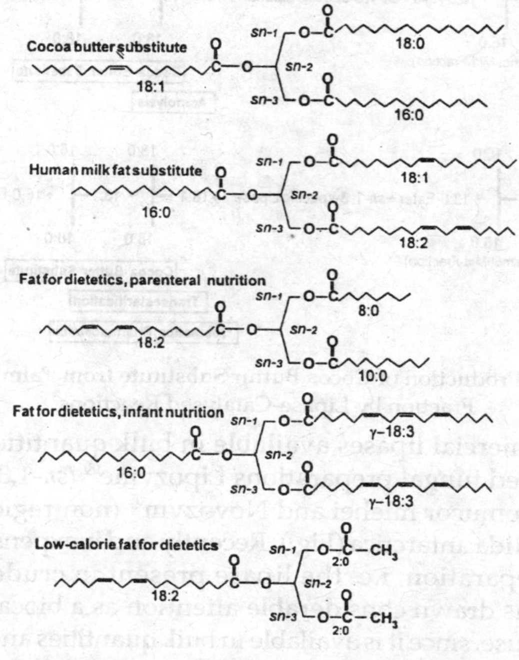

Fig. 10.17 Enzymatically Produced Structured Triglycerides

infants, because intestinal lipolysis yields 2-monopalmitin that is directly absorbed and unsaturated fatty acids cleaved from the *sn*-1, 3-positions do not form insoluble calcium salts, thus readily absorbed; moreover calcium absorption is not impaired.

Structured triglycerides containing medium-chain or short-chain acids esterified at the *sn*-1,3 -positions and unsaturated fatty acids esterified at the *sn*-2-position should be useful in parenteral nutrition and dietetics due to ease of absorption, direct utilization of medium- or short-chain fatty acids as energy source, and better availability of essential fatty acids. Such structured triglycerides can be obtained, as shown in Fig. 10.19, by interesterification of a plant oil, such as sunflower or Canola-type rapeseed oil, with short- or medium-chain fatty acids catalyzed by sn-1,3-specific lipases.

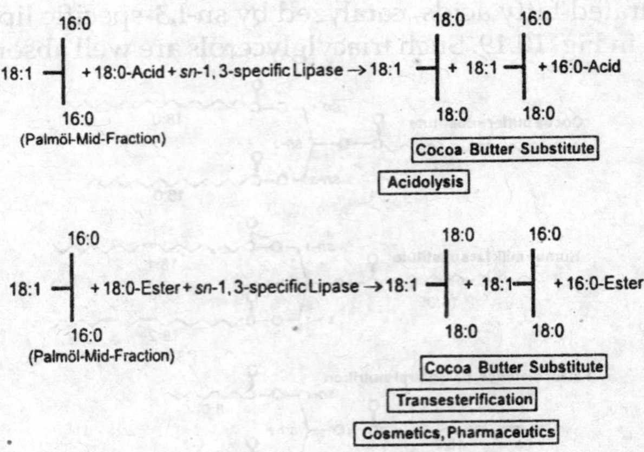

Fig. 10.18 Production of Cocoa Buttur Substitute from Palm Oil Mid-Fraction by Lipase-Catalysed Reactions

Commercial lipases available in bulk quantities are the immobilized fungal preparations Lipozyme® (*sn*-1,3 specific) from Rhizomucor miehei and Novozym® (non-regiospecific) from Candida antarctica (Fig.). Recently and inexpensive plant lipase preparation, i.e. the lipase present in crude latex of papaya has drawn considerable attention as a biocatalyst for industrial use, since it is available in bulk quantities and because papaya latex is being used since long in food and beverage

Industries due to the proteolytic enzymes papain and chymopapain contained in the crude latex. Papaya lipase has a strong *sn*-3-regiopreference and a selectivity towards unsaturated fatty acids/acyl moieties which is similar to many microbial lipases (Fig. 10.20).

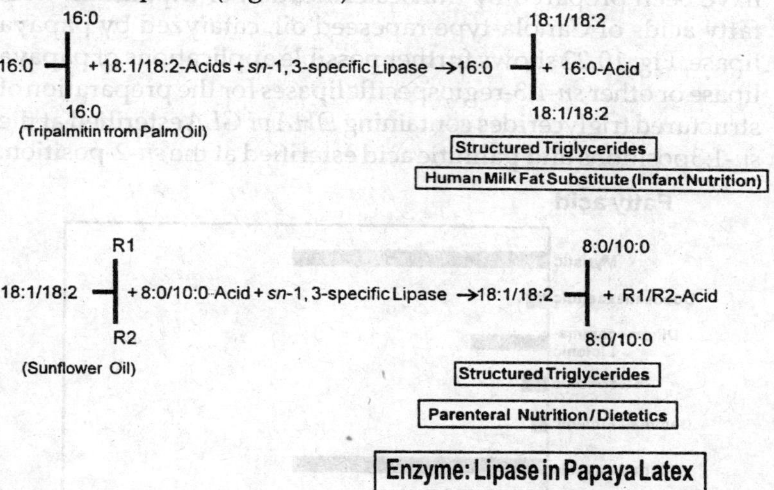

Fig. 10.19 Production of Structured Triglycerides by Lipase-Catalyzed Acidolysis Reactions

Fig. 10.21 shows, for example, that in esterification of fatty acids with n-butanol, catalyzed by papaya lipase, oleic and α-linolenic acids are the preferred substrates, whereas the fatty acids having a cis-4 double bond (docosahexaenoic acid), cis-6 double bond (petroselinic, γ-linolenic and stearidonic acids) or a cis-8 double bond (dihomo-γ-linolenic acid) are discriminated against as substrates.

Lipozyme (*Rhizomucor miehei*) — sn-1,3-specific
Novozym (*Candida antarctica*) —— unspecific

Lipase in Latex from Papaya (*Carica papaya*)
* sn-3-specific
* fatty acid selectivity similar to other commercial lipases
* inexpensive enzyme used in food products

Fig. 10.20 Lipase Preparation for Lipid Designing

The above properties of papaya lipase can be utilized to prepare designer lipids, such as structured triglycerides or triglycerides enriched with specific *Pufa* of interest. For example, structured triglycerides resembling human milk fat have been prepared by transesterification of tripalmitin with fatty acids of Canola-type rapeseed oil, catalyzed by papaya lipase. Fig. 10.22 shows further possible applications of papaya lipase or other *sn*-1,3-regiospecific lipases for the preparation of structured triglycerides containing *DHA or GLA* esterified at the *sn*-1,3positions and palmitic acid esterified at the *sn*-2-position.

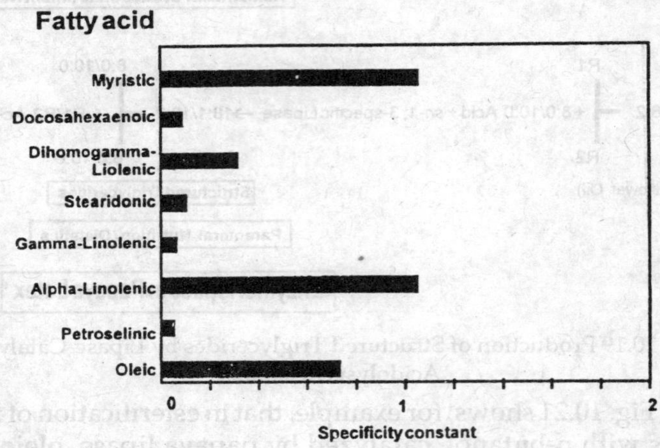

Fig. 10.21 Substrate selectivity of lipase in papaya (Carica papaya) latex towards individual fatty acids in esterification with n-Butanol

Such structured triglycerides should find application in infant nutrition, parenteral nutrition and dietetics. Further possible applications of *sn*-1,3-regiospecific lipases is the preparation of low-calorie structured triglycerides containing short-chain acyl moieties at the *sn*-1,3-positions and C 18 fatty acids esterified at the sn-2-position (Fig. 10.22).

PERSPECTIVES

With the advances in biomedical research further lipids and lipid constituents, than hitherto known, with beneficial health effects, such protection against cardiovascular diseases and cancer, are likely to become known. Moreover, specific nutritional benefits achieved by structuring fatty acyl moieties in triacylglycerols and other lipids in a different manner than

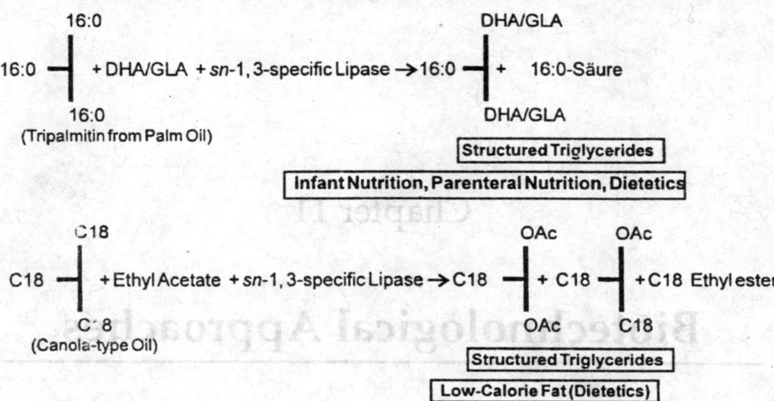

Fig. 10.22 Production of Structured Triglycerides by Lipase-Catalyzed Acidolysis (DHA = 22:6 ω3; GLA = 18:3 ω6)

those occurring naturally are envisaged to be unravelled in the future. It is conceivable, for example, that the calories available from animal fats and their hyperlipidemic effects can be reduced by structuring the saturated fatty acids at the *sn*-1,3-positions of the glycerol backbone from where they are less absorbed than from the *sn*-2-position. In this scenario it is left to the imagination of the food designer to apply conventional technology and novel biotechnology to design lipids for specific health use.

Chapter 11

Biotechnological Approaches

Increasing global human population and specific consumer demands have and will continue to have a direct impact on food supplies. As a major component of many diets definite concern needs to be directed toward oils and fats. This class of compounds not only serves as a major source of dietary energy but also has nutritional implications associated with specific lipids. Particular biological roles exist, for example, for essential fatty acids, sterols (especially cholesterol) and fat-soluble vitamins. Consequently it is imperative that oils and fats be available in both the desired quantity and quality.

There is an essentially universal reliance on plants as the major supplier of human food. In particular, certain herbaceous oilseed crops and cultivated trees are the primary source of edible oils and fats. Projections have been made, however, that greater supplies will be necessary to meet increasing future demands. In addition, health concerns are directing more immediate attention toward the actual composition of dietary oils and fats as related to nutritional requirements.

Several approaches may be adopted to meet the projected supply and demand for edible oils and fats. More efficient agronomic practices can be adopted to ensure greater crop productivity. Use of improved developed plant strains (crop cultivars) can be made to obtain increased oil yields. These improvements in plants can be sought not only from conventional breeding programmes, but also from the adaptation of many of the new techniques of biotechnology and molecular biology. Applications of cell culturing, genetic engineering and modification of enzyme systems have been made with varying degrees of success.

Table 11.1 The source and range of oil and protein yields in major crops (as percentage dry mass)

CROP	SOURCE	OIL	PROTEIN
Cottonseed	seed	18-20	19 - 24
Groundnut	seed	45-50	21 - 36
Linseed	seed	40-46	22 - 26
Rapeseed/cano la	seed	40-45	25 - 29
Safflower	seed	30-35	14 - 17
Sesame	seed	50-55	18 - 22
Soybean	seed	19-23	36 - 44
Sunflower	seed	35-45	14 - 19
Coconut	endosperm	65-68	
Oilpalm	mesocarp	>60	
Olive	mesocarp	vanous	

DESIRED GOALS

Several major areas for possible crop improvements may be considered.

Plant

Application of better practices of agronomy should result in increased crop productivity. Cultivation of strains with optimal desired characteristics is, however, ultimately dependent upon the availability of seeds at a reasonable cost. In addition, a definite need exists for the development of further cultivars which can tolerate abnormal soil conditions of high acidity, alkalinity and salinity. Frost tolerance, drought resistance and other environmental conditions are also factors to be considered. Furthermore, scope exists for improvements in plant viability in such areas as germination, maturation and seed-pod shattering propensity.

High crop production is frequently governed by the availability of fertilizers that is accompanied by a significant cost component. A further cost can be incurred from the need to apply pesticides and herbicides which, in turn, can introduce problems of potential environmental concern.

Post-harvest practices can have a major impact on oilseed storage. Protection from conditions of extreme temperatures and moisture, fungal activity and rodent infestation is of prime importance and can be accomplished by procedures that do not require sophisticated approaches such as biotechnology.

Oil Yield

Most of the oilseed crops as presently grown appear to have reached a plateau in oil production. Crops such as rapeseed, sunflower, groundnut and linseed routinely yield seed with an oil content of approximately 40% with a general range of 38-48%. Soybean has a relatively low oil productivity of 19-23%. The very high oil output associated with the oil palm and coconut is associated with the mesocarp of the fruit of these trees.

Subject to a better fundamental understanding of crop physiology, plant genetics and the biochemistry involved with oil production, some definite improvement in the quantity of oil obtainable from oilseeds may be anticipated. Cell culturing techniques and various approaches of biotechnology can be fully expected to be applied in the relevant studies for the development of new crop strains.

Oil Quality

The fatty acid composition of a vegetable oil has a definite significance. Primary concern exists with nutritional requirements and their impact on human health. A second, but none the less important, consideration is with the keeping quality of the oil. Based on the general understanding of the involvement of saturated fatty acids in the aetilogy of cardiac vascular disease interests definitely exist in oils which are deficient or low in this class of fatty acids. Although olive oil and canola oils are characterized by very small contents of saturated fatty acids deliberate lowering of these acids in other commercial edible oils is a major topic of study.

Manipulation of the unsaturated fatty acid component of vegetable oils is variously sought to improve their nutritional value. Oils enriched in oleic acid and certain polyunsaturated fatty acids such as essential fatty acids as well as gamma linolenic acid are considered to be highly desirable.

The polyunsaturated fatty acid content of oils, on the other hand, due to their oxidative instability can present problems of storage and keeping qualities. As a consequence there is a demand by the food industry for the availability of oils with good keeping qualities obtainable from a designed reduction of the polyunsaturated fatty acid content of edible oils.

PLANT ENGINEERING

Trait Detection

Recognition of the occurrence within a plant of a desirable characteristic is essential if the intention is to incorporate that trait into some host crop. The older practice involving selective breeding is handicapped by an inefficient success rate and the length of time generally associated with this approach.

Plant stock evaluation, characterization of the diversity of germplasm and identification of quantitatively inherited traits can be made by one of several techniques of molecular biology. The practice of restriction fragment length polymorphisms (RFLPs) is being replaced by the faster procedures of randomly amplified polymorphic DNAs (RAPDs) or amplified fragment length polymorphisms (AFLPs). Application of expressed sequence tags (ESTs) generated by rapid-throughput automated cDNA sequencing has been made to facilitate the isolation of certain genes associated with the biosynthesis of plant oils.

The basis of these procedures is associated with electrophoretic separation of variously sized fragments of the plant DNA, obtained by the use of restriction endonucleases, and their visualisation by ethidium bromide. The procedures lend themselves to automation. Use of the polymerase chain reaction (PCR) is now being made routinely and at low cost to increase the number of copies of a specific DNA region under investigation. 'Finger-printing' or probing of these segments is then made through hybridisation with DNA fragments recognized to be related to some desired trait.

Particular applications of RAPDs and ESTs are anticipated to be most useful for locating and manipulating particular genes as well as in the mapping of polygenic sources of variation such as plant characteristics and oil yields.

There has also been an increasing use of genomic and molecular techniques to assist in the rapid domestication of presently non-commercial plants such as coriander and cuphea.

Cell Culturing

Use of modern cell culturing techniques is continuing to have a major impact on the development of new plant strains. In particular, the time required for the introduction of new commercial cultivars can be shortened significantly from a minimum of approximately fifteen years to about five years. Some further delays, however, may be experienced in obtaining legislative approval for use as a food.

The ready totipotency of individual living cells when isolated in a suitable medium allows for cultured cells to be regenerated into whole plants possessing the desired genomic composition with the ultimate development of a new stable homozygous variety. The general practice is to culture explants (plant cells, tissue or organs) on a chemically defined medium under sterile conditions. Either microspore (immature pollen) or anther culture techniques are commonly used and are so effective that many plants can be produced eventually at a time. Manipulation of the genetics of these isolated haploid (single set of chromosomes) cells is relatively simple and can be attempted by one of several procedures. Diploidisation (creation of a double set of chromosomes) is assisted by colchicine addition or heat treatment and further cell fusion encouraged by electrical stimulation or by the presence of polyethylene glycol. Cell growth and division is then regulated to produce individual cells and callus (clumps of cells) showing little differentiation. Subsequent organised growth is accompanied by embryogenesis and organogenesis of shoots and roots resulting in the ultimate development of plantlets. Following a hardening period the plantlets may be planted in the open field.

The ease of regeneration of plants from cell cultures varies, but a particularly useful system has been established for rapeseed; the system for soybean is more labour-intensive while that for sunflower continues to be developed.

Genetic Modification

Use of haploid cell systems allows for simpler manipulation of the single set of chromosomes than would be the case with diploid cells. As a general practice protoplasts are created by the enzymatic removal of the cell wall to facilitate the introduction of the desired foreign DNA.

Several approaches have been undertaken to induce changes in the plant genome and the incorporation of novel traits. Commercially valuable successes have been limited and varied.

Mutagenesis

An earlier practice with some noted successes in mutation breeding involves the use of a variety of mutagenic agents. The process is associated with point mutations and the attempt to unmask traits normally hidden in the native strain. Compared to targeted procedures, the process te1ds to have a low success rate due to its random approach.

Exposure of explants, seeds or whole plants can be made to ionizing irradiation or chemical mutagens such as ethyl methanesulphonate (EMS) and sodium azide. The practice is heavily dependent upon the availability of good screening methods. For example, when modification of the fatty acid composition of an oilseed is the aim, good use can be made of gas liquid chromatography to analyse a portion of a developing cotyledon and the remainder can be allowed to develop into a mature plant if found to possess the desired characteristic.

Gene Transfer Technology

Several systems are available for the introduction of selected foreign DNA into plant cells or protoplasts.

1. **Transfection.** Certain bacteria and viruses can be

used as vectors in which a desired DNA fragment has been incorporated and then used to infect host cells leading to alteration of the host genome. Frequent application has been made with the Ti plasmid of *Agrobacterium tumefaciens* as vector to transfect dicots—rapeseed, soybean, sunflow'er flax and cotton; this system is ineffective with monocot plants such as corn and oil palm.

2. **Electroporation.** The desired foreign DNA can be introduced into host cells where a transient increase in membrane permeability due to a brief voltage pulse allows for its uptake from the medium.

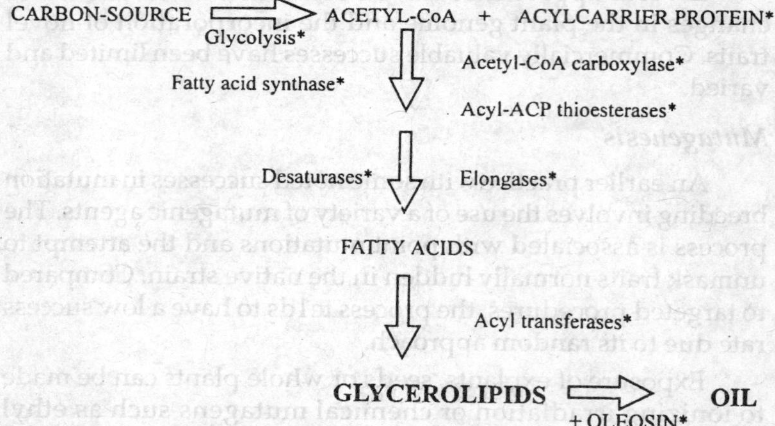

Fig. 11.1 Major target systems * for fatty acid, glycerolipid and oil formation in oilseeds.

3. **Microprojection.** Very small particles of tungsten or gold coated with the desired DNA can be shot into host cells at very high speed using the 'gene gun.'

These procedures are generally limited in their efficiency. Selection of protoplasts and cells where successful uptake of the foreign DNA has occurred frequently involves a known characteristic of the vector system where e.g. the gene for resistance to some antibiotic such as ampicillin or tetracycline resides. Only the transformed systems will survive and grow in medium treated with the appropriate antibiotic.

Successful uptake of the desired genetic material resulting in the creation of a new allele does not guarantee that it will be either expressed or inherited.

cDNA

Specific isolation and purification of a particular gene is seldom undertaken although the successful cloning of many of the targeted enzymes of fatty acid biosynthesis and glycerolipid formation has been obtained. Two general practices can be used to prepare a fragment of DNA that contains the desired genetic information. Re-combinant DNA libraries representing all of the mRNAs in a given cell can be prepared. Following the isolation and purification of the desired individual single-stranded mRNA double-stranded cDNA can be produced by the action of reverse transcriptase.

Alternatively the desired cDNA may be prepared from synthesis of the mRNA based on the knowledge of the full or partial amino acid sequence of the target protein. A frequently used approach involves the purification of the target protein (enzyme) and raising antibodies against them. Expression gene libraries are then screened, preferably with probes of plant origin, and the desired DNA located. Increased quantities of the desired cDNA can be obtained by cloning in some established system such as *Escherichia coli* or, more effectively, through application of the polymerase chain reaction (PCR). Genetic materials including partial and complete cDNAs which encode particular targeted enzymes have been obtained from several systems derived from seed tissue.

The particular procedure to be used is governed by the ease of isolation, purification and solubilisation of the desired protein. The enzymes systems associated with fatty acid biosynthesis are generally membrane-bound, difficult to isolate and are unstable; the enzymes involved in glycerolipid formation are usually easier to isolate and purify.

Targeted Systems

Several approaches. are available for the modification of a single enzyme system. Exact knowledge of the critical involvement of the enzyme in a particular aspect of plant lipid formation is essential. There are at least 50 enzyme systems associated with the conversion of carbohydrate carbon-source to oil droplets in the seed. Some of the systems that have been targeted are given in Fig. 11.1.

1. **Site-directed mutagenesis.** A selected enzyme is targeted for modification of its activity by introduction of chemically synthesised deoxynucleotides having discrete changes in sequence from that normally present in the genome. Expression of this altered gene in the host plant is then expected to yield the required protein generally modified at its active site.
2. **Anti-sense RNA technology.** Gene expression for a particular system is effectively reduced by providing a chemically synthesised RNA fragment which is complementary to the mRNA for the targeted enzyme. Subsequent hybridisation of the sense and anti-sense RNA molecules is expected to reduce the quantity of sense mRNA available for translation. Application of this technology is particularly useful in obtaining mutants for specific genes when they are members of multigene families e.g. the fatty acid synthase complex and hence more difficult to engineer. Problems, however, may arise when the desired reduction of targeted traits is accompanied with unpredicted accentuation of others. None the less it should be possible, at least in theory, to 'fine tune' the overall fatty acid composition of plant oils.
3. **Sense gene/co-suppression technology.** Introduction of a homologous gene into the host plant system can result in suppression of subsequent genetic expression rather than an expected increase in the quantity of the targeted enzyme. Further application of this procedure may prove to be useful for the modification of the unsaturated fatty acid com-position of oils with improved nutritional

characteristics.

All of these approaches to modification of the plant genome are heavily dependent upon accurate knowledge of plant lipid biochemistry, genetics and factors influencing genetic expression. Introduction of foreign DNA into a host plant with the creation of a new allele does not guarantee its expression or its inheritance. Particular requirement exists for the availability of a strong promoter to initiate the process of transcription.

TRANSGENIC OILSEED CROPS

Successful modifications of oilseed crops have been made in two general areas and have allowed the introduction of cultivars with desired characteristics.

Table 11.2 Some available transgenic crops and their trait modification.

CROP	ALTERED TRAIT
Canola *(Brassica napus)*	glufosinate ammonium tolerance *
	glyphosate tolerance **
	high-lauric
	high-stearate/low-polyunsaturate
	high-oleic
	high-medium chain
Soybean *(Glycine max)*	glyphosate tolerance
	high -oleic/low-saturate
Corn *(Zea mays)*	Bt -resistant ***
	glufosinate ammonium tolerance
	glyphosate tolerance
Cotton *(Gossypiumhirsutum)*	Bt-resistant
	glyphosate tolerance
Flax (Linum usitatissimum)	sulphonylurea tolerance

* Liberty Link * * Roundup Ready ***Bt = Bacillus thuringiensis

Plant Modifications

Plant protection has been achieved through genetic engineering to provide both tolerance to selected herbicides and resistance to insect infestation.

Strains of corn (maize), flax, rapeseed and soybean are commercially available which are herbicide-tolerant. The presence of weeds can have a major impact on crop yield where growth inhibition of e.g. untreated canola can be as much as 35%. Theoretically plant growth of the crop can be expected to be favoured when weeds sprayed with the herbicide die off. The anticipation is that there will be a lower need to apply the herbicide with an accompanying reduction of costs and potential environmental problems. Preliminary results suggest that herbicide-tolerant cultivars show slightly enhanced crop yield with a lower need for spraying.

Similar increases in crop yields have been obtained with both corn and cotton which have been engineered to contain the gene from Bacillus thuringiensis responsible for the production of a toxin fatal to certain insects.

Selective breeding of cultivars and the application of good agronomic practices have resulted in an apparently optimal range of total oil content for the major oilseed crops.

At present there does not appear to have been any success in developing transgenic crops with enhanced oil productivity. Based on the fact that certain (non-domesticated) plants are capable of accumulating up to 70% (w/w) oil in the seed an understanding of the mechanism(s) involved might conceivably allow development of higher oil-yielding crops.

Oil Modifications

Some very notable successes have been obtained in the development of transgenic plants in which the seed oil has been modified to yield increased and/or decreased contents of particular fatty acids (Table 11.2). Certain of these cultivars are available for commercial use, subject to legislative approval, while others are still in field trials. The impact, if any, of modified fatty acid composition of seed oil on the viability of the plant is not generally known. An engineered high stearic acid content in Brassica rapa seeds resulted in low oil content and low germination.

Further progress in the tailoring of seed oils can be fully anticipated.

COMMENTARY

Successful use of genetic engineering has been made with the incorporation of some desirable traits into commercial crops, particularly canola. The techniques and protocols are avai lable, but more definite applications are heavily dependent upon the availability of more advanced knowledge of plant lipid metabolism — total fatty acid synthesis,unsaturated fatty ac id synthesis and glycerolipid (triacylg lycerol formation) — gene expression and regulation, and all factors that impact on plant viability.

The first generation of transgenic crops has been primarily associated with input traits associated with protection — herbicide-tolerance and insect-resistance and with certain modifications of the fatty acid composition of the seed oil. Rapeseed (canola) has proved itself to be particularly amenable to manipulation and further advantages lie in the fact that it is an established crop.

The second generation it is believed will focus on output traits associated with the use of crops modified to produce a wide variety of consumer-desirable products such as nutraceuticals, pharmaceuticals and renewable feedstocks for the oleochemical industry.

Great care will need to be exercised with the use of crop engineering to avoid any possibly negative impacts on the environment and human health. A Luddite philosophy exists in certain segments of society that will have to be overcome by good solid reasoning if further successes with the technology is to be achieved.

Claims that transgenic plants will go a long way to solving the global lack of available food have yet to be substantiated. Emerging nations which generally lack adequate supplies of food may look toward biotechnology and genetic engineering as a solution, but the cost may be prohibitive. It may well be prudent for them to give greater attention to other approaches such as better agronomic practices and post-harvest technologies.

Chapter 12

Recommended Techniques

ISO is the International Standards Organisation and is made up of National Standards Institutes from countries large and small, industrialised and developing, in all regions of the world. ISO develops voluntary technical standards which "add value" to all types of business operations. They contribute to making the development, manufacture and supply of products and services more efficient, safer and cleaner and make trade between countries easier and fairer. ISO standards also serve to safeguard consumers (and users in general) of products and services, as well as making their lives simpler.

ISO standards are voluntary; ISO does not enforce their implementation. A certain percentage of ISO standards—mainly those concerned with health, safety or the environment—have been adopted in some countries as part of their regulatory framework or are referred to in legislation for which they then serve the technical basis. However, such adoptions are sovereign decisions by the regulatory authorities or governments of the countries concerned. ISO itself does not regulate or legislate.

In the United Kingdom the work of running the Secretariat of the United Kingdom National Oils and Fats Committee and the Secretariat of the ISO Oils and Fats Committee is shared between the British Standards Institute and the Federation of

Oils, Seeds and Fats Associations Ltd (FOSFA International).

The job of the secretary is to that deal with all the scientific aspects of the Federation's work, such as the Standard Contractual Methods of Analysis, laboratory accreditation, collaborative trials, research and day-to-day superintending matters on behalf of the trade. Secondly, the author is the Secretary of the ISO Technical Sub-Committee on Animal and Vegetable Fats and Oil (ISO/TC 34/SC 11). In this capacity, the secretary totally neutral and the ISO Directives govern my work of secretary.

A few words of interest about the historic relationship that has existed for well over 120 years between the continent of India, the Royal Botanic Gardens at Kew and the international oilseeds trade in London. With the help ofMs Lesley Price, the Kew Archivist, we have been able to find a letter dated 19 July 1881 from the Office of the Secretary of State for India to Professor Church of the Royal Botanic Gardens requesting a supply of Indian white linseed for experimental purposes. In the records of the United Oilseeds Association, the earliest international contract was dated 1862.

FOSFA International is an international contract issuing and arbitral body based in the City of London. It is not a national or even a regional association – it is an international global federation with over 650 members in 70 countries around the world. Its contracts are governed by English law and the largest proportion of oilseeds, oils and fats traded internationally are bought and sold of FOSFA contracts, of which there are over 50. Many FOSFA contracts contain specifications and references to analytical methods. The Federation also tests and accredits independent laboratories around the world to conduct analytical work under FOSFA contracts – provided these laboratories meet well-defined criteria. The fact that there are other Member Analysts in overseas laboratories carrying out analytical work means that it is essential that all analysts use the same methods of analysis. Some time ago, the Council ofFOSFA made the conscious policy decision to use ISO methods wherever possible because they had been internationally accepted.

In its oils and fats contracts, FOSFA stipulates the use of 30 different ISO standards and, in its oilseeds contracts, the use of 13 different ISO standards. It may interest you to know that the Codex Alimentarius Committee on Fats and Oils has endorsed and named about 20 different ISO standards for use in laboratories world-wide for sampling and analysis of fats and oils.

ISO only develops those standards for which there is a market requirement. In this sense, ISO standards are trade-driven. This work is carried out by experts "on loan" from the industrial, technical and business sectors which have asked for the standards and which subsequently put them to use. These experts may be joined by others with relevant knowledge, such as representatives of government agencies and testing laboratories. Published under the designation of "International Standard", ISO standards represent an international consensus on the state of the art in the technology concerned.

THE STRUCTURE OF ISO/TC 34/SC 11

The ISO Sub-Committee for Fats and Oils is made up, firstly, of those countries in "Participating Membership" who have a formal obligation to vote on all questions that are formally submitted for voting. These 16 "participating countries" are shown on this viewfoil.

The second group of countries on the Committee are the "Observer Members" who follow the work of the Committee as observers. They receive Committee documents, have the right to submit comments and can attend meetings. The 30 "observer countries" are shown on this viewfoil.

The Committee also has 6 organisations who have been granted "Category A" Liaison Status. They do not have a vote. They are shown on this viewfoil.

Among the 16 Participating Members of the Committee is the Member Body for India, represented by the Bureau ofIndian Standards in New Delhi. They have the right to vote upon the acceptance or rejection of any new ISO project work that is proposed for adoption by other countries, as well as the right to vote upon and make comments and technical recommendations about the various drafts of international oils and fats standards at each progressive stage of their development. This voting right is not extended to Observer Members or to organisations in Liaison Membership, such as FOSFA, the International Olive Oil Council or the International Fishmeal and Oil Manufacturers Association.

PUBLISHED ISO STANDARDS

Shown in VIEWFOIL 6 are the countries currently holding the Secretariats of some of the ISO Agricultural Food Product Technical Committees, together with the number of standards (as at 1998) that each of these Technical Committees have had published by ISO.

Technical Committee Name	Secretariat	Number of Published Standards
Oleaginous Seeds and Fruits	France	26
Cereals and Pulses and Derived Products	Hungary	49
Spices and Condiments	India	66
Tea	United Kingdom	12
Animal Feeding Stuffs	Netherlands	20
Animal and Vegetable Fats and Oils	United Kingdom	40

HOW THE WORK IS DONE

Show on the next three viewfoils are details of how the work in ISO is progressed.

Each Technical committee is encouraged by the ISO Central Secretariat to develop its own way of working, provided it abides by the ISO Directives. The Secretariat must have the means to process the texts electronically. If a country is going to suggest to the ISO Committee the adoption of a new standard, then it must appoint a Project Officer to see the project through to its final conclusion.

There are 6 stages prior to publication of a new ISO standard and the "participating countries" are able to vote at the proposal stage to accept or reject the proposal, as well as the CD and DIS stages. Any country is entitled to make comments and suggested amendments to the text. Once the standard reaches the FDIS stage, no further technical changes may be made.

The ISO Central Secretariat has set new time limits for the progression of work. They are shown on this viewfoil .

PROJECT LEADERS AND ACTIVE OILS AND FATS PROJECTS

The technical content of all 19 of the current active work projects being undertaken by various countries in the Animal and Vegetable Fats and Oils Technical sub-Committee are not described here. However, shown on this viewfoil is a complete list of the active projects that are being worked upon at the moment, together with the country that has nominated the Project Leader and is co-ordinating the work:

	Project	Country
1.	Ultra Violet Absorbance - Precision Data	United Kingdom
2.	Sampling - Amendment to the existing standard	United Kingdom
3.	Refractive index - Precision Data	Germany
4.	Alkalinity - Precision Data	France
5.	Phosphorus -Colorimetric Method - Precision Data	USA
6.	Phosphorus - GFAAS Method - Precision Data	USA
7.	Phosphorus - ICP Method	Canada
8.	Individual and Total Sterols Content—GC Method	Germany
9.	Trans Isomers Content - IR Method	USA
10.	Trans Isomers Content - GC Method	Netherlands
11.	Sediments and Foots	Netherlands and United Kingdom

12.	Volatile Organic Contaminant - GCMS Method	United Kingdom
13.	Chlorophyll in Crude Vegetables Oils	Canada
14.	Polycyclic Aromatic Hydrocarbons	Germany
15.	Cadmium Content - GFAAS Method	France
16.	Stigmastadienes - CCGC Method	France
17.	Stigmastadienes - HPLC Method	Germany
18.	Low boiling Halogenated Hydrocarbons	Germany
19.	Polymerized Triglycerides - HPLC Method	France

FOUR SELECTED PROJECTS

Four projects from among the 19 illustrated on the viewfoil to best illustrate how an ISO standard comes into being, what national and international agencies may be involved and how some of the costs of research and collaborative trials are met, as well as giving an idea of who contributes to the actual work.

ISO/CD 15301 - Determination of Visible Foots and Centrifuged Sediment in Sunflowerseed Oil

In 1992, the United Kingdom and The Netherlands submitted a proposal to the ISO Oils and Fats Committee to commence a project to find a new method of analysis for the Determination of Visible Foots in Crude Oils and Fats. Many of you will be aware that crude oils contain matter which is insoluble and settles out at the bottom of storage tanks. This material is called "foots" and, for trade purposes, it is important to know the level of "foots" in oil as the material normally represents a loss for the buyer. Previously, the only available method was to be found in ISO 150. Unfortunately, this was not suited to all oils. For example, some oils may need warming to make them fully fluid – an aspect not covered in ISO 150 which was the standard for linseed oil.

This standard will specify two methods for the determination of insoluble matter in crude oil which can be separated by force of gravity -

Method A: In this method, the amount of separated material, called "visible foots" is separated from the bulk of the oil by leaving a sample of the crude oil to settle by natural gravity differences during 96 hours and volumetrically measuring it in a prescribed measuring device. The authors of this draft international standard have defined "visible foots" as the matter insoluble in oil with occluded oil that settles out at 20°C or 10°C above the melting point of the oil or fat. It is quantified by storage of a sample of the homogenised oil or fat for a period of96 hours at 20°C or 10°C above the melting point.

Method B: In this method, the amount of separated material, called "sediment", is separated from the bulk of the oil by centrifugal forces in a prescribed procedure and volumetrically measured in a prescribed measuring device. The authors of the standard defined "sediment" as that part of the insoluble matter in a crude oil sample" which can be centrifugally separated. It can be quantified according to Method B of the standard. Sediment is the total amount of unclear layer of components collected at the bottom of the measuring cell after centrifuging. It contains phospholipids, impurities and dirt dispersed in a water-containing phase. White crystalline components deposing on top of and with the dark materials are regarded as part of the" foots" or of the "sediment".

It should be noted that Methods A and B usually give different values. Method A is the reference method. All measurements should be made on a sufficient homogenous and representative oil sample.

The final conclusions of the trial were that, although, spiked samples of sunflowerseed oil were used for the ring test, it can be concluded, by combining the analysis results of both ring tests, that there is an acceptable linear relationship between the visible foots (96 hours test) and sediment (as determined by centrifuging) for levels of sediment above 0.5ml/100ml.

For samples of sunflowerseed oil with sediment below 0.5ml/100ml, the results of the ring test shows a large deviation between Methods A and B. The results, however, show that the centrifuge sediment method can be considered as a good alternative to the visible foots method for levels of foots above 0.5ml/100ml, where a rapid analysis method is required.

ISO/CD 10540- Determination of Phosphorus Content - AAS - Colorimetric and IPC

FOSFA and the American Oil Chemists Society are funding and co-ordinating 3 international collaborative trials which will compare results for the Determination of Phosphorus Content using 3 different methods of analysis. The overall aim of the series is to publish a new ISO standard with 3 parts.

Part One: This method of analysis is the Colorimetric Method and, in principle, the test portion is carbonised in the presence of magnesium hydroxcarbonate and then ashed . The ash is dissolved in dilute hydrochloric acid, after which the phosphorus is determined colorimetrically by the molybdenum blue method. Fourteen laboratories in 7 countries are taking part.

Part Two: This method of analysis uses the Graphite Furnace Atomic Absorption Spectrometric Method. In principle, a test portion is mixed with a matrix modifier and vaporized in a graphite furnace atomic absorption spectrometer, previously calibrated using standard dilutions of soya lecithin. The phosphorus content is calculated from the absorption at a wave length of213 .5nm. Seven laboratories in 5 countries are taking part.

Part Three: This method of analysis uses a determination by Inductively Coupled Plasma Emission Spectroscopy. This procedure is new and uses expensive equipment. Depending on dilution solvent, nearly all forms of vegetable oils may be analysed from crude oil to degummed, refined, bleached, deodorised and hardened and nearly all types of lecithins and phosphatides. This procedure is only suitable when the elements are present in a solubilized form. When they are present as fine particles, such as bleaching earth, catalysts or rust, ICP-ES analysis results in poor recovery due to nebulization and atomization problems. The principle is that liquid samples are nebulized and carried into the excitation source by a flowing gas. Atoms are quantitated by measuring the specific emission lines produced by atoms decaying from high energy levels. Eleven laboratories in 4 countries are taking part and the results should be known by the end of March 1999.

ISO/DIS 15304- Determination of the Content of Trans Fatty Acid Isomers- Gas Chromatographic Method

In 1995, the Unilever Research and Engineering Division in Vlaardingen (The Netherlands) organised and conducted an international collaborative trial to determine the amount of trans- isomers formed in oils and fats during high temperature refining. They focused on low levels of trans fatty acids as they occurred in liquid oils. This new GLC method is not only suitable for low trans levels in refined oils, but also for trans in hydrogenated materials. Dr Duchateau and his team developed, by careful column selection and optimization procedures, one single GLC method for all types of vegetable oils; refined and hydrogenated. The method was presented at the AOCS Annual Meeting in 1995 and the manuscript was accepted for publication by the lAOCS in 1996. The hydrogenated oils were not incorporated in the ring test as the levels were higher and there was no real need to have figures in two decimal digits. The method was validated with isolated cis - and trans -fractions being the ultimate proof of accuracy. The ring test was of a double split-level design with three oil types (sunflower, soybean and rapeseed oil), with 12 samples in total.

ISO 5555 (1991) - Sampling of Oils and Fats -Document Revision

ISO 5555 is the international standard that describes methods of sampling crude or processed animal and vegetable fats and oils, whatever their origin and whether liquid or solid. It also describes the apparatus used for this process. Methods of sampling milk and milk products, including milk fats, are specified in a different standard, namely ISO 707.

The object of sampling and preparing samples is to obtain, from a consignment (which may be in lots), a manageable quantity of the fat (the properties of which correspond as closely as possible to the properties of the consignment sampled). The methods of taking samples described in this standard are intended for the guidance of expert superintendents and can be used for bulk consignment in land tanks, ships' tanks, tank wagons and tank cars. The standard can also be used for consignments consisting of a number of packages, such as barrels, drums, cases, tins, bags and bottles.

Sampling instruments, ancillary apparatus and sample containers (including caps) are required to be made of materials which are chemically inert to the fat being sampled and they must not catalyse chemical reactions. For sampling instruments, stainless steel is the most suitable material. Copper and copper alloys or any toxic material is banned. When the standard was written in 1991, plastics were also banned.

In 1994 a number of international supervision companies and internationally recognised independent analytical laboratories began to write to their national trade associations in various parts 6f the world saying that some of their clients who were involved in the trade of oils and fats under FOSFA contracts had advised them that they wished to maintain their refineries and food production sites as, "glass free sites" There was increasing presure to eliminate all glass from being used in any area where edible oils were present, from the point of production of the crude oil to the final sale ofthe food containing the oil.

Over the last 18 months a number of tests have been carried out in the Netherlands and the United Kingdom on the suitability of sample bottles made of high IV PET copolymer for the packaging of samples of animal and vegetable fats and oils in comparison with sample bottles made of glass. The following tests were carried out as part of this programme.

Migration of Chemical Bottle Components into the Oil

Storage of Contaminated Oil Samples

Carotene Content and Colour Tests

Physical Tests on PET Bottles

Hot - Fill Investigation

Migration of Chemical Bottle Components into the oil

250ml PET sample bottles were filled with refined coconut oil, crude sunflower oil and crude fish oil. The bottles were then stored for 30 days at 50°C (I 1°C). After storage, the oil in the bottles was analysed for the presence of mono - ethylene glycol, di -ethylene glycol, terephtalic acid and acetaldehyde. GC -MS analysis showed that none of the named chemicals had migrated from the bottle into the oil. The lower detection limit of the analysis method applied was 0.01 ppm for all 4 chemicals.

Storage of Contaminated Oil Samples

Refined coconut oil, crude sunflowerseed oil and crude fish oil were contaminated with trace amounts of hexane (isomers), benzene, vinyl acetate monomer (VAM) and phenol. PET, as well as plass sample bottles, were filled with contaminated oil and stored for 25 days at 25°C (I 1°C). Before and after storage, the oils in the respective bottles were analysed for the amount of contaminant present and for free fatty acid and peroxide value.

Carotene Content and Colour Tests

Tests were carried out on the influence of light on the carotene content and the colour of crude palm oil packed in PET bottles, compared to white glass bottles as packaging material. Both bottle types were filled with crude palm oil. The carotene content and the colour in a 1 inch L:ovibond cell of the samples were measured immediately after filling the bottles. The samples were placed in a storage room for 25 days at an ambient temperature of 25°C C ± 5°C). The storage room was fitted with a glass window facing north. The samples remained in natural daylight and in the dark at night. The results of the carotene and colour are shown on this viewfoil (VIEWFOIL 18).

Results of the Carotene and Colour Tests

Bottle Type	Carotene Content Initial	After Storage	Colour Lovibond 1 Inch Cell Initial		After Storage	
Glass	275.9	275.0	20Y	14.9R	20Y	14.7R
PET	276.4	274.8	20Y	14.9R	20Y	14.8R

Physical Tests on the PET Bottles

PET bottles were filled with crude sunflowerseed oil. The bottles and contents were heated to 60°C and kept at that temperature for 24 hours. The bottles were then cooled down to minus 18°C and kept at that temperature for 24 hours. For the next 24 hours they were stored at ambient temperature (20°C). After this treatment the bottles showed no sign of leakage, even when they were held upside down and pressure exerted by hand. A slight deformation was found to have taken place during the cold test.

Hot Fill Investigation

In a separate test for FOSFA, 500ml PET bottles were filled with hot palm fatty acid distillate at 75°C, capped and allowed to cool; one sample in the upright position, the other inverted. Both samples were then left on an inclined plane (neck down) overnight to check for leaks. Wadded tamper evident closures were used throughout this test. The wad was wood pulp with an aluminium facing. No evidence of leakage or distortion of the cap or bottle thread was observed.

Test Conclusions

On the basis of these tests, it is beginning to look as if "food grade" PET bottles fitted with tamper evident caps will be authorised as suitable sample containers for use in the international oils and fats trade.

ISO 5555 (1991) contains an Informative Annex about temperature limits for a range of oils and fats. The Secretariats of ISO/TC 34/SC 11, Codex, the IASC and FOSFA are actively seeking to harmonise this table for the benefit of the

(c) a world in which electronic communications dramatically increase human collaboration in every field and between all countries.

In this rapidly evolving scenario, globally applicable standards will play a key role. Such standards, whether developed by ISO or by others, will become one of the primary driving forces to support international commerce. In this context ISO intends to be globally recognized as an influential and innovative leader and as an effective and responsive producer in the development of globally applicable international standards which meet or exceed the expectations of the community of nations.

ISO Constantly strives to perfect the application of consensus and transparency principles in standardization and, in this way, promote the values of rationality, quality, safety and environmental protection for the benefit of all people of all races.